Untethered Miniature Soft Robots

# Untethered Miniature Soft Robots

## Materials, Fabrications, and Applications

*Li Zhang, Jiachen Zhang, Neng Xia, and Yue Dong*

WILEY-VCH

**Authors**

***Prof. Li Zhang***
The Chinese University of Hong Kong
William M.W. Mong Engineering
Building
Shatin, N.T., Hong Kong SAR
China

***Dr. Jiachen Zhang***
City University of Hong Kong
Yeung Kin Man Academic Building
Kowloon, Hong Kong SAR
China

***Dr. Neng Xia***
The Chinese University of Hong Kong
William M.W. Mong Engineering
Building
Shatin, N.T., Hong Kong SAR
China

***Dr. Yue Dong***
Guangdong Provincial Key Laboratory
of Intelligent Morphing Mechanisms
and Adaptive Robotics
School of Mechanical Engineering and
Automation
Harbin Institute of Technology
Shenzhen (HITSZ)
China



**Library of Congress Card No.:** applied for

**British Library Cataloguing-in-Publication Data**
A catalogue record for this book is available from the British Library.

**Bibliographic information published by the Deutsche Nationalbibliothek**
The Deutsche Nationalbibliothek lists this publication in the Deutsche Nationalbibliografie; detailed bibliographic data are available on the Internet at <http://dnb.d-nb.de>.



**Print ISBN:** 978-3-527-35177-0
**ePDF ISBN:** 978-3-527-84089-2
**ePub ISBN:** 978-3-527-84090-8
**oBook ISBN:** 978-3-527-84091-5

**Typesetting:** Straive, Chennai, India
**Printing and Binding:** CPI Group (UK) Ltd, Croydon, CR0 4YY

C9783527351770_191023

# Contents

# Preface

Soft robots with large deformation capabilities have received extensive attention in both academia and industry. Owing to the outstanding maneuverability and environmental adaptability, they are expected to perform complex tasks in tortuous or unstructured spaces and present great application prospects in industrial monitoring, space exploration, biomedical fields, and many others. However, with the increasingly complex application scenarios and the demand for *in vivo* medical treatment, the development of miniaturized soft robots shows great potential. Due to the limited onboard space, conventional components and control methods are difficult to adapt to these miniaturized soft robots. The development of smart materials, especially stimuli-responsive materials, provides a powerful solution for the construction of miniaturized soft robots, which derives the topic of this book: *Untethered Miniature Soft Robots: Materials, Fabrications, and Applications.*

In recent years, advanced polymer materials and their fabrication technologies have developed rapidly. Tough materials such as silicon-based elastomers and double-network hydrogels facilitate the construction of robust robotic structures. Widely available stimuli-responsive materials, such as magnetic materials, liquid crystal elastomers, responsive hydrogels, and shape-memory polymers, enable miniaturized robots to operate under various external stimuli (e.g. magnetic fields, light fields, and chemical stimuli). The application of biocompatible and biodegradable materials makes it possible for robots to evade immune system attacks or carry therapeutic cells to perform precision medical tasks. Self-healing polymers based on dynamic networks or non-covalent bonding enable soft robots to repair themselves autonomously after damage and are widely used to develop modular robotic systems. Meanwhile, advances in additive manufacturing techniques (such as two-photon polymerization and direct ink writing) and assembly approaches could satisfy the fabrication of 3D robotic structures on a range of scales from micron to centimeter scale, and even integrating different materials in 3D space. Therefore, a large number of achievements in materials and manufacturing have greatly enriched the design space of miniaturized soft robots and enhanced their performance.

This encouraging prospect has inspired a great deal of fundamental research and application exploration of miniaturized soft robotic systems. Existing soft robotic systems have achieved remarkable progress in independent control, selective manipulation, multimodal locomotion, and precise prediction due to sophisticated regulation and modeling techniques. It provides powerful tools for many application scenarios in the field of biomedicine, such as micro-manipulation, targeted drug delivery, and biological scaffolds, and is expected to improve the quality of life of patients. In addition, combined with flexible electronics technology, miniaturized flexible robots play an important role in applications including environmental monitoring and sensing.

The goal of this book is to convey the latest progress of untethered miniaturized soft robotic systems, focusing on the materials, preparation, and applications related to soft robotic systems. It is suitable for students, engineers, and scientists who are engaged in the research of soft robotics and soft materials. The authors will systematically introduce the research progress of miniaturized soft robots as well as the opportunities and challenges in the future, hoping to inspire more excellent ideas from researchers and enhance general readers' interest in miniature soft robots.

This book consists of seven chapters. The first chapter mainly introduces various soft robots developed in recent years. Chapter 2 presents the recent research outcome of silicone elastomer-based untethered miniature robots. Chapter 3 introduces botanical-inspired strategies for constructing multi-stimuli-responsive robots using carbon-based materials. Chapter 4 discusses 3D-printed environmental-responsive hydrogels for constructing programmable miniature robots. Chapter 5 is devoted to the application of liquid crystal networks and elastomers to fabricate monolithic or bimorph-material-based robots with programmable shape-morphing and self-adaptation capabilities. Chapter 6 presents ferrofluid droplets used to form miniature soft robots. Finally, Chapter 7 summarizes the recent key progress made by researchers on the investigation of miniature soft robots and discusses the prospects and challenges toward intelligent and autonomous soft robots.

The book is written with the kind help of many organizations and individuals. The authors would like to acknowledge support from the Hong Kong Research Grants Council with project Nos. RFS2122-4S03 (RGC Research Fellow Scheme), R4015-21 (RGC Research Impact Fund), E-CUHK401/20 (EU/RGC Research and Innovation Cooperation Co-funding Mechanism by RGC), 14300621 (RGC GRF), 14301122 (RGC GRF); the HKSAR Innovation and Technology Commission (ITC) with project No. MRP/036/18X (ITC Midstream Research Programme for Universities); the Croucher Foundation with project No. CAS20403; the Chinese University of Hong Kong (CUHK) internal grants; the SIAT-CUHK Joint Laboratory of Robotics and Intelligent Systems; the Multiscale Medical Robotics Center (MRC) at the Hong Kong Science Park; the CUHK Chow Yuk Ho Technology Centre for Innovative Medicine; and the CUHK T Stone Robotics Institute.

We are very grateful to Prof. Bradley Nelson from ETH Zurich, Switzerland; Prof. Joseph Sung Jao-yiu, Dean of Lee Kong Chian School of Medicine and Senior Vice President (Health and Life Sciences), at the Nanyang Technological University; Prof. Yangsheng Xu, President of CUHK (Shenzhen); Prof. Carmel Majidi, at the Carnegie Mellon University; Prof. Liu Wang, at the University of Science and Technology of China; Prof. Philip Chiu Wai Yan and Prof. Samuel Au Kwok Wai, Co-Directors of MRC, the other collaborators, as well as the colleagues from the

Faculty of Engineering and Faculty of Medicine at CUHK, for their long-term guidance and/or generous support to our group's research. We appreciate all the lab ex- and current members, including Dr. Tiantian Xu, Dr. Xiaohui Yan, Dr. Jiangfan Yu, Dr. Yabin Zhang, Dr. Ben Wang, Dr. Tony Chan Kai Fung, Dr. Lidong Yang, Dr. Mengmeng Sun, Dr. Dongdong Jin, Dr. Xingzhou Du, Dr. Fengtong Ji, Jialin Jiang, and many other researchers, for contributing their excellent research work to this book. We would also like to thank Jiaqi Zhu, Yanfei Cao, Xurui Liu, Xinyu Hou, and Bonan Sun in our group for their assistance in drawing figures and proofreading. The editorial and production staff of Wiley are gratefully acknowledged for the kind arrangement in writing this book that enables its publication. Thanks to the great and fascinating efforts and discoveries numerous researchers keep pushing the forefront of untethered miniature soft robots, and we sincerely hope the prospect of intelligent soft robots presented in this book can come into reality in the near future.

Finally, we would like to use this chance to send our best greeting for celebrating the 60th anniversary of the Chinese University of Hong Kong in 2023.

Hong Kong SAR, China
May 2023

*Li Zhang*
(On behalf of all authors)
The Chinese University
of Hong Kong

# 1

# Introduction to Untethered Miniature Soft Robots

## 1.1 Introduction

Miniature soft robots with inherent compliance could exhibit dynamic interaction with the real world [1, 2]. These controllable microdevices attracted growing attention because of their promises in a wide spectrum of applications, e.g. biomimetic study, environmental monitoring, precision medicine, as well as minimally invasive surgery [3, 4]. With a proper design of miniaturized structure and selection of material, the actuation and locomotion of miniature soft robots in tortuous and unstructured environments such as artificial vascular networks and animal tissues have been verified [5–8]. Advanced control techniques of physical fields allow us to elaborately tune the transformation of the robots, subsequently resulting in the generation of a variety of locomotion modes inspired by their natural counterparts, e.g. earthworm, inchworm, midge larvae, starfish larvae, bacteria, as well as jellyfish [9–14].

Various functional polymers have been introduced to endow robotic structures with intelligent properties, including self-healing property [15–17], degradability or stimuli-responsive deformation [18–21]. Due to the limited onboard space, conventional control units or driven parts can be hardly integrated into the miniature soft robots. In this respect, diverse stimuli-responsive materials (e.g. liquid crystal elastomers [LCE], shape memory polymers [SMP], and hydrogels) have been adopted for the construction of micromachines, so that they can present controllable deformation under external stimuli. In addition, to ensure the service life of soft robots, such as avoiding the influence of cracks caused by sharp parts in the physical environment or fatigue damages induced by multiple cycles of large deformations, self-healing polymers based on noncovalent interaction mechanisms or dynamic covalent networks have been proposed. Biomedical application is one of the most important development directions of miniature soft robots. The *in vivo* environments usually require the biodegradability or biocompatibility of robotic materials such as gelatin methacryloyl (GelMA) as well as zwitterionic materials [22, 23]. For instance, miniature soft machines have been applied for targeted cell delivery. The machines that load cells should be biocompatible to facilitate the adhesion and growth of therapeutic cells. Moreover, for robotic structures that can be hardly retrieved, they should

*Untethered Miniature Soft Robots: Materials, Fabrications, and Applications*, First Edition.
Li Zhang, Jiachen Zhang, Neng Xia, and Yue Dong.

be biodegradable to avoid the adverse effect of long-term accumulation. It is worth noting that sophisticated application scenarios may bring high requirements for the adaptability and versatility of a robotic system, which are often difficult to meet by robotic structures made of single-component materials difficult to meet. Therefore, the seamless integration of multiple functional modules or material compositions in the robotic system becomes pivotal, albeit a challenge [24, 25].

## 1.2 Working Mechanisms of Untethered Soft Robots

### 1.2.1 Magnetic Actuation

Soft robots actuated by external magnetic field have been intensively investigated due to their operation capability in large and enclosed workspaces, e.g. human body, and their great potential in minimally invasive surgery [26]. Magnetic field forces and torques could be used for the actuation of magnetic small-scale robots based on the interaction between the magnetic properties of the robot and the externally exerted magnetic fields (Figure 1.1a,b) [36, 37]. Assuming that there is no current

**Figure 1.1** Working mechanisms of untethered soft robots. (a,b) Magnetic actuation [27, 28]. (a) A hexapod soft structure with programmable ferromagnetic domains was fabricated by direct ink writing (DIW) printing technique. The 2D–3D morphological transformation is realized under the action of magnetic torque. Source: Kim et al. [27]/Reproduced from Springer Nature. (b) A microneedle robot with a magnetic base navigates over obstacles and penetrates the wall of the small intestine under the control of permanent magnets. Source: Zhang et al. [28]/John Wiley &Sons. (c,d) Acoustic actuation [11, 29]. (c) Ultrasonic-powered microrobots inspired by starfish larvae. Under ultrasonic stimulation, the cilia array in the robot body produces a swinging motion, which in turn induces complex vortices flows. Source: Dillinger et al. [11]/Reproduced from Springer Nature/CC BY 4.0. (d) Ultrasound-actuated structures with arrays of bubbles. Under ultrasonic stimulation, the bubbles resonate and meanwhile form a thrust force acting on the surface of the structure. Source: Adapted from Qiu et al. [29]. (e,f) Light actuation [9, 30]. (e) The photo-responsive liquid crystal elastomer (LCE) generates oscillatory behavior under the stimulation of two mutually perpendicular laser beams, as well as the photothermal distribution on the surface of the LCE structure. Source: Deng et al. [30]/Reproduced from John Wiley & Sons, Inc./CC BY 4.0. (f) LCE-based soft robot driven by dynamic light field. The beam distribution in the space is adjusted in real time through a digital micromirror device (DMD), and then the deformation control of the soft robot is realized. Source: Palagi et al. [9]/Reproduced from Springer Nature. (g) Propulsion driven by Marangoni effect. A rove beetle-inspired hydrogel rotor achieves efficient rotation and energy transfer by consuming hexafluoroisopropanol (HFIP fuel). Source: Wu et al. [31]/Springer Nature/CC BY 4.0/Public domain. (h) Thermal actuation. The self-propelled rolling of a robot with a bilayer structure (ferroelectric polyvinylidene fluoride (PVDF) and polydopamine-modified reduced graphene oxide-carbon nanotube (PDG-CNT)) was realized via a mechano-thermal feedback loop. Source: Wang et al. [32]/Springer Nature/CC BY 4.0/Public domain. (i) Deformation driven by humidity stimuli. Source: Adapted from Dong et al. [33]. (j) Chemical actuation. Source: Hu et al. [34]/John Wiley & Sons. (k) Biohybrid actuation. Source: Wang et al. [35]/American Association for the Advancement of Science.

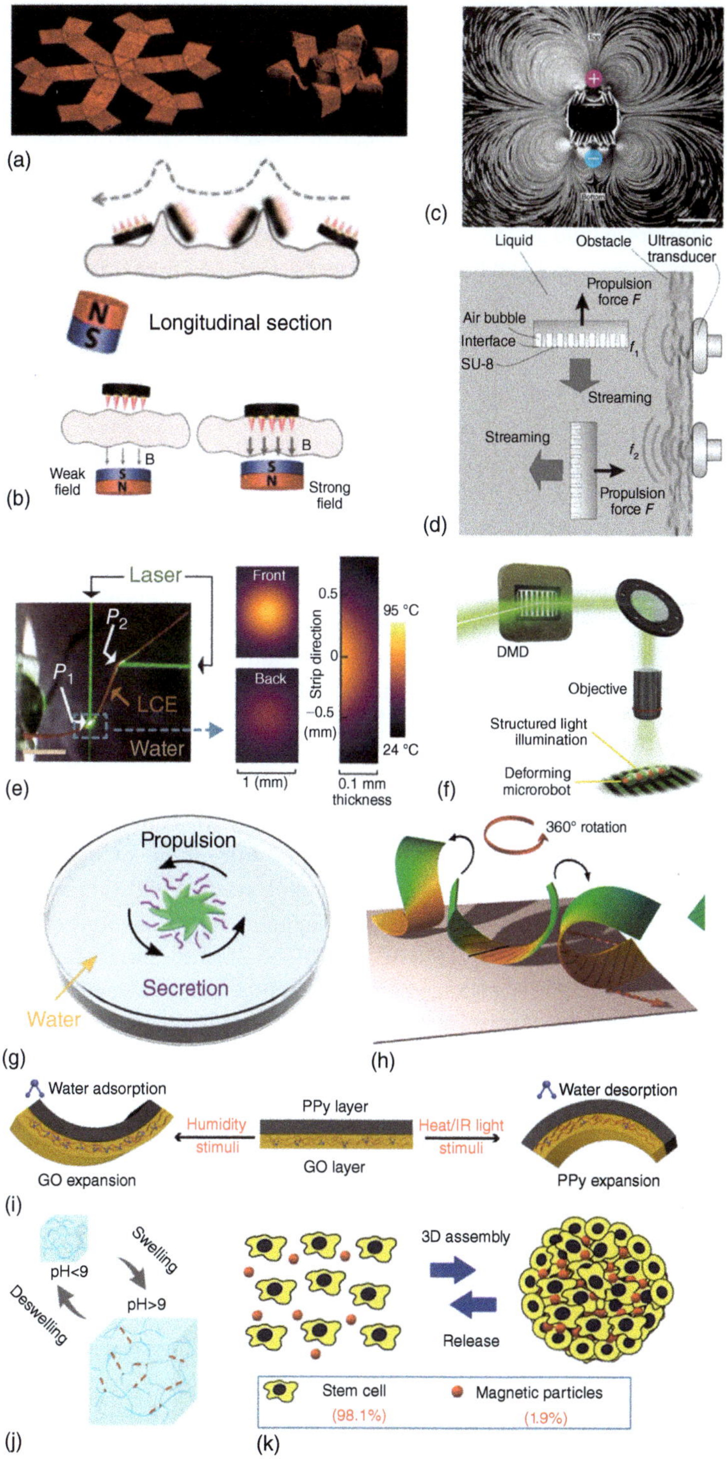

(a)
Longitudinal section
Weak field
B
Strong field
(b)
(c)
Liquid
Obstacle
Ultrasonic transducer
Propulsion force F
Air bubble
Interface
SU-8
$f_1$
Streaming
Streaming
$f_2$
Propulsion force F
(d)
Laser
$P_2$
$P_1$
LCE
Water
Front
Back
0.5
0
−0.5
Strip direction (mm)
95 °C
24 °C
1 (mm)
0.1 mm thickness
(e)
DMD
Objective
Structured light illumination
Deforming microrobot
(f)
Propulsion
Secretion
Water
(g)
360° rotation
(h)
Water adsorption
GO expansion
Humidity stimuli
PPy layer
GO layer
Heat/IR light stimuli
Water desorption
PPy expansion
(i)
pH<9
Swelling
pH>9
Deswelling
(j)
3D assembly
Release
Stem cell (98.1%)
Magnetic particles (1.9%)
(k)

in the workspace, according to Maxwell's equations, it can be derived that the static magnetic field satisfies Eq. (1.1) [38]:

$$\nabla \cdot \boldsymbol{B} = 0 \nabla \times \boldsymbol{B} = 0 \tag{1.1}$$

where $\boldsymbol{B} = [B_x, B_y, B_z]$ represents the applied magnetic field, whose gradient matrix is traceless and symmetric.

The magnetic force ($f$) and torque ($\tau$) applied to a magnetic device that are induced by a nonuniform field and the misalignment of directions of magnetic field and magnetization, respectively, can be calculated as Eq. (1.2) and Eq. (1.3):

$$\tau = \boldsymbol{m} \times \boldsymbol{B} = \begin{bmatrix} 0 & B_z & -B_y \\ -B_z & 0 & B_x \\ B_y & -B_x & 0 \end{bmatrix} \begin{bmatrix} m_x \\ m_y \\ m_z \end{bmatrix} \tag{1.2}$$

$$f = (m \cdot \nabla)B = \begin{bmatrix} \frac{\partial B_x}{\partial x} & \frac{\partial B_x}{\partial y} & \frac{\partial B_x}{\partial z} \\ \frac{\partial B_x}{\partial y} & \frac{\partial B_y}{\partial y} & \frac{\partial B_y}{\partial z} \\ \frac{\partial B_x}{\partial z} & \frac{\partial B_y}{\partial z} & -\left(\frac{\partial B_x}{\partial x} + \frac{\partial B_y}{\partial y}\right) \end{bmatrix} \begin{bmatrix} m_x \\ m_y \\ m_z \end{bmatrix} \tag{1.3}$$

The programmable parameters of magnetic field include the direction, amplitude, and gradient. Thanks to the high temporal resolution control and the capability of deep tissue penetration, multimodal locomotion including biomimetic modes on various terrains, sophisticated functionalities, and shape-morphing behaviors could be achieved by the magnetic machines. For instance, a variety of agile movements are developed for magnetic robots with fruitful inspirations from multilegged animals, zebrafish larvae, midge larvae, scallops, and jellyfish [10, 39–42]. Moreover, advanced functionalities including self-adaptation, shape memory, logic circuits, and mechanical tunability are achieved by the integration of magnetic control properties with structural designs and intelligent materials. Encoded with heterogeneous 3D magnetization profiles, magnetic robots could exhibit multiple complex deformations, e.g. 2D-to-3D and 3D-to-3D structural changes, upon magnetic stimulation. The arrangement of heterogeneous magnetization inside robotic structures is extensively studied, such as with the assistance of photolithography, modular assembly using bonding agents or dynamic covalent bonds, template-assisted magnetic programming, 3D printing techniques, and laser heating [16, 24, 27, 43–47]. In addition, the magnetothermal effect that could remotely generate heat using high-frequency magnetic fields is also developed for the activation of magnetic soft robots.

### 1.2.2 Light Actuation

Light is another commonly adopted actuation source for untethered miniature soft robots with the advantages of high spatial and temporal resolutions, enabling the precise and selective control [48]. Sophisticated optical equipment including photomasks, optical choppers, and lenses have been developed for the precise light actuation (Figure 1.1e,f). For instance, a single material component of LCE in cilia shape developed by Li et al. could present diverse complex deformation behaviors including photophobic, phototropic motions, bending, and twisting using tunable light source [49]. The photo-responsive properties of soft robots could adapt to a wide spectrum or selective wavelength based on the optical absorptive and chemical properties of the material components. Photothermal effect and photochemical reactions are widely adopted as the working mechanisms for the synthetic light-responsive actuators. For example, through photonic–thermal energy conversion or photoisomerization of azobenzene derivatives, the ordering change in liquid crystal networks is often activated. A variety of other mechanisms including water desorption, change of surface tension, hydrophobicity, and magnetic properties, inequivalent thermal strain, and shape memory effect are also developed for the actuation of light-responsive materials [33, 50, 51]. For photo-responsive actuators, a two-layer structure design including an active layer and a passive layer has been widely adopted. The bending deformation of the bimorph actuators can be calculated according to the Timoshenko theory [50]:

$$\rho = \frac{(h_1 + h_2)\left[3(1+m)^2 + (1+mn)\left(m^2 + \frac{1}{mn}\right)\right]}{6(Le_2 - Le_1)(1+m)^2} \tag{1.4}$$

$$m = \frac{h_1}{h_2} \tag{1.5}$$

$$n = \frac{E_1}{E_2} \tag{1.6}$$

$\rho$ represents the curvature radius. $h_i$ and $E_i$ ($i = 1, 2$) are the thickness and elastic modulus of the materials, respectively. $Le_i$ ($i = 1, 2$) represent the expansion or shrinkage of the materials.

### 1.2.3 Acoustic Actuation

Acoustic actuation owns a wide range of applications in miniature robots and biomedical fields due to its good environmental adaptability and deep penetration into biological tissues [52, 53]. Acoustic radiation force and bubble assistance vibration have been developed for acoustic-actuated microrobots (Figure 1.1c,d).

The acoustic radiation force ($F_A$) applied to microrobots and the resonance frequency of a microbubble ($f_{re}$) can be obtained by Eqs. (1.7) and (1.8), respectively [54].

$$F_A = \oint_{\partial\Omega} \langle\sigma\rangle \cdot n\,\mathrm{d}A - \oint_{\partial\Omega} \rho\langle v_1 v_2\rangle \cdot n\,\mathrm{d}A \tag{1.7}$$

$$f_{re} \approx \frac{1}{2\pi}\sqrt{\frac{3\pi\kappa P_0}{8RL\rho}} \tag{1.8}$$

$\Omega$ is the surface of the microrobot. $\sigma$ is the stress applied to the microrobot's surface and $\langle\sigma\rangle$ represents the time averaging value of $\sigma$. $\rho$ and $n$ refer to fluid density and the normal direction of the microrobot surface, respectively. $v_1$ is the vibration velocity. $v_2$ is the streaming velocity. $R$ represents the bubble's radius and $L$ represents the cavity's length. $P_0$ and $\kappa$ are the hydrostatic liquid pressure and adiabatic index, respectively.

Mechanical resonance generated by integrated oscillatory units, e.g. sharp structures or bubbles in robot body, is needed to efficiently transform acoustic energy into mechanical energy and produce propulsion. Activated by acoustic energy, diverse locomotion modes including rotation, sliding movement, vertical climbing, starfish larvae-like motions, and pull-type motions have been achieved for microrobots [11, 53, 55]. A large thrust force can be generated via bubble assistance vibration in liquid environment even upon a low-amplitude acoustic field, while this type of actuation would be limited by the long-term stability of bubble. In comparison, the vibration of flexible structures is free from the long-term stability problem, but requires the input of high acoustic energy. One of the important properties of acoustic actuation is selective activation based on the design of robotic structures with well-separated resonance frequencies. To generate precise acoustic fields, strategies including time-lapse Fourier synthetic harmonics, acoustic holography methods, and phased-array acoustic waves have been proposed [56, 57].

### 1.2.4 Thermal Actuation

Thermo-responsive soft materials can be actuated by the change of environmental temperature in liquids or air (Figure 1.1h). However, at the small-scale size, the direct conversion of thermal energy to robots is relatively difficult. Strategies including electrical heating, electromagnetic heating, and photothermal conversion have been developed for miniature robots made of temperature-responsive materials, e.g. SMP, LCE, and poly(*N*-isopropylacrylamide) (PNIPAAm) hydrogel [58]. To generate heterogeneous deformation or execute target functions with better output performance, diverse methods have been proposed including the integration of materials with distinct thermal response behaviors, anisotropic alignment or patterning of thermal-responsive units, the optimization of robotic structures, and the pre-treatment of the thermal-responsive robots [59, 60]. For instance, soft robots that could exhibit self-oscillating behaviors upon temperature gradient generated by a hot plate were developed by Wang et al. and Dong et al. [32, 33]. Through the optimization of robotic structures, the thermal-responsive robots could achieve continuous crawling and rolling locomotion on the hot plate.

### 1.2.5 Chemical Actuation

Stimuli-responsive soft materials can also be actuated by other environmental stimuli including Marangoni effect (Figure 1.1g), humidity, chemical fuels, and ionic strength [54]. Chemical reactions that could generate bubbles have been widely adopted to propel miniature robots. The volumetric swelling or deswelling behaviors of water-absorbing materials have been exploited for hygroscopic robots to achieve programmable shape-morphing and movements (Figure 1.1i,j). Made of smart materials, soft robots could also exhibit responsive behaviors to the variations in pH, ionic strength, selective DNA sequences, surface tension gradient, and the presence of solvent or solvent vapor [61, 62]. Due to the advantages of ease in downscaling and wireless control, these robots could present high mobility and execute tasks in hard-to-reach environments. Nevertheless, for some robots that rely on chemical reactions, the continuous supply of chemical fuels is challenging in confined spaces and the fuels may induce adverse effects to the real environment.

### 1.2.6 Biohybrid Actuation

Chemical energy could be efficiently transformed into mechanical work by biological microorganisms and cells. Due to the limitations of materials and fabrication techniques, it remains challenging to develop artificial robots with comparable control and actuation performance at microorganism and cell scales. Soft robotic systems that integrate flexible materials and biological components exhibit the capability of generating high power density and high output force for small-scale robotic actuation and promise practical application prospects in organ-on-a-chip, tissue engineering, drug delivery, minimally invasive surgery, and cell manipulation (Figure 1.1k) [63–66]. For instance, different substrates, e.g. elastomers and hydrogels, could be integrated with modified muscle cells which could respond to chemical fuels such as ATP, electrical stimulation, or light via optogenetic modification. Diverse functions including sensing, pumping, or artificial muscles could be executed via the contractile locomotion of these cells. In addition, organisms such as sperm cells, invertebrates, zebrafish, and *Escherichia coli* integrated with control components could enable the achievement of steerable movements of these biohybrid machines along targeted trajectories. Moreover, biological units in these artificial micromachines could carry functional dopants such as drugs and cells for therapeutic use to execute biomedical tasks.

## 1.3 Fabrication Methods of Untethered Soft Robots

### 1.3.1 Molding

Molding is a commonly used fabrication method that cures polymeric precursors to fabricate robots with specific shapes in molds (Figure 1.2a,b). For complex 3D structures, molds formed by 3D printing techniques (such as two-photon polymerization [TPP] or fused deposition modeling [FDM]) are usually used for molding or casting processes. On the one hand, molds made of sacrificial materials such as PVA materials could be used. After the curing of the polymeric precursors, the mold is removed, and the formed 3D structure is retained. On the other hand, the mold can also be used as a part of the robotic structure. For example, magnetically responsive mechanical metamaterials are fabricated by injecting magnetorheological fluids or elastomers into 3D-printed lattice structures [72, 73]. When using the molding method to form robotic structures, it is usually necessary to generate internal heterogeneous orientations in polymers to fulfill the deformation or actuation requirements, especially for structures based on magnetic composites and liquid crystal materials. When polymeric precursors doped with hard magnetic or soft magnetic particles are molded, a uniform external magnetic field can be applied

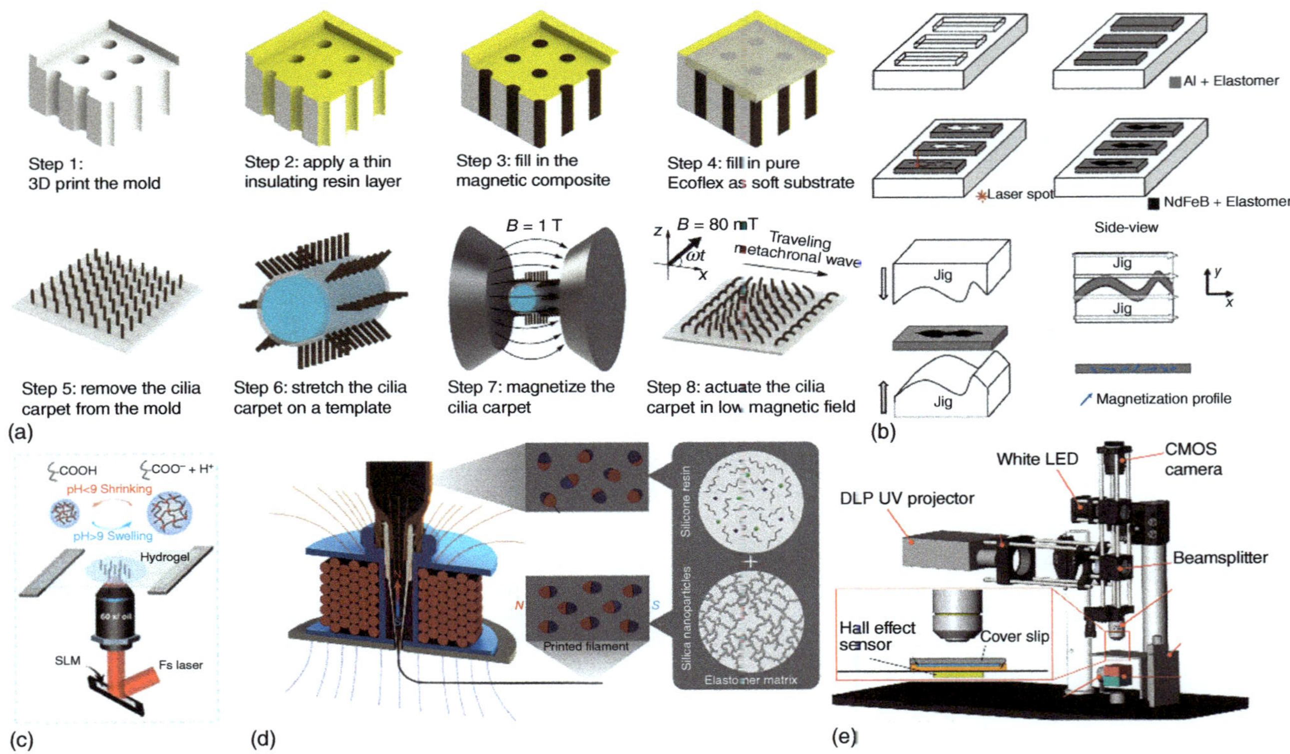


**Figure 1.2** Fabrication methods of untethered soft robots. (a,b) Molding method. Source: Gu et al. and Lum et al. [67, 68]/Springer Nature/CC BY 4.0/Public domain/Pnas. (c–e) Additive manufacturing. Source: Adapted from Kim et al., Xu et al., and Li et al. [27, 45, 69]/Springer Nature/American Association for the Advancement of Science. (f–h) Semiconductor and microelectronic techniques. Source: Bandari et al., Kim et al., and Miskin et al. [61, 70, 71]/Reproduced from Springer Nature. (i,j) Modular assembly based on bonding agents. Source: Zhang et al. and Dong et al. [24, 25]/American Association for the Advancement of Science (AAAS)/© The Authors, some rights reserved; exclusive licensee AAAS. Distributed under a CC BY-NC 4.0 license http://creativecommons.org/licenses/by-nc/4.0/.

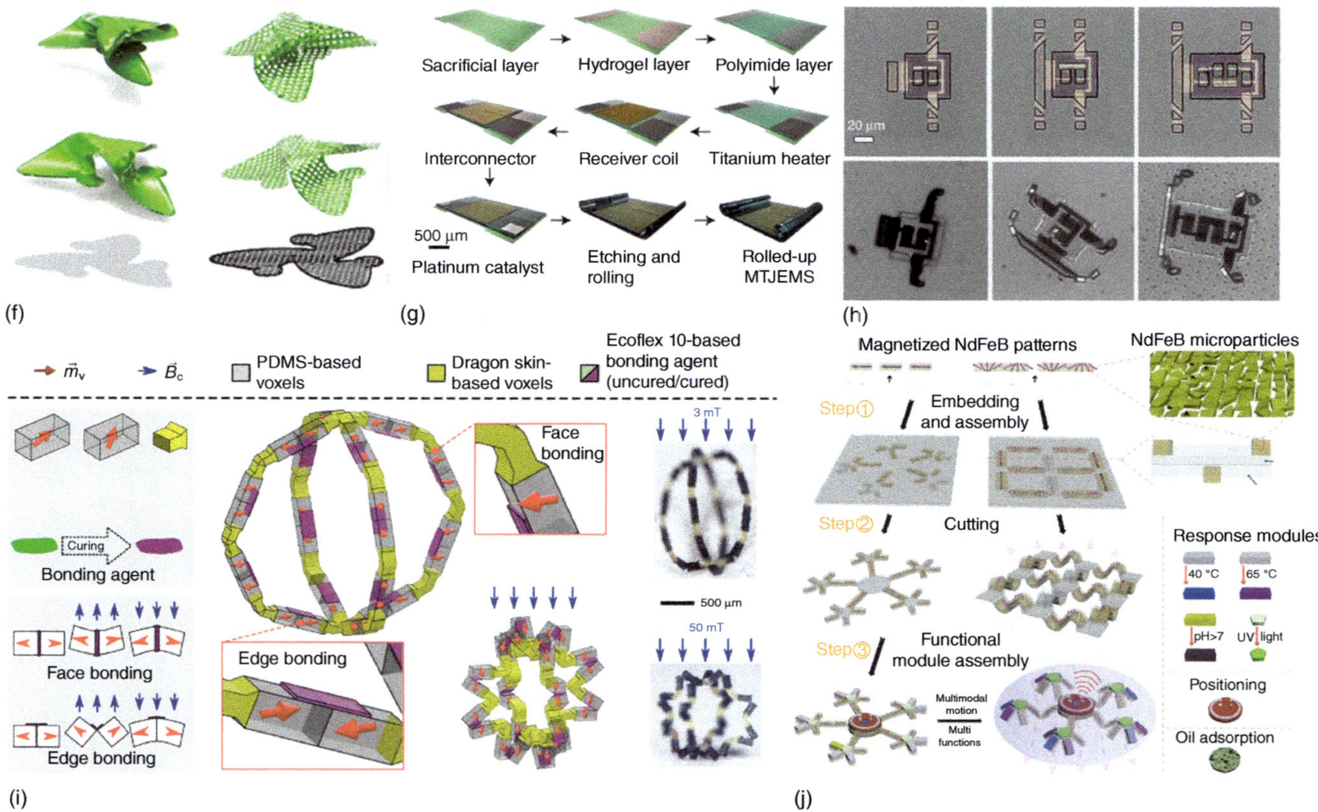


**Figure 1.2** (*Continued*)

during the preparation process to control the orientation of the magnetic particles. In addition, after the curing of hard magnetic soft materials, the soft structure can be deformed into a targeted shape with the assistance of template and magnetized by applying a strong magnetic field under the deformed state, which results in a nonuniform magnetization profile formed in the fabricated soft materials. Compared with other fabrication techniques, e.g. stereolithography (SLA) and DIW, the molding method does not impart high requirements for material properties including optical and rheological properties. Thus, the molding method exhibits procedural simplicity and wide applicability, and is most commonly adopted for the preparation of robotic prototypes.

### 1.3.2 3D Printing Techniques

The progress in the development of miniature soft robots with more sophisticated structures and functions has been accelerated by the advances in diverse 3D printing techniques. Classified based on their working mechanisms, widely used 3D printing techniques for the fabrication of miniature soft robots include material extrusion, light-based 3D printing, and material jetting (Figure 1.2c–e) [74].

Extrusion-based 3D printing techniques usually deposit uncured material through a nozzle via mechanical force. 3D structures are printed by stacking consecutive layers of 2D patterns with the assistance of motile extrusion nozzles or a three-axis motion platform. Fused filament fabrication (FFF) and DIW are two powerful extrusion 3D printing techniques. For the FFF method, a heater near the nozzle is used to liquefy thermoplastic polymers, which would be solidified again after extruding out the nozzle with an ambient temperature lower than the glass transition temperature. For the DIW method, with the assistance of piston, screw, or air pressure, pseudoplastic polymers that exhibit reversible rheological behavior are extruded out of the nozzle. With such a flexible filament-by-filament printing strategy, complex structures made of diverse materials such as elastomers, hydrogels, and shape memory polymers could be fabricated. Compared with the printing speed of FFF (40–480 mm $s^{-1}$), DIW presents a relatively slow speed (~10 mm $s^{-1}$). Nevertheless, due to a large variety of printable materials and multi-material ability, DIW is more attractive for broad application prospects in robotics [75–77]. A customized DIW printing system developed by Zhao et al. exhibits superior capability of in situ magnetization programming which is achieved by controlling the magnetizations of hard magnetic particles upon the localized magnetic field generated by coils near the nozzle [27]. Complicated structures with desired magnetization distributions can be prepared by tuning the printing direction to achieve functional shape transformation under magnetic stimulation.

Light-based 3D printing is a kind of technique using focused laser light or patterned ultraviolet to solidify photopolymerizable polymer resins. Printing methods including digital light processing (DLP), computed axial lithography (CAL), continuous liquid interface production (CLIP), and direct laser writing (DLW) have been developed. During the DLP printing, a supporting plate would move according to the given commands to generate a sequential pattern on each layer whose resolution could reach 7 μm. Compared with voxel-based printing methods such as selective laser sintering (SLS) and DLW, the printing speed of DLP is relatively high (40–400 mm $h^{-1}$) and DLP shows good compatibility with diverse photocurable composites including polymeric precursors doped with micro–/nano magnetic particles or conductive fillers [78–80]. For example, during the selective curing of the photopolymer resin with patterned ultraviolet, electromagnetic coils or a permanent magnet are adopted in the printing system to tune the alignment or magnetization of magnetic particles, leading to the formation of soft-magnetic and hard-magnetic soft robotic structures. By initiating two-photon or multi-photon polymerization with femtosecond laser pulses, the DLW method could achieve a printing resolution of 100 nm, promising great potential in the microscopic fabrications of robots. Nevertheless, the printing speed (~2 mm $h^{-1}$) and allowable printing dimensions (~$2.2 \times 2.2 \times 0.25$ $mm^3$) of the DLW method are usually limited [81].

### 1.3.3 Semiconductor and Microelectronic Techniques

Semiconductor processing technology offers a powerful platform for the etching and deposition of diverse materials (Figure 1.2f–h). Due to the advantages of easy integration with electronic components and mass fabrication, they have been widely used for the development of flexible electronic devices. Compressive buckling and self-rolling strategies have been developed to form complex 3D structures [61, 82–84]. The control of internal stress in the processed materials enables the formation of robotic structures with hierarchical 3D layouts integrated with electronic components. As shown in Figure 1.2f, 3D microfliers that carry electronic components including near-field communication (NFC) chips and silicon nanofilm transistors are developed by Kim et al. By applying metal hard mask and oxygen plasma reactive ion etching, patterned SMP films were obtained, followed by depositing a multilayer of Ti/Mg/Ti/$SiO_2$ [70]. There are hydroxyl groups generated on the surface of the ozone-treated $SiO_2$ layer, which could ensure strong covalent bonding between a pre-strained silicone elastomer substrate and the treated SMP film at specific locations. By using transfer printing technique, electronic components encapsulated by the polyimide can be integrated with the SMP film. A morphological transformation of the film from 2D structure to 3D wind-dispersed seeds shape structure was performed via the release of pre-strain and mechanical buckling process. These flying devices released in the air can present controllable falling speeds and execute tasks such as environmental monitoring via the loaded wireless electronic devices. In addition, a flexible motile microsystem was proposed by Bandari et al. shown in Figure 1.2 [61]. A wireless energy transfer module in the microsystem facilitates the control of the propulsion direction of the microsystem and the power supplier of the integrated electronic modules. After the depositing of multiple layers of polymers on a substrate, self-rolling operation activated by the removal of sacrificial layer was adopted for the generation of tubular shape structures at the edges of the proposed microsystem. In the micro-robotic system, there are heating wires and a square coil capable of heating the catalytic engines to tune the production of oxygen and wirelessly transfer energy. Recently, Miskin et al.

developed a new class of electrochemical-actuated microrobot systems that are compatible with silicon processing technology, which allows the formation of thousands of microrobots in a chip via an integrated process [71]. With the photovoltaics integrated into the robotic structure, the legs of the microrobots would deform upon laser stimulation, leading to the locomotion of the micro-robotic system.

### 1.3.4 Modular Assembly Based on Bonding Agents

The bottom-up assembly strategy allows a facile integration of heterogeneous components in a robotic system by using bonding agents including tapes, glues, and uncured materials (Figure 1.2i,j). For instance, uncured elastomers were adopted by Zhang et al. to connect heterogeneous micro-components in small-scale magnetic soft machines and develop micro-robotic systems with arbitrary material compositions and 3D magnetization profiles (Figure 1.2i) [24]. With the assistance of TPP-printed micro-molds, edge connection or surface connection between magnetic modules and skeleton modules could be formed by the bonding agents. Microrobotic systems developed by this bottom-up assembly strategy exhibit great promises for diverse biomedical applications including anchoring machines, soft capsules, and peristaltic pumps. In addition, Dong et al. developed multifunctional and multimodule magnetic soft robots using adhesive sticker network (Figure 1.2j) [25]. The adhesive sticker allows the transfer of hard magnetic particles with different 3D magnetization patterns to the substrate and the easy integration of diverse functional modules including photosensitive modules, oil absorption modules, and electronic components.

## 1.4 Applications of Miniature Soft Robots

### 1.4.1 Biomedical Application

Recent advances in miniature robotic techniques, e.g. medical guidewire and catheter, greatly promote the practical applications of minimally invasive surgery with shortened recovery times and reduced surgical risks [85–87]. The catheter-based interventions have been widely adopted for the cardiovascular disease treatment [88, 89]. Nevertheless, further advancements in catheter technology require them to perform more functional and even challenging tasks in addition to delivering catheters to designated locations, such as the capability of monitoring diverse physiological environments, delivering therapeutic cells or drugs, performing thermal or electrical stimulation to soft tissue, embolization, and clot removal. To fulfill these requirements, the catheter structures need to integrate with various functional components, e.g. soft sensors and manipulator tools. However, the introduction of functional components would impede the miniaturization of catheter structures, which is important for the application of medical catheters in narrow spaces. A feasible solution for the trade-off between miniaturization and functionalization is provided by semiconductor processing technology. A hollow microcatheter with a diameter of 0.1 mm carrying magnetic sensor and electrically actuated gripper was developed by Rivkin et al. via the self-rolling method [90]. The fabricated microcatheter exhibited the locomotion capability in a thin curved channel (only 0.2 mm diameter), and the fluid delivery function in the stomach and esophagus of mouse. The end of the catheter is integrated with a conductive polymer film actuated microgripper, which could capture 0.1 mm particles in a tube. In addition, the magnetic sensor based on the anisotropic magnetoresistive effect in the microcatheter could execute *in vivo* localization and navigation with a resolution of 0.1 mm. Han et al. proposed the integration strategy of multilayer configurations of soft electronic arrays and actuators on commercialized endocardial balloon catheters [89]. The integrated flexible electronic arrays include a pressure sensor array, a temperature sensor array, and an electrode array with electrical stimulation and electrophysiology measurement functions, and are capable of withstanding 10 000 cycles of uniaxial stretching, the deflation and inflation of the balloon on the catheter. Irreversible electroporation and programmable radiofrequency ablation for the treatment of arrhythmias could be achieved by the selective powering of electrode arrays. Moreover, ventricular action can be simultaneously monitored by the integrated pressure sensors for surgical improvement.

Wirelessly actuated microrobots have been widely adopted for minimally invasive surgery, e.g. targeted drug and cell delivery. To execute the delivery function, therapeutic units need to be integrated into the robotic system. For instance, Zhang et al. proposed a magnetic anchoring device with a 3D lattice shape equipped with TPP printed cell cages [24]. The fabricated anchoring device could shrink and restore its radial dimension upon magnetic stimulation and stem cells can be carried by the cell cage as a cell scaffold. After reaching the targeted position, the proliferation, migration, and differentiation of stem cells would be performed to achieve treatment functions such as vascular regeneration. The separation and retrieval of the robotic body from the therapeutic units is a key issue for the practical application of miniature robots to avoid the adverse impact of the robotic body. In this respect, a multifunctional magnetic soft robot was developed by Dong et al. (Figure 1.3a) [25]. Magnetic components with 3D heterogeneous magnetization profiles and a therapy patch for gastric ulcer treatment were integrated into the robotic system by an adhesive sticker network. The connection between the therapy patch and the robotic body was realized by a soluble tape. The robot body wrapped the therapy patch to avoid contact with gastric fluids during the movement. After reaching a gastric ulcer position, the release of the therapy patch was performed by the dissolution of the soluble tape in the gastric mucosa.

### 1.4.2 Environmental and Proprioceptive Sensing

Natural organisms could exhibit sensing and adaption functions to the physical environment with developed intelligence. To mimic the intelligence of biological systems, various robotic systems with embedded perception and sensing capabilities have been developed, enabling the closed-loop control of deformation and locomotion (Figure 1.3b). For example, a magnetic soft robotic system with

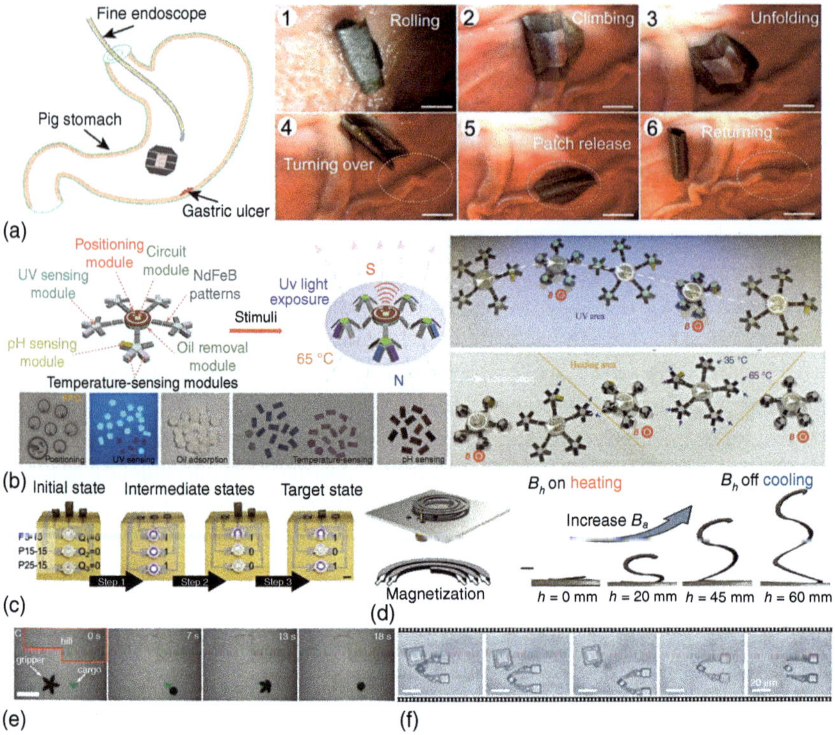


**Figure 1.3** Applications of miniature soft robots. (a) Magnetic soft robot actuated in stomach for gastric ulcer treatment. Source: Dong et al. [25], © The Authors, some rights reserved; exclusive licensee AAAS. Distributed under a CC BY-NC 4.0 license http://creativecommons.org/licenses/by-nc/4.0/. (b) Multifunctional multilegged soft robot integrated with environmental sensing modules. Source: Dong et al. [25], © The Authors, some rights reserved; exclusive licensee AAAS. Distributed under a CC BY-NC 4.0 license http://creativecommons.org/licenses/by-nc/4.0/. (c) Magnetic shape memory material applied for logic circuit. Source: Ze et al. [59]/Reproduced from John Wiley & Sons, Inc. (d) Magnetically actuated helical antenna. Source: Ze et al. [59]/Reproduced from John Wiley & Sons, Inc. (e) Magnetic responsive microgripper. Source: Xu et al. [45]/American Association for the Advancement of Science (AAAS). (f) pH-responsive microgripper. Source: Ma et al. [91]/Reproduced from Springer Nature/CC BY 4.0.

seamless integration of multiple functional units was developed by Dong et al. to achieve environmental sensing. Through an adhesive network, pH sensing paper, temperature, and UV sensing particles could be easily loaded by the developed soft robot, and these sensor modules can be replaced after use via the repeatability of the stickers [25]. Recently, Zhang et al. proposed a frog-inspired origami robot made of laser-scanned PDMS sheet which could selectively absorb thermochromic ink, photosensitive ink, and quantum dot solutions [92]. Upon thermal and light stimuli, the origami robot could switch skin color. In addition, to construct a biomimetic intelligent drive system, the integration of the actuation and proprioception units is required. In this respect, somatosensory sensors, e.g. contact sensor, curvature sensor, and inflation sensor, that could monitor the deformation type of the robot body and surface roughness of contacted objects were developed and integrated with a somatosensitive pneumatic actuator via multi-material embedded 3D printing technique [93]. Different types of movements such as flicking, upward bending, and downward bending could induce different resistance changes for the somatosensory feedback. Furthermore, with the variation of contact pressure and local conductivity, the temperature and surface texture of the manipulated object could be monitored by the pneumatic actuator. Different from the multi-material printed robotic structures, an interpenetrating polymer network proposed by Zhao et al. consists of conductive polyaniline and thermally responsive PNIPAAm, which allow the seamless integration of piezoresistive sensing and photothermal actuation [58]. Upon near-infrared light stimulation, the hydrogel soft robot could achieve a series of locomotion, e.g. bending and contraction. Meanwhile, with the intrinsic conductivity, the external force applied to the robot body or structural deformation activated by remote stimulation could induce the variation of resistance. Therefore, the fabricated somatosensory soft robot is capable of executing closed-loop control and recognizing the manipulated unknown object.

### 1.4.3 Intelligent Electronics

Advances in miniature soft robots promise great potential in reconfigurable electronic devices, e.g. morphable antennas and logic circuits (Figure 1.3c–d). For instance, Kim et al. developed an annular-ring-shaped magnetic film integrated with a soft electronic circuitry [27]. Via magnetic stimulation, the magnetic film exhibits two distinct deformation modes which cause the selective contact of electrodes and light up different micro-LEDs. To realize more complicated shape transformation modes, a magnetic SMP-based robotic structure developed by Ze et al. could present tunable stiffness by magnetothermal effect, resulting in different deformation capabilities under magnetic stimulation [59]. By using an actuation magnetic field and a high-frequency magnetic field as input, and the off or on states of LED as output signal, the developed magnetic SMP-based robots could achieve logical functions such as three-bit memory and D-latch. In addition to the 2D thin-film structure, 3D origami robotic structures are also applied to the logic circuit. Kresling origami robots consisting of conductive components and magnetic discs with different magnetization patterns were fabricated by Novelino et al. [94]. Upon magnetic stimulation, bistable modes including folded and deployed states of robotic structures were exhibited, and results in the connection of different circuits. Moreover, three Kresling origami robots were assembled to demonstrate the function of digital computing of three-bit information. The shape-morphing capabilities of soft robots enable the antennas to exhibit reconfigurable frequency responses. For example, a tapered helical antenna made of magnetic SMP was developed by Ze et al. [59]. Under magnetic stimulation, the morphable antenna could present controllable deformation height, leading to tunable resonant frequency from 2.15 to 3.26 GHz. In addition, Bai et al. utilized a buckling-guided assembly strategy to fabricate a 3D reconfigurable antenna [95]. Nine deformation modes of the morphable structure could be induced by the sequential release of an elastic substrate with pre-strain and endow the antenna with widely tunable radiation directions.

### 1.4.4 Micromanipulation

Miniature soft robots have been widely used in micromanipulation to execute the tasks of in situ analysis and active delivery (Figure 1.3e,f). Recently, a micropillar array that could perform reversible bending upon the variation of pH was fabricated by Li et al. using the asymmetric DLW method and adopted as a pH-responsive microgripper [69]. Neural stem cells and microparticles with diameters of 10–15 μm could be in situ captured by the developed microgripper showing great manipulation resolution. A microgripper made of pH-responsive protein and rigid SU-8 resin was developed by Ma et al. using a multi-material TPP printing strategy [91]. Assisted by a microfluidic chip, a photopolymerization chamber was filled with printable polymers that can be replaced by different polymer precursors to achieve the sequential printing of multi-materials in one microstructure. The integration of a 3D moving stage and the adjustment of the pH of surrounding media allow the microgripper with a side length of ~30 μm to precisely deliver and capture a micro cube (10 μm length). Apart from the pH-responsive behavior, magnetic actuation provides a remotely controlled strategy. For instance, a multi-arm magnetic gripper was developed by Xu et al. through the DLP printing method [45]. The gripper was made of photocurable polymer doped with ferromagnetic particles. With the arrangement of heterogeneous magnetization patterns and external magnetic field stimulation, the gripper could achieve wireless transporting, releasing, and grasping of millimeter-scale cargo with ~2 mm length in an unstructured environment.

## 1.5 Scope and Layout of the Book

### 1.5.1 Scope of the Book

Miniature soft robots refer to controllable soft devices with maximum characteristic sizes in millimeters. Untethered miniature soft robots are able to perform reversible deformations and actively interact with the environment due to their inherent compliance. These robots have attracted great attention because of their broad potential in various fields, including intelligent electronics, smart grippers, biomedicine, and environmental applications. Due to their small-scale sizes, miniature robots may access diverse tortuous and confined spaces such as eustachian tubes and cerebrovascular networks. In addition, due to the ability to conduct micromanipulation with high precision, microactuators have the potential to perform in situ cell analysis. It demonstrates that micro-robots and micro-actuators have great potential on the important research topic of biomedical applications. Other meaningful applications are also explored in previous research. Over 5000 articles related to soft robots and their applications were published in the period from 2017 to 2021, and there has been an exponential growth of published works about small-scale soft robots over the past decades. However, there is a lack of a specialized and dedicated book on this topic. We aim to fill this gap and present the latest achievements in research on untethered miniature soft robots.

This book is focused on the emerging field of untethered miniature soft robots. It introduces fundamental understanding of various small-scale soft robots, including actuation mechanisms, soft matter, fabrication strategies, actuation, and locomotion principles. This book also demonstrates applications of miniature soft robots in different fields, such as intelligent electronics, smart grippers, biomedicine, and environmental applications. The detailed analysis of new materials and fabrication strategies, as well as experimental demonstrations in this book, provide readers ranging from students to researchers from diverse communities of robotics, materials, and biomedical engineering with a realistic understanding of progress achieved recently in the field of miniature soft robots.

### 1.5.2 Layout of the Book

This book introduces readers to the emerging field of miniature soft robots. From the perspective of fundamental research, we describe different types of functional materials to build miniature soft robots, such as silicone elastomer, carbon-based materials, hydrogels, liquid crystal polymer, flexible ferrofluid, and liquid metal. The material properties, fabrication strategies, and functionalities in soft robots are presented in detail, together with the underlying mechanisms. The description in this book is concise and explicit, which is easy to understand even for readers who have not been exposed to related fields. Due to the limitation of space, we emphatically introduce magnetically, thermal, light, and chemically actuated soft robots in this book. Furthermore, various specific applications of miniature soft robots in biomedical, environmental, and electrical fields are demonstrated. Finally, we summarize the opportunities and challenges faced by miniature soft robots in the future of this field. For researchers in this academic field, this book can serve as a reference for their research on soft robots, which may inspire more excellent ideas. For non-expert readers, such as undergraduate students, the attractive contents presented in this book can intrigue their interest in miniature robotics by showing them the amazing behaviors of tiny robots.

The layout of this book is briefly introduced as follows:

Chapter 1 Introduction to untethered miniature soft robots

This chapter introduces different kinds of miniature soft robots and provides a brief description of their development history and recent progress of miniature soft robots in terms of actuation strategies, materials, fabrication, control, and applications.

Chapter 2 Silicone elastomers-based miniature soft robots

This chapter presents the recent research outcome of silicone elastomers-based untethered miniature robots. Silicone elastomers have good biocompatibility and mechanical properties. They are widely used in forming soft robots, especially magnetic-responsive robots. By using 3D printing techniques, mold-assisted fabrication, or bottom-up assembly strategy, silicone elastomers-based miniature robots with 3D programmable magnetization profiles have been developed and exhibited fast shape-morphing and multimode locomotion capabilities.

Chapter 3 Carbon-based miniature soft robots with rolled-up concept

This chapter describes botanical-inspired strategies for constructing multi-stimuli-responsive robots. Some plants adopt rolled-up strategies (e.g. bending and curling) by sensing surrounding environmental changes. By learning from them, a series of carbon-based robots are fabricated to display dexterous structural changes once activated by external stimuli. Among these robots, graphene oxide (GO)-based films have been widely adopted. The programmable deformation of GO film is achieved by precisely patterned oriented wax or polymers on the GO film. The fabricated GO-based robots exhibit responses to light, humidity, and temperature, and are applied as shape-adaptation grippers and autonomous crawling robots.

Chapter 4 Hydrogels-based miniature soft robots

This chapter presents 3D-printed environmentally responsive hydrogels for constructing programmable miniature robots. Stimuli-responsive hydrogels feature broadly tunable chemical and mechanical properties and have shape-morphing capability in response to diverse environmental stimuli such as pH, temperature, and ion concentration. By using direct laser writing or digital light processing printing strategies, the hydrogel-based robots exhibit revisable and programmable 2D-to-3D or 3D-to-3D shape transformation by applying external stimuli.

Chapter 5 Liquid crystal network and elastomer-based miniature soft robots

This chapter discusses the application of liquid crystal networks and liquid crystal elastomers (LCNs and LCEs) to fabricate monolithic or bimorph material based robots with programmable shape-morphing and self-adaptation capability. By embedding magnetic particles into LCN or LCE films or integrating LCN or LCE and magnetic-responsive elastomers, the fabricated materials could integrate magnetic responsiveness with environmental awareness by responding to environmental cues like light and temperature, while simultaneously being remotely steered by the external magnetic field.

Chapter 6 Flexible ferrofluid as soft robotic agents

This chapter reports ferrofluid droplets used to form miniature soft robots. By the integration of elastomeric structures with ferrofluids, such as selective modification of elastomeric surfaces with laser scanning method or the injection of droplets into polymer structure with internal channels, the fabricated robots could exhibit diverse complex and controllable deformation and are applied as soft robots with multimode locomotion and multi-functionality.

Chapter 7 Conclusions and future prospects

This chapter summarizes the recent key progress made by researchers in the investigation of miniature soft robots and discusses the prospects and challenges of intelligent and autonomous soft robots. Section 7.1 introduces other functional materials including shape memory materials and bio-hybrid materials and discusses the comparison between these functional materials regarding their properties and applications. Section 7.2 introduces the integration of different types of functional materials. Multi-material integration strategies could build miniature soft robots with different functional modules, enabling them to respond to different external stimuli or execute multiple tasks. Sectionr 7.3 discusses multifunctional integration strategies used to fabricate soft robots with powerful perception capabilities that enable the feedback and adaptive control of the miniature soft robots. Section 7.4 discusses the development and challenges of miniature soft robots with intelligence and autonomy. The development of functional materials continuously improves the performance of miniature soft robots. The envisioned positive influences of these intelligent miniature robots in different aspects of our real world and perspectives for future miniature soft robots will be discussed.

## Abbreviations

| | |
|---|---|
| CAL | computed axial lithography |
| CLIP | continuous liquid interface production |
| DIW | direct ink writing |
| DLP | digital light processing |
| DMD | digital micromirror device |
| FDM | fused deposition modeling |
| FFF | fused filament fabrication |
| GelMA | gelatin methacryloyl |
| LCE | liquid crystal elastomer |
| NFC | near-field communication |
| PNIPAAm | poly(*N*-isopropylacrylamide) |
| SLA | stereolithography |
| SLS | selective laser sintering |
| SMP | shape memory polymer |
| TPP | two-photon polymerization |

## References

1 Li, M., Pal, A., Aghakhani, A. et al. (2022). Soft actuators for real-world applications. *Nature Reviews Materials* 7: 235–249.

2 Wallin, T., Pikul, J., and Shepherd, R.F. (2018). 3D printing of soft robotic systems. *Nature Reviews Materials* 3: 84–100.

3 Cianchetti, M., Laschi, C., Menciassi, A., and Dario, P. (2018). Biomedical applications of soft robotics. *Nature Reviews Materials* 3: 143–153.

4 Sitti, M. (2018). Miniature soft robots—road to the clinic. *Nature Reviews Materials* 3: 74–75.

5 Kim, Y., Parada, G.A., Liu, S., and Zhao, X. (2019). Ferromagnetic soft continuum robots. *Science Robotics* 4: eaax7329.

6 Ren, Z., Zhang, R., Soon, R.H. et al. (2021). Soft-bodied adaptive multimodal locomotion strategies in fluid-filled confined spaces. *Science Advances* 7: eabh2022.

7 Zhang, Z., Wang, L., Chan, T.K. et al. (2022). Micro-/Nanorobots in antimicrobial applications: recent progress, challenges, and opportunities. *Advanced Healthcare Materials* 11: 2101991.

8 Wang, Q., Chan, K.F., Schweizer, K. et al. (2021). Ultrasound Doppler-guided real-time navigation of a magnetic microswarm for active endovascular delivery. *Science Advances* 7: eabe5914.

9 Palagi, S., Mark, A.G., Reigh, S.Y. et al. (2016). Structured light enables biomimetic swimming and versatile locomotion of photoresponsive soft microrobots. *Nature Materials* 15: 647–653.

10 Xia, N., Jin, B., Jin, D. et al. (2022). Decoupling and reprogramming the wiggling motion of midge larvae using a soft robotic platform. *Advanced Materials* 34: 2109126.

11 Dillinger, C., Nama, N., and Ahmed, D. (2021). Ultrasound-activated ciliary bands for microrobotic systems inspired by starfish. *Nature Communications* 12: 6455.

12 Dong, X., Lum, G.Z., Hu, W. et al. (2020). Bioinspired cilia arrays with programmable nonreciprocal motion and metachronal coordination. *Science Advances* 6: eabc9323.

13 Ji, F., Wu, Y., Pumera, M., and Zhang, L. (2023). Collective behaviors of active matter learning from natural taxes across scales. *Advanced Materials* 35: 2203959.

14 Wang, Q. and Zhang, L. (2021). External power-driven microrobotic swarm: from fundamental understanding to imaging-guided delivery. *ACS Nano* 15: 149–174.

15 Cheng, Y., Chan, K.H., Wang, X.Q. et al. (2021). A fast autonomous healing magnetic elastomer for instantly recoverable, modularly programmable, and thermorecyclable soft robots. *Advanced Functional Materials* 31: 2101825.

16 Kuang, X., Wu, S., Ze, Q. et al. (2021). Magnetic dynamic polymers for modular assembling and reconfigurable morphing architectures. *Advanced Materials* 33: 2102113.

17 Zhu, G., Hou, Y., Xia, N. et al. (2023). Fully recyclable, healable, soft, and stretchable dynamic polymers for magnetic soft robots. *Advanced Functional Materials* 2300888. https://doi.org/10.1002/adfm.202300888.

18 Heiden, A., Preninger, D., Lehner, L. et al. (2022). 3D printing of resilient biogels for omnidirectional and exteroceptive soft actuators. *Science Robotics* 7: eabk2119.

19 Llacer-Wintle, J., Rivas-Dapena, A., Chen, X.Z. et al. (2021). Biodegradable small-scale swimmers for biomedical applications. *Advanced Materials* 33: 2102049.

20 Wang, B., Zhang, B., Tan, Y. et al. (2022). Leech-inspired shape-encodable liquid metal robots for reconfigurable circuit welding and transient electronics. *Advanced Intelligent Systems* 4: 2200080.

21 Xin, C., Jin, D., Hu, Y. et al. (2021). Environmentally adaptive shape-morphing microrobots for localized cancer cell treatment. *ACS Nano* 15: 18048–18059.

22 Jeon, S., Kim, S., Ha, S. et al. (2019). Magnetically actuated microrobots as a platform for stem cell transplantation. *Science Robotics* 4: eaav4317.

23 Wang, X., Qin, X.H., Hu, C. et al. (2018). 3D printed enzymatically biodegradable soft helical microswimmers. *Advanced Functional Materials* 28: 1804107.

24 Zhang, J., Ren, Z., Hu, W. et al. (2021). Voxelated three-dimensional miniature magnetic soft machines via multimaterial heterogeneous assembly. *Science Robotics* 6: eabf0112.

25 Dong, Y., Wang, L., Xia, N. et al. (2022). Untethered small-scale magnetic soft robot with programmable magnetization and integrated multifunctional modules. *Science Advances* 8: eabn8932.

26 Ji, F., Li, T., Yu, S. et al. (2021). Propulsion gait analysis and fluidic trapping of swinging flexible nanomotors. *ACS Nano* 15: 5118–5128.

27 Kim, Y., Yuk, H., Zhao, R. et al. (2018). Printing ferromagnetic domains for untethered fast-transforming soft materials. *Nature* 558: 274–279.

28 Zhang, X., Chen, G., Fu, X. et al. (2021). Magneto-responsive microneedle robots for intestinal macromolecule delivery. *Advanced Materials* 33: 2104932.

29 Qiu, T., Palagi, S., Mark, A.G. et al. (2016). Wireless actuation with functional acoustic surfaces. *Applied Physics Letters* 109: 191602.

30 Deng, Z., Zhang, H., Priimagi, A., and Zeng, H. (2022). Light-fueled non-reciprocal self-oscillators for fluidic transportation and coupling. *Advanced Materials* 2209683. https://doi.org/10.1002/adma.202209683.

31 Wu, H., Chen, Y., Xu, W. et al. (2023). High-performance Marangoni hydrogel rotors with asymmetric porosity and drag reduction profile. *Nature Communications* 14: 20.

32 Wang, X.-Q., Tan, C.F., Chan, K.H. et al. (2018). In-built thermo-mechanical cooperative feedback mechanism for self-propelled multimodal locomotion and electricity generation. *Nature Communications* 9: 1–10.

33 Dong, Y., Wang, L., Xia, N. et al. (2021). Multi-stimuli-response programmable soft actuators with site-specific and anisotropic deformation behavior. *Nano Energy* 88: 106254.

34 Hu, Y., Wang, Z., Jin, D. et al. (2020). Botanical-inspired 4D printing of hydrogel at the microscale. *Advanced Functional Materials* 30: 1907377.

35 Wang, B., Chan, K.F., Yuan, K. et al. (2021). Endoscopy-assisted magnetic navigation of biohybrid soft microrobots with rapid endoluminal delivery and imaging. *Science Robotics* 6: eabd2813.

36 Ji, F., Liu, W.S., Yang, L. et al. (2021). On-demand assembly and disassembly of a 3D swimming magnetic mini-propeller with two modules. *IEEE Robotics and Automation Letters* 6: 6008–6015.

37 Yang, Z., Yang, L., and Zhang, L. (2020). 3-D visual servoing of magnetic miniature swimmers using parallel mobile coils. *IEEE Transactions on Medical Robotics and Bionics* 2: 608–618.

38 Yang, Z. and Zhang, L. (2020). Magnetic actuation systems for miniature robots: a review. *Advanced Intelligent Systems* 2: 2000082.

39 Ren, Z., Hu, W., Dong, X., and Sitti, M. (2019). Multi-functional soft-bodied jellyfish-like swimming. *Nature Communications* 10: 2703.

**40** Wang, T., Ren, Z., Hu, W. et al. (2021). Effect of body stiffness distribution on larval fish-like efficient undulatory swimming. *Science Advances* 7: eabf7364.

**41** Qiu, T., Lee, T.C., Mark, A.G. et al. (2014). Swimming by reciprocal motion at low Reynolds number. *Nature Communications* 5: 5119.

**42** Lu, H., Zhang, M., Yang, Y. et al. (2018). A bioinspired multilegged soft millirobot that functions in both dry and wet conditions. *Nature Communications* 9: 3944.

**43** Deng, H., Sattari, K., Xie, Y. et al. (2020). Laser reprogramming magnetic anisotropy in soft composites for reconfigurable 3D shaping. *Nature Communications* 11: 6325.

**44** Hu, W., Lum, G.Z., Mastrangeli, M., and Sitti, M. (2018). Small-scale soft-bodied robot with multimodal locomotion. *Nature* 554: 81–85.

**45** Xu, T., Zhang, J., Salehizadeh, M. et al. (2019). Millimeter-scale flexible robots with programmable three-dimensional magnetization and motions. *Science Robotics* 4: eaav4494.

**46** Cui, J., Huang, T.Y., Luo, Z. et al. (2019). Nanomagnetic encoding of shape-morphing micromachines. *Nature* 575: 164–168.

**47** Alapan, Y., Karacakol, A.C., Guzelhan, S.N. et al. (2020). Reprogrammable shape morphing of magnetic soft machines. *Science Advances* 6: eabc6414.

**48** Ji, F., Jin, D., Wang, B., and Zhang, L. (2020). Light-driven hovering of a magnetic microswarm in fluid. *ACS Nano* 14: 6990–6998.

**49** Li, S., Lerch, M.M., Waters, J.T. et al. (2022). Self-regulated non-reciprocal motions in single-material microstructures. *Nature* 605: 76–83.

**50** Li, C., Xue, Y., Han, M. et al. (2021). Synergistic photoactuation of bilayered spiropyran hydrogels for predictable origami-like shape change. *Matter* 4: 1377–1390.

**51** Lahikainen, M., Zeng, H., and Priimagi, A. (2018). Reconfigurable photoactuator through synergistic use of photochemical and photothermal effects. *Nature Communications* 9: 4148.

**52** Ahmed, D., Baasch, T., Blondel, N. et al. (2017). Neutrophil-inspired propulsion in a combined acoustic and magnetic field. *Nature Communications* 8: 770.

**53** Ren, L., Nama, N., McNeill, J.M. et al. (2019). 3D steerable, acoustically powered microswimmers for single-particle manipulation. *Science Advances* 5: eaax3084.

**54** Chen, X.Z., Jang, B., Ahmed, D. et al. (2018). Small-scale machines driven by external power sources. *Advanced Materials* 30: 1705061.

**55** Aghakhani, A., Pena-Francesch, A., Bozuyuk, U. et al. (2022). High shear rate propulsion of acoustic microrobots in complex biological fluids. *Science Advances* 8: eabm5126.

**56** Melde, K., Mark, A.G., Qiu, T., and Fischer, P. (2016). Holograms for acoustics. *Nature* 537: 518–522.

**57** Yang, S., Tian, Z., Wang, Z. et al. (2022). Harmonic acoustics for dynamic and selective particle manipulation. *Nature Materials* 21: 540–546.

**58** Zhao, Y., Lo, C.Y., Ruan, L. et al. (2021). Somatosensory actuator based on stretchable conductive photothermally responsive hydrogel. *Science Robotics* 6: eabd5483.

**59** Ze, Q., Kuang, X., Wu, S. et al. (2020). Magnetic shape memory polymers with integrated multifunctional shape manipulation. *Advanced Materials* 32: 1906657.

60 Zhang, J., Guo, Y., Hu, W. et al. (2021). Liquid crystal elastomer-based magnetic composite films for reconfigurable shape-morphing soft miniature machines. *Advanced Materials* 33: 2006191.

61 Bandari, V.K., Nan, Y., Karnaushenko, D. et al. (2020). A flexible microsystem capable of controlled motion and actuation by wireless power transfer. *Nature Electronics* 3: 172–180.

62 Maeda, S., Hara, Y., Sakai, T. et al. (2007). Self-walking gel. *Advanced Materials* 19: 3480–3484.

63 Sun, L., Yu, Y., Chen, Z. et al. (2020). Biohybrid robotics with living cell actuation. *Chemical Society Reviews* 49: 4043–4069.

64 Striggow, F., Medina-Sánchez, M., Auernhammer, G.K. et al. (2020). Sperm-driven micromotors moving in oviduct fluid and viscoelastic media. *Small* 16: 2000213.

65 Fu, F., Shang, L., Chen, Z. et al. (2018). Bioinspired living structural color hydrogels. *Science Robotics* 3: eaar8580.

66 Magdanz, V., Khalil, I.S., Simmchen, J. et al. (2020). IRONSperm: sperm-templated soft magnetic microrobots. *Science Advances* 6: eaba5855.

67 Gu, H., Boehler, Q., Cui, H. et al. (2020). Magnetic cilia carpets with programmable metachronal waves. *Nature Communications* 11: 2637.

68 Lum, G.Z., Ye, Z., Dong, X. et al. (2016). Shape-programmable magnetic soft matter. *Proceedings of the National Academy of Sciences* 113: E6007–E6015.

69 Li, R., Jin, D., Pan, D. et al. (2020). Stimuli-responsive actuator fabricated by dynamic asymmetric femtosecond bessel beam for in situ particle and cell manipulation. *ACS Nano* 14: 5233–5242.

70 Kim, B.H., Li, K., Kim, J.T. et al. (2021). Three-dimensional electronic microfliers inspired by wind-dispersed seeds. *Nature* 597: 503–510.

71 Miskin, M.Z., Cortese, A.J., Dorsey, K. et al. (2020). Electronically integrated, mass-manufactured, microscopic robots. *Nature* 584: 557–561.

72 Lee, H., Jang, Y., Choe, J.K. et al. (2020). 3D-printed programmable tensegrity for soft robotics. *Science Robotics* 5: eaay9024.

73 Jackson, J.A., Messner, M.C., Dudukovic, N.A. et al. (2018). Field responsive mechanical metamaterials. *Science Advances* 4: eaau6419.

74 Kim, Y. and Zhao, X. (2022). Magnetic soft materials and robots. *Chemical Reviews* 122: 5317–5364.

75 Skylar-Scott, M.A., Mueller, J., Visser, C.W., and Lewis, J.A. (2019). Voxelated soft matter via multimaterial multinozzle 3D printing. *Nature* 575: 330–335.

76 Sydney Gladman, A., Matsumoto, E.A., Nuzzo, R.G. et al. (2016). Biomimetic 4D printing. *Nature Materials* 15: 413–418.

77 Cheng, Y., Chan, K.H., Wang, X.Q. et al. (2019). Direct-ink-write 3D printing of hydrogels into biomimetic soft robots. *ACS Nano* 13: 13176–13184.

78 Han, D., Farino, C., Yang, C. et al. (2018). Soft robotic manipulation and locomotion with a 3D printed electroactive hydrogel. *ACS Applied Materials & Interfaces* 10: 17512–17518.

79 Zhang, Y.F., Ng, C.J.X., Chen, Z. et al. (2019). Miniature pneumatic actuators for soft robots by high-resolution multimaterial 3D printing. *Advanced Materials Technologies* 4: 1900427.

**80** Zhang, B., Li, H., Cheng, J. et al. (2021). Mechanically robust and UV-curable shape-memory polymers for digital light processing based 4D printing. *Advanced Materials* 33: 2101298.

**81** Saha, S.K., Wang, D., Nguyen, V.H. et al. (2019). Scalable submicrometer additive manufacturing. *Science* 366: 105–109.

**82** Xu, S., Yan, Z., Jang, K.I. et al. (2015). Assembly of micro/nanomaterials into complex, three-dimensional architectures by compressive buckling. *Science* 347: 154–159.

**83** Zhang, Y., Zhang, F., Yan, Z. et al. (2017). Printing, folding and assembly methods for forming 3D mesostructures in advanced materials. *Nature Reviews Materials* 2: 1–17.

**84** Miao, L., Song, Y., Ren, Z. et al. (2021). 3D temporary-magnetized soft robotic structures for enhanced energy harvesting. *Advanced Materials* 33: 2102691.

**85** Piskarev, Y., Shintake, J., Chautems, C. et al. (2022). A variable stiffness magnetic catheter made of a conductive phase-change polymer for minimally invasive surgery. *Advanced Functional Materials* 32: 2107662.

**86** Zhang, T., Yang, L., Yang, X. et al. (2021). Millimeter-scale soft continuum robots for large-angle and high-precision manipulation by hybrid actuation. *Advanced Intelligent Systems* 3: 2000189.

**87** Wang, L., Zheng, D., Harker, P. et al. (2021). Evolutionary design of magnetic soft continuum robots. *Proceedings of the National Academy of Sciences* 118: e2021922118.

**88** Gopesh, T., Wen, J.H., Santiago-Dieppa, D. et al. (2021). Soft robotic steerable microcatheter for the endovascular treatment of cerebral disorders. *Science Robotics* 6: eabf0601.

**89** Han, M., Chen, L., Aras, K. et al. (2020). Catheter-integrated soft multilayer electronic arrays for multiplexed sensing and actuation during cardiac surgery. *Nature Biomedical Engineering* 4: 997–1009.

**90** Rivkin, B., Becker, C., Singh, B. et al. (2021). Electronically integrated microcatheters based on self-assembling polymer films. *Science Advances* 7: eabl5408.

**91** Ma, Z.-C., Zhang, Y.L., Han, B. et al. (2020). Femtosecond laser programmed artificial musculoskeletal systems. *Nature Communications* 11: 4536.

**92** Zhang, S., Ke, X., Jiang, Q. et al. (2021). Programmable and reprocessable multifunctional elastomeric sheets for soft origami robots. *Science Robotics* 6: eabd6107.

**93** Truby, R.L., Wehner, M., Grosskopf, A.K. et al. (2018). Soft somatosensitive actuators via embedded 3D printing. *Advanced Materials* 30: 1706383.

**94** Novelino, L.S., Ze, Q., Wu, S. et al. (2020). Untethered control of functional origami microrobots with distributed actuation. *Proceedings of the National Academy of Sciences* 117: 24096–24101.

**95** Bai, K., Cheng, X., Xue, Z. et al. (2020). Geometrically reconfigurable 3D mesostructures and electromagnetic devices through a rational bottom-up design strategy. *Science Advances* 6: eabb7417.

# 2

# Silicone Elastomers-Based Miniature Soft Robots

## 2.1 Introduction

Silicone elastomers typically consist of a catalyst, a crosslinker, reinforcing fillers, and a silicone polymer. The crosslinking of silicone elastomers is often performed by radical reaction, hydrosilylation, and condensation (Figure 2.1a) [1]. According to the required curing temperature, silicone elastomers can be divided into high-temperature vulcanizing (HTV) and room-temperature vulcanizing (RTV). Silicone elastomers are non-flammable, electrically insulating, as well as biocompatible materials. Furthermore, they could maintain electrical and mechanical properties among a wide range of temperatures (−50~300 °C) [6–8]. Due to these unique properties, silicone elastomers are valuable for broad application prospects, e.g. medical implants, minimally invasive surgery, sealants, and stretchable electronics.

One of the most common fabrication strategies of silicone elastomer is molding or casting method, which is suitable for elastomers with relatively simple geometries [9, 10]. By pouring uncured elastomer polymer into a mold and providing the required temperature environment, the elastomer structure can be obtained. It is highly desirable to fabricate complex silicone elastomers structures, e.g. medical implants for healthcare applications, via additive manufacturing methods. Recently, various additive manufacturing techniques have been developed for silicone elastomers, including direct ink writing (DIW) [2, 3, 11–17], direct laser writing [18, 19], vat photopolymerization and inkjet printing [20–30], which could provide better freedom in design space and the capability of high throughput.

To actively control silicone elastomer-based structures, a variety of strategies have been developed, including pneumatic/fluidic actuation, magnetic actuation, chemical actuation, thermal and electrical actuation, as shown in Figure 2.1 [31–39]. A soft millipede walker composed of soft and rigid elastomers was fabricated by a multimaterial multinozzle 3D printing (Figure 2.1b) [2]. Pneumatic actuators were embedded in the 3D printed walker with complex channel networks. By applying a vacuum to these channels, a vertical and lateral displacement of the robotic leg will be induced by the buckling of the soft silicone “muscle.” Magnetic components (e.g. hard magnetic particles) are often added into the

*Untethered Miniature Soft Robots: Materials, Fabrications, and Applications*, First Edition.
Li Zhang, Jiachen Zhang, Neng Xia, and Yue Dong.

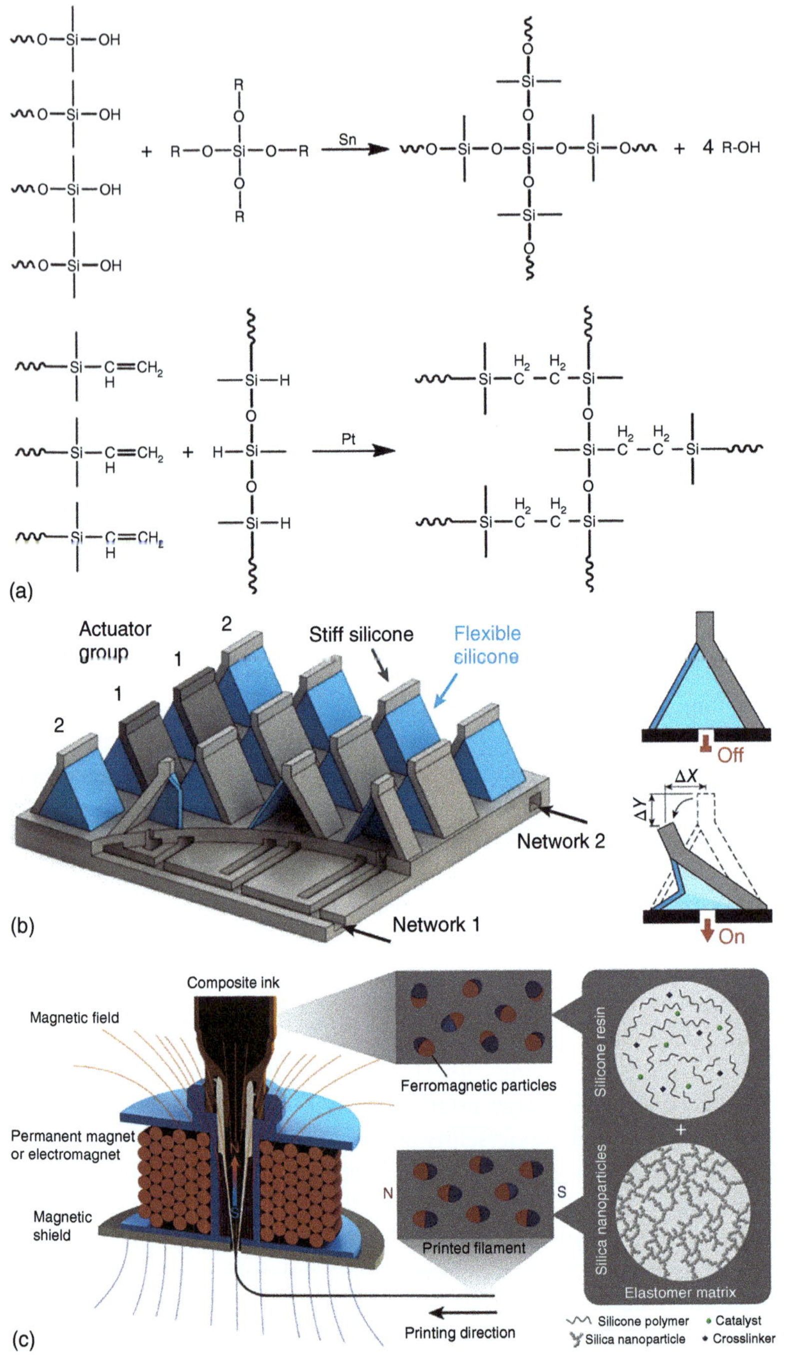


**Figure 2.1** Crosslinking and actuation of silicone elastomer-based soft structures. (a) Pt-catalyzed and Sn-catalyzed reactions for the crosslinking of silicone elastomer. Source: Mazurek et al. [1]/Royal Society of chemistry/CC BY 3.0/Public domain. (b) Pneumatic actuation. Source: Skylar-Scott et al. [2]/Springer Nature. (c) Magnetic actuation. Source: Kim et al. [3]/Springer Nature. (d) Thermal actuation. Source: Zhou et al. [4]/Reproduced from American Chemical Society. (e) Chemical actuation. Source: Kang et al. [5]/Reproduced from John Wiley & Sons, Inc.

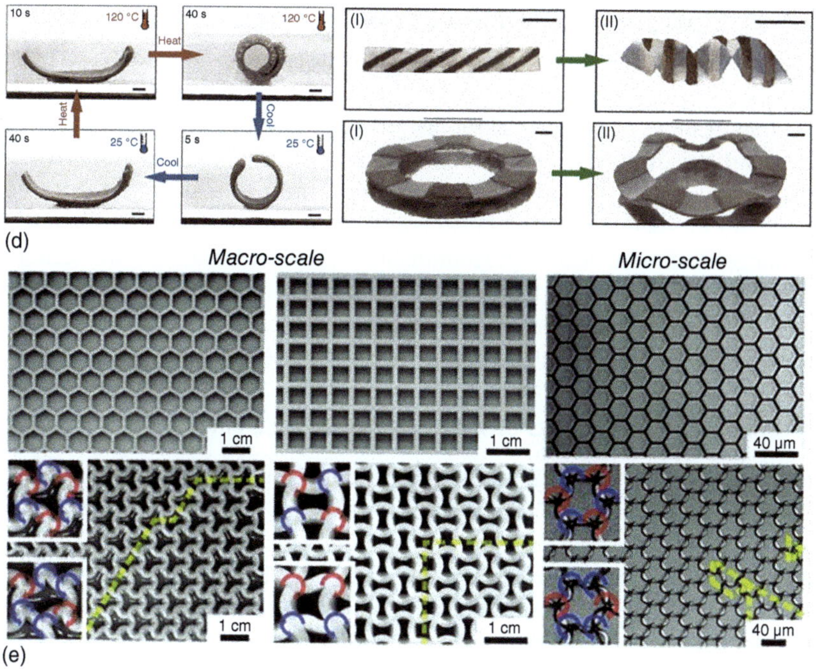


**Figure 2.1** (*Continued*)

matrix of silicone elastomers (e.g. Ecoflex [Young's modulus: 0.01~0.1 MPa], poly(dimethylsiloxane) (PDMS) [Young's modulus: 1.2~4.2 MPa]) to endow the soft structure with magnetic-responsive capability [40, 41]. Different types of deformation, e.g. bending, twisting, elongation, and compression, have been presented by the activation of embedded magnetic components in the silicone elastomer. Moreover, with advanced magnetization programming techniques, more complex shape-morphing behaviors can be achieved [3, 10, 40, 42–45]. For example, Kim et al. utilized an external magnetic field to reorient magnetized magnetic particle inside composite ink during the direct ink printing process [3]. Such strategy facilitates the formation of various fast-response and reconfigurable soft robots. The printing of soft materials with heterogeneous responsive swellable properties has been widely adopted for building shape-morphing architectures [46]. To endow silicone elastomer with thermal-responsive swellable property, Zhou et al. added ethanol (low boiling point liquid) and liquid metal to the elastomer matrix (Figure 2.1d) [4]. The liquid metal could improve thermal conductivity of the composite and the inflation of silicone matrix can be induced by the vaporization and expansion of ethanol after heating. The printability of the elastomer ink can be improved by mixing a rheology modifier (e.g. fumed silica). A multimaterial printing technique with thermal-responsive elastomer ink and non-responsive silicone ink was developed to fabricate diverse bilayer structures, which could exhibit complex shape-morphing behaviors once heating.

## 2.2 Soft Elastomer-Based Robots with Programmable Magnetization Profiles

The programmability of magneto-elastomers is realized by patterning the doped magnetic particles in the polymer matrix, such as the orientation of soft magnetic particles or the direction of magnetic domains of hard magnetic particles. Due to the ability of hard magnetic particles to maintain their magnetization, more complex shape changes and more flexible programmability can be achieved. In this section we discuss four programming methods for elastomers containing hard magnetic particles, including template-assisted magnetization, buckling instability-encoded magnetization, 3D printing, and bottom-up heterogeneous assembly.

### 2.2.1 Untethered Soft Robots Based on Template-Aided Magnetizing

Template-assisted magnetization is a widely used approach to construct magneto-elastomers with 3D continuous magnetization. As shown in Figure 2.2, a magnetic soft robot developed by Hu et al. was composed of soft active materials (Ecoflex 00-10) and NdFeB magnetic particles mixed in a mass ratio of 1 : 1 [10]. The doped magnetic microparticles (MMPs) will be demagnetized under the application of an external magnetic field of 600 mT.

A film structure with the thickness of 185 μm was obtained by pouring the magnetic slurry to a flat plate made by poly(methyl methacrylate) followed by thermal curing. Laser cutting was performed to obtain the desired robot shape (the height, width, length of the robot are 0.185, 1.5, 3.7 mm, respectively) before removing it from the poly(methyl methacrylate) plate. Thereafter, the magneto-elastomer based robot was wrapped around a rod-shaped structure (circumference: 3.7 mm) and experienced a strong magnetic field (1.65 T) to form a harmonic magnetization distribution. The angle $\beta R$ shown in Figure 2.2 could be applied to tune the phase shift of the harmonic magnetization distribution. The properties of the fabricated robot including Young's modulus and remanent magnetization were measured as $8.45 \times 10^4 \pm 2.5 \times 10^3$ Pa and $62\,000 \pm 10\,000\ \text{A m}^{-1}$, respectively. The experimental results show that increasing the content of magnetic particles has a more obvious effect on the remanent magnetization than on Young's modulus, so a stronger magnetic torque will be generated, and the robot will be more easily deformed. But the presence of a large amount of magnetic particles would hinder the curing of the silicon elastomer; thus, a 1 : 1 mass ratio was adopted in this work to fabricate the robot.

For a robot with the above-encoded magnetization profile, it will exhibit various deformation and locomotion modes under the control of an external magnetic field, as shown in Figure 2.2. When the magnetic field strength is small, the robot generated a single-wavelength shape deformation (e.g. cosine and sine shape). When a large magnetic field strength is applied, the robot will produce large deformations such as letter "C" or "V" shape. If the direction of the external magnetic field strength is inconsistent with the direction of the main axis in Figure 2.2, a large net magnetic moment will be formed, causing the robot to rotate around the $z$-axis, and then the angular displacement of the robot can be adjusted to enable it to achieve various locomotion modes such as roll and jump.

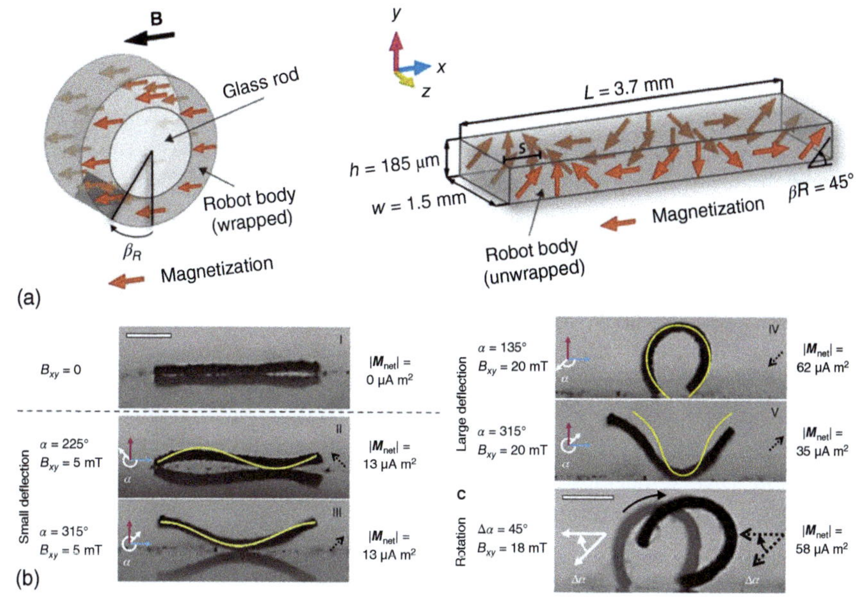


**Figure 2.2** Template-assisted magnetization process and magnetic response behavior of elastomers doped with hard magnetic particles. (a) A robot with size $L \times w \times h$ is wrapped on a cylindrical rod to generate deformation, and placed in a 1.65 T magnetic field (generated by a vibrating magnetometer) for magnetization, and the angle $\beta R$ is used to adjust the phase shift in the magnetization distribution. (b) The initial state of the robot and the deformed state under different magnetic field parameters. $\boldsymbol{B} = [B_{xy}\ B_z]$ represents the applied external magnetic field, and the parameter $\alpha$ is used to describe the direction of the magnetic field strength, that is, the angle between the external magnetic field and the $x$-axis along the clockwise direction. Source: Hu et al. [10]/Reproduced from Springer Nature.

### 2.2.2 Small-Scale Soft Machines Based on Buckling Instability-Encoded Magnetization

The active transformation of 3D soft architectures attracts growing attention because of the capability of dynamic interaction with environment and geometry-determined functionality [47–54]. Highly controllable geometrical transformations have been widely adopted in natural systems with enhanced adaptability to physical environments. For instance, the hydration of plants could be regulated by changing surface areas of wrinkled surfaces, the stability and maneuverability of the swimming of glass knifefish could be modulated by its wavy ribbon fin [55, 56]. Inspired by these natural transformation behaviors, active soft devices across different scales have been developed using stimuli-responsive materials and advanced fabrication strategies. The investigation of the active transformation behavior facilitates the design of optimized soft robots and intelligent systems, as well as unveiling related physical mechanisms.

Soft architectures made of magnetoactive materials open possibilities due to the excellent programmability and controllability of magnetic actuation strategy. The fast and reversible transformation of magnetoactive materials can be achieved by programmed magnetic inputs and magnetization profiles. The application of time-varying magnetic field allows the dynamic modulation of magnetoactive materials. Metamaterial or mechanical systems used in preceding studies were often made of magneto-elastomers encoded by uniform magnetization (UM) patterns which may present limited deformation freedom [57, 58]. Different magnetization programming techniques have been developed to explore complex morphological transformation including template-assisted approach and point-by-point programming approach. However, because of the lack of reverse design strategy toward targeted transformations and the difficulty in the fabrication of sophisticated 3D molds for encoding arbitrary magnetization profiles, there are still some problems confronting researchers to achieve complex transformation behaviors (e.g. wrinkling or buckling) of 3D magnetoactive structures.

In this section, morphable architectures made of magneto-elastomers with buckling instability-encoded heterogeneous magnetization patterns are developed, allowing the formation of multimodal anisotropic shape-morphing and dynamic modulation. Diverse buckling transformations of non-freestanding structures are activated by the absorption of organic solvents and geometric constraints. The active transformation process offers a template-free manner for encoding 3D heterogeneous magnetization profiles at different scales.

As shown in Figure 2.3, silicone elastomers with embedded ferromagnetic particles, e.g. NdFeB microparticles, were used to fabricate the 3D morphable structures. Silicone-based materials consisting of Ecoflex 00-30 parts A and B were mixed vigorously and then NdFeB MMPs were added at a mass ratio of 1 : 1. After completely mixing, air bubbles were eliminated by vacuum degassing in a vacuum chamber for 1 min. Diverse organic solvents including $n$-hexane, ethyl acetate, and toluene can be absorbed by the magneto-elastomers leading to the swelling of the structure (Figures 2.4 and 2.5). The volume expansion of the silicone elastomers was caused by the diffusion of these organic solvents into the elastomeric network [59, 60]. In this chapter, we use toluene as the organic solvent for the activation of the stimuli-responsive behavior of the magneto-elastomers because it is relative stable at room temperature and exhibits good absorption rate with a swelling ratio of 1.58. Due to the fast diffusion of toluene solution in ethanol, the magneto-elastomer could recover to its undeformed state by changing the liquid environment to ethanol. Glass substrates that experience hydrophilic treatment was adopted to attach the fabricated elastomeric structures. After submerging the magneto-elastomer in toluene solution, the toluene diffused into the polymer matrix leading to the swelling of the elastomer structure. While the constraint of substrate would affect the swelling behavior of the bottom section of the structure. The generated axial compressive forces facilitate the buckling transformation of the swelled elastomer, as the buckled state shown in Figure 2.3. A variety of factors including different connectivity types, geometrical parameters, the change of organic solvents, and the arrangement of artificial defects can be adopted to control the shape of the buckled state.

Magnetic actuation provides a strategy to dynamically control the shape-morphing of the magnetic elastomers. The magnetic elastomers were magnetized in the buckled state by using a ~2 T pulse magnetic field generated by a magnetizer. The

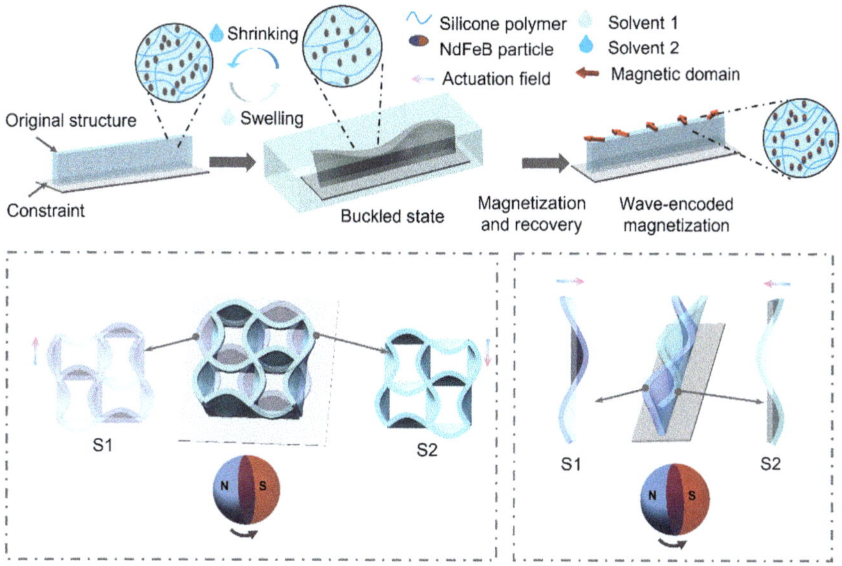


**Figure 2.3** Schematic illustration of the morphological transformation of the buckling instability encoded magneto-elastomers. Reversible deswelling and expansion of the magneto-elastomer could be achieved by the addition of solvents 1 and 2, respectively. The buckling transformation of the elastomeric structure was generated due to the existence of geometric constraints, and provided a template-free strategy for encoding 3D heterogeneous magnetization profile, which allows the formation of dynamic transformation upon magnetic stimulation. Source: Xia et al. [41]/Springer Nature/CC BY 4.0/Public domain.

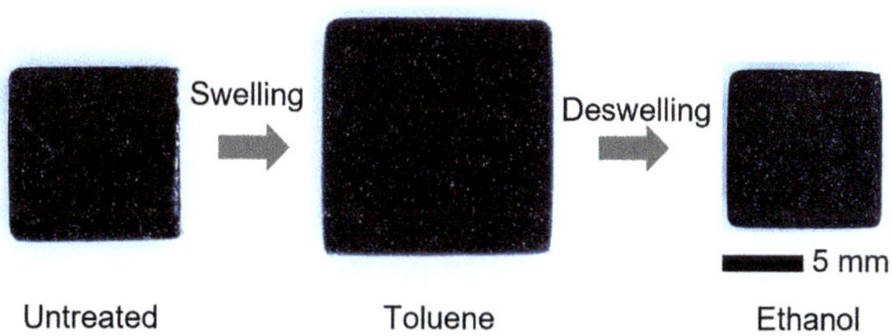


**Figure 2.4** Demonstration of the swelled and recovered magneto-elastomer by adding toluene and ethanol. Scale bar: 5 mm. Source: Xia et al. [41]/Reproduced from Springer Nature/CC BY 4.0.

remanent magnetic moment of the magnetic material was 37.5 emu $g^{-1}$ (Figure 2.6). After the magnetization and recovering to the initial state, an anisotropic magnetization pattern would be formed in the elastomeric structures. As shown in Figure 2.3, we focus on the transformation of non-freestanding strip and cellular structures which consist of multiple strips interlaced together. Upon the stimulation of external magnetic field, magnetic torques caused by the misalignment between the external field direction and anisotropic magnetization profile could deform these structures to various geometric configurations.

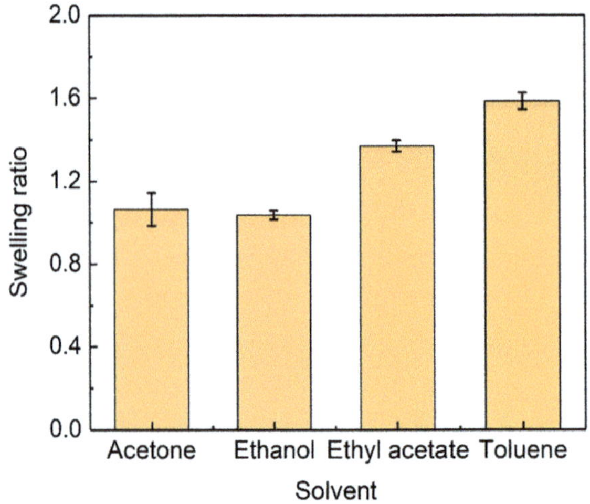


**Figure 2.5** Measurement of the swelling ratios of the material in four kinds of solvents. Source: Xia et al. [41]/Springer Nature/CC BY 4.0/Public domain.

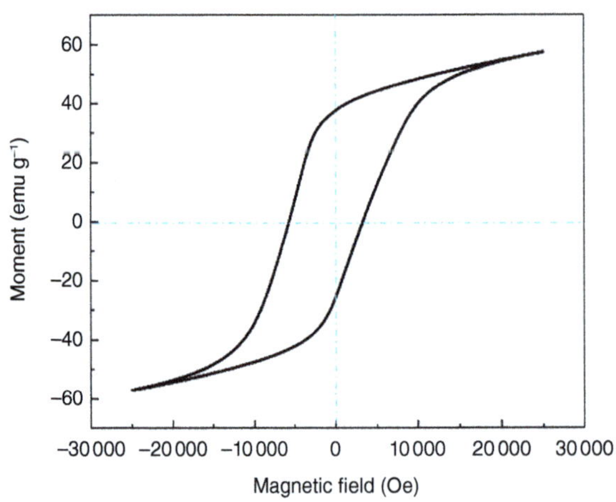


**Figure 2.6** The characterization of the magnetic hysteresis of the magneto-elastomer using a Quantum Design VSM. Source: Xia et al. [41]/Springer Nature/CC BY 4.0/Public domain.

### 2.2.3 Small-Scale Soft Machines Based on 3D Printing of Ferromagnetic Materials

Kim et al. fabricated magneto-elastomers by DIW and demonstrated their programmable and complex deformation capabilities, as shown in Figures 2.1 and 2.7. The ink used for DIW printing is composed of NdFeB magnetic particles, fumed silica particles, silicon-based polymers containing catalysts and crosslinking agents. Fumed silica is used to adjust the rheological properties of the magnetic ink, such as forming shear-thinning and shear-yielding properties, so that it can meet the requirements of extrusion printing, and then can maintain a complete 3D shape after multi-layer stacking. After the particles were thoroughly mixed with the polymer matrix, they were magnetized under a strong pulsed magnetic field of ~2.7 T for subsequent use.

During the DIW printing, the orientation of the magnetic particles was adjusted in real time by applying an external magnetic field, such as using an adjustable

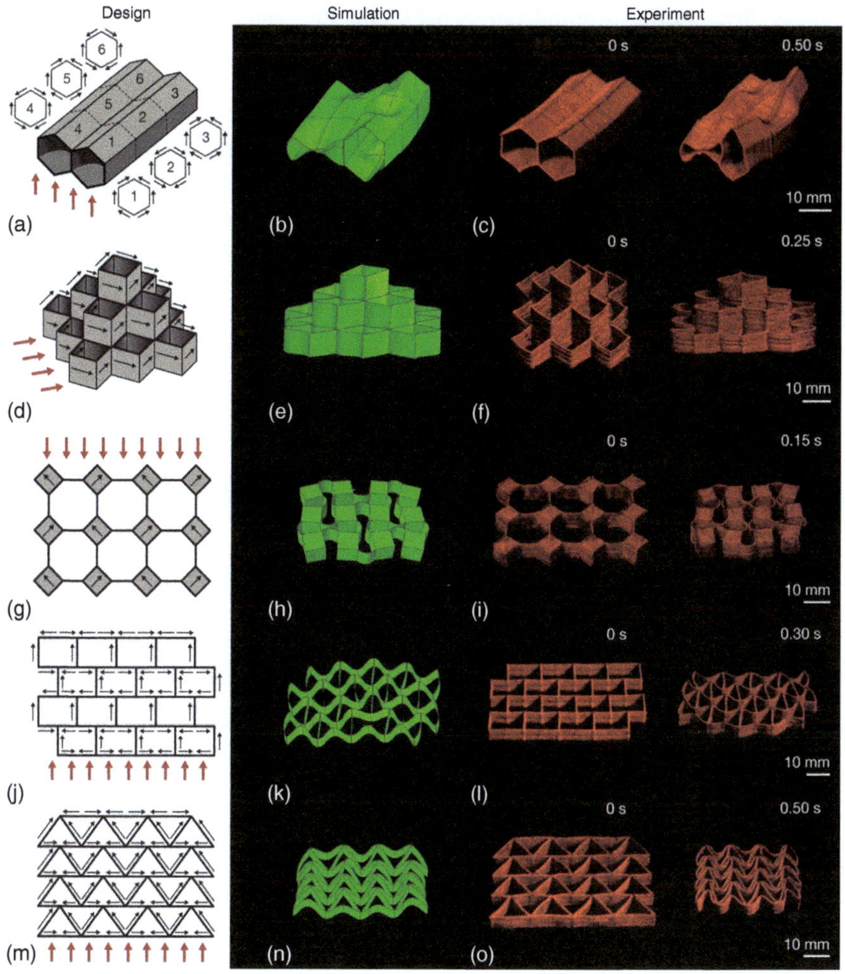


**Figure 2.7** Using DIW to print silicon elastomers containing 20 vol % hard magnetic particles, including the design of its magnetic domain pattern and structure, deformation simulation, and deformed shape under the stimulation of a magnetic field. (a–c) Elastomeric structure with hexagonal tube shape. (d–f) Elastomers exhibiting pyramidal shapes. (g–o) Elastomeric structures with metamaterial properties. Source: Kim et al. [3]/Reproduced from Springer Nature.

electromagnetic coil or permanent magnet near the nozzle. The application of a time-varying magnetic field provides an efficient means to control the magnetic domain orientation of the extruded ink, thereby forming programmable discrete magnetic domain patterns in the 3D structure. In addition, in order to prevent the adverse effect of the applied external magnetic field on the printed structure, a magnetic shielding device was installed at the nozzle to enhance the attenuation of the magnetic field. The printed 3D structure needs to be cured in a 120 °C environment for one hour. The remanent magnetization of printed structures can be affected by the diameter of the nozzle (50–1190 μm), the strength of the external magnetic field during printing (20–50 mT), and the concentration of magnetic particles in the ink (5–20%).

Figure 2.7 demonstrates that various ferromagnetic 3D structures can be fabricated using the proposed DIW printing. In order to ensure the stability of ink filaments, when they are stacked into complex 3D structures, it is necessary to use supporting ink to build temporary supports on both sides of the magnetic ink. The support ink is made from a catalyst-containing silicone rubber mixed with fumed silica nanoparticles, and the support structure can be removed with a solvent after printing. Thin-walled structures with high aspect ratios are shown in Figure 2.7. They can exhibit fast 3D shape transformation actuated by a magnetic field, and mechanical metamaterial properties such as simultaneous shrinkage in both width and length directions under magnetic stimulation.

### 2.2.4 3D Miniature Soft Machines Based on Bottom-up Heterogeneous Assembly

Silicone elastomers are rubbery materials whose Young's modulus are on the order of MPa and deformations are reversible under external loads. Soft robots based on silicon elastomers have been widely adopted due to their excellent toughness and ability to exhibit large deformations. Miniature soft robots actuated by external stimuli have broad application prospect in robotic manipulation and biomedical engineering. The magnetic actuation strategy provides a fast, dexterous, precise, and safe choice for untethered miniature machines [42, 61–65]. Although recent advances in the fabrication of magnetic soft robots (e.g. molding and casting, ultraviolet lithography, extrusion-based 3D printing, chemical self-assembly, and laser cutting), it still remains challenging to integrate multiple material compositions with arbitrary geometries and 3D magnetization patterns [66, 67]. For example, the extrusion-based methods are limited by the extrudate swell of soft materials and the size of magnetic soft fibers, as well as the magnetization programming of ultraviolet lithography methods is restricted by the local interaction of magnetic components inside polymer precursors [3, 42, 68].

The magnetic components and encoded magnetization patterns determine the functionality of a magnetic-responsive structure. However, most reported magnetic devices consisted of only one type of material. Existing fabrication approaches are often limited in geometry design, the programmability of magnetization profile, and the material selection, which prevent the improvement of functionalities and design space of magnetic soft machines.

A 3D heterogeneous assembly approach assisted by jig is proposed to tackle these challenges. With the integration of micrometer-scale building blocks or voxels, miniature magnetic soft machines composed of magnetic and nonmagnetic materials (e.g. PDMS and Ecoflex), with 3D magnetization profiles and complex geometries are fabricated, as shown in Figure 2.8. Thus, the design space and functionality of magnetic soft machine are significantly improved, enabling the development of tailored machines for specific biomedical applications.

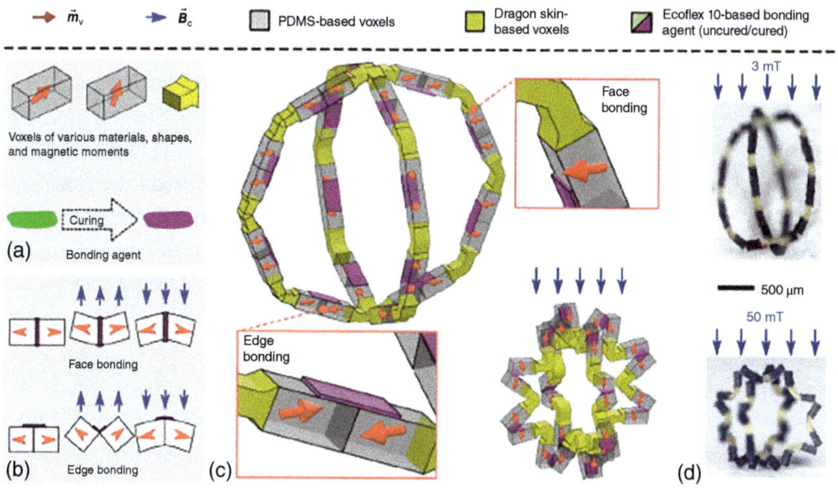


**Figure 2.8** Miniature reconfigurable devices fabricated by bottom-up assembly assisted by jigs. (a) Building blocks with various magnetization domains, geometries, and materials, as well as bonding agents. (b) Building blocks were connected by bonding agent in two manners including edge bonding and face bonding. (c) Schematic illustration of 3D ring structure assembled by different kinds of building blocks. (d) Experimental results of the developed miniature device which exhibited the property of negative Poisson's ratio. Source: Zhang et al. [40]/American Association for the Advancement of Science (AAAS).

#### 2.2.4.1 Voxel Fabrication and Magnetization

Polymer precursors doped with MMPs were poured into designed molds and cured to fabricate voxels with different materials and structures. Firstly, as shown in Figure 2.9, two photon polymerization (TPP) (Photonic Professional GT, Nanoscribe

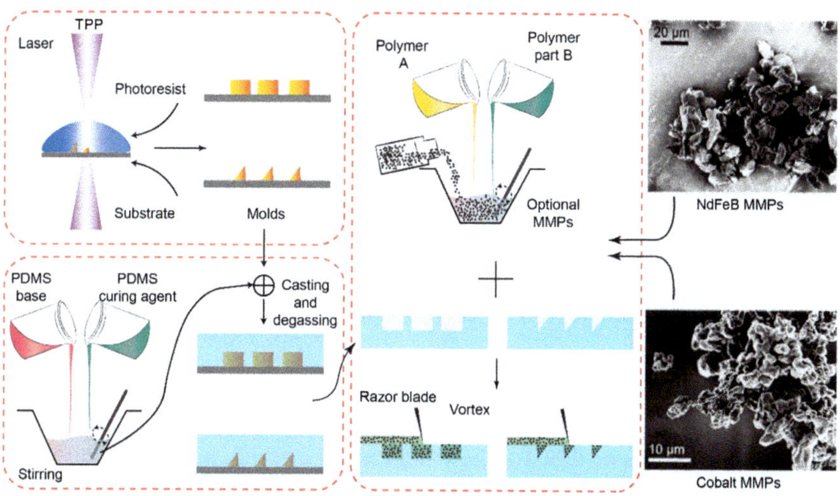


**Figure 2.9** Schematic illustration of the process of voxel fabrication and magnetization. TPP, two-photon polymerization. Source: Zhang et al. [40]/American Association for the Advancement of Science (AAAS).

GmbH) and photoresists including IP-Q and IP-S (Nanoscribe GmbH) were used to prepare positive molds. Due to its high precision shaping capability for 3D structures, TPP has been widely adopted for quick iterations and fast prototyping. Negative molds with concave features were fabricated by adding PDMS with a mass ratio between curing agents and base at 1 : 5 (Dow Corning, Sylgard 184) to the positive molds. Air inside the polymer precursor was cleared by vacuum degassing. To improve the crosslinking of polymer precursor and the release of building block, the silanization of these molds was performed by Trichloro(1H,1H,2H,2H-perfluorooctyl)silane (PFOCTS). Fluidic mixtures consisting of selected base materials mixed with magnetic particles such as NdFeB particles (MQP-15-7, Magnequench) was poured into the negative molds, followed by vacuum degassing to remove trapped air. The mass ratio between magnetic particles and base materials can be tuned for different remnant magnetizations. Razor blades were applied to remove excess mixture. After curing at specific temperature and time, voxels with targeted geometric features can be obtained and taken out by needles or tweezers from the negative molds.

Individual voxels with various geometries are shown in Figure 2.10. The facile fabrication of voxels with diverse geometries is capable of saving cost and time in fabrication, as better geometric profiles are provided for a device and lots of generic voxels can be replaced by a larger non-generic voxel in a soft machine to improve efficiency. For example, a smooth surface of a peristaltic pump is hard to be created by generic cubic voxels, and in the fabrication of a capsule, ring voxels would be a better choice to save time compared with generic cubic voxels. Some of the voxels can also be fabricated by other reported fabricating methods [3, 42, 43], but they are usually compatible with one kind of recipes of material. The bottom assembly approach exhibits a key advantage that is multimaterial integration.

Assistant jigs were used to position magnetic voxels with targeted 3D orientations, which would be magnetized using a magnetizing field ($\boldsymbol{B}_{\mathrm{m}}$) created by a vibrating sample magnetometer (VSM, EZ7, Microsense) for the formation of magnetization profile ($\boldsymbol{m}_{\mathrm{v}}$) with 3D orientation and strength, as shown in Figure 2.11. The strength of the magnetization $\boldsymbol{m}_{\mathrm{v}}$ can be tuned by $\boldsymbol{B}_{\mathrm{m}}$ and the concentration of magnetic particles. $\boldsymbol{B}_{\mathrm{m}}$ was generated between the two parallel circular plates shown in Figure 2.11, with the direction perpendicular to the circular plates. The terminal of the VSM could tune the strength of $\boldsymbol{B}_{\mathrm{m}}$. Jigs were used to control the orientation of the voxels, which were fixed at the center of the VSM. The strength of $\boldsymbol{B}_{\mathrm{m}}$ increases at a speed of $0.2\,\mathrm{T\,s^{-1}}$ and stays at 1.8 T for 2s, after which would decrease to 0.

#### 2.2.4.2 Jig-Assisted Assembly Approach

To ensure the repeatability and fabrication precision, reusable assistant jigs were used as reference and guideline during the assembly process. These jigs were fabricated by mold casting or TPP. For the mold-casting method, negative molds were formed by casting PDMS with the mass ratio between the base and curing agent at 5 : 1 to the TPP printed structures. The out-of-plane features of the 3D jigs could help to support the 3D parts. The designs of jigs derived from trial and error as well as experiences.

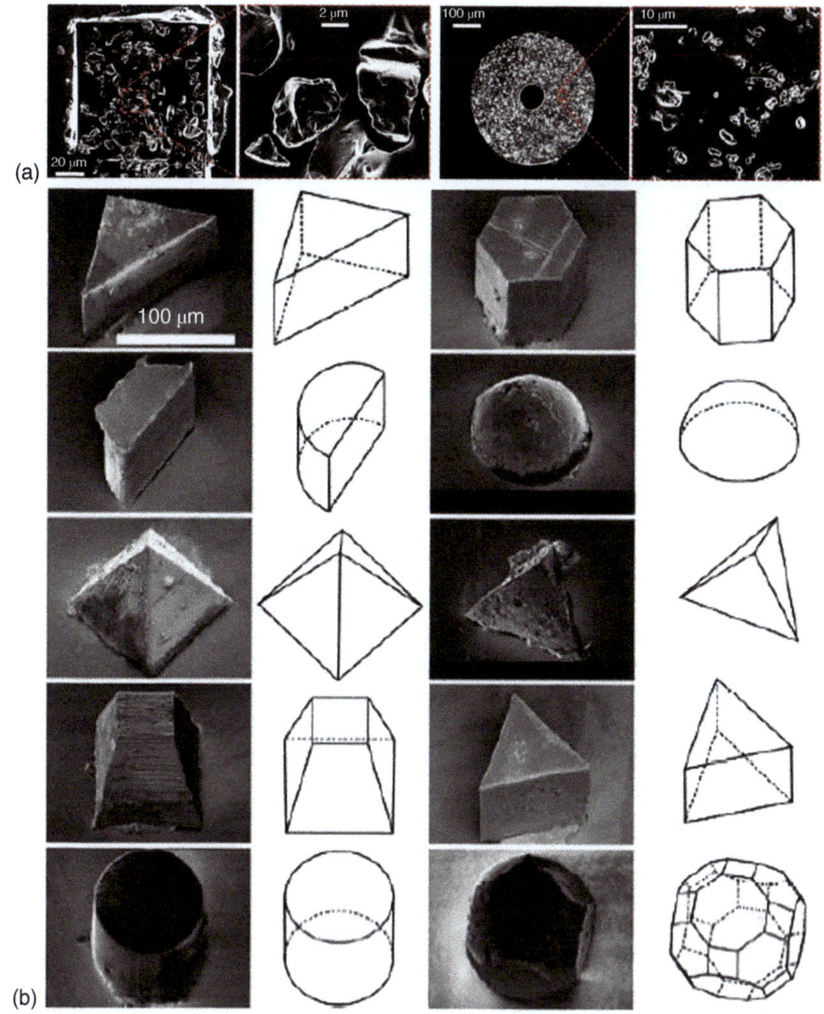


**Figure 2.10** SEM images and design of voxels with various geometries. Source: Zhang et al. [40]/American Association for the Advancement of Science (AAAS).

Bonding agent was applied to connect and assemble voxels together. The utilization of assistant jigs enables the control of the orientation and the position of each voxel, leading to the formation of final machine with targeted 3D geometry and magnetization profile. The mechanical fixing provided by the jigs could help to avoid the magnetic interaction between voxels, which may cause the collapse of the structure.

Face-bonding and edge-bonding strategies shown in Figures 2.12 and 2.13 were used to form 3D soft machines. Uncured liquid soft elastomers were used as bonding agents and cured under the contact of surrounding voxels. Other kinds of glues, e.g. optical glues and universal glues, can also be adopted as alternatives. A stereomicroscope with a magnification varying from ×6.3 to ×50 (ZEISS Stemi 508, Carl Zeiss Microscopy GmbH) was used to assist the assembly process. Rotary

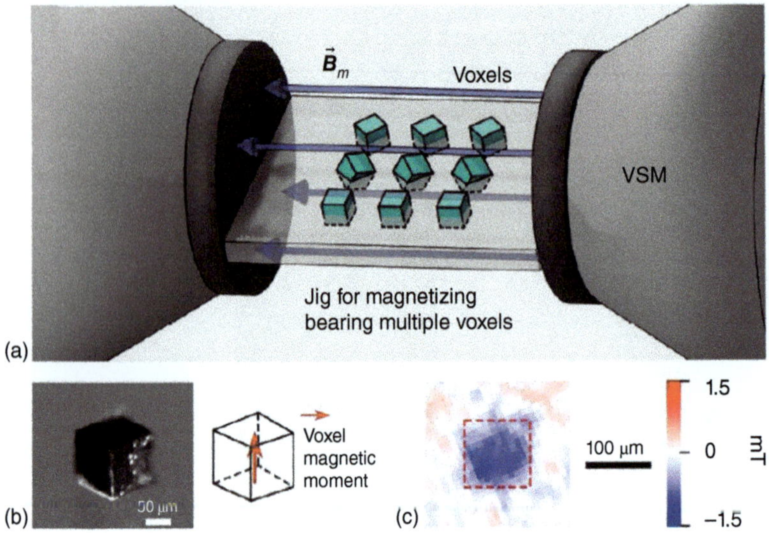


**Figure 2.11** The magnetization of magnetic voxels. (a) A jig inside the VSM was used to control the orientation of multiple voxels during the magnetization process. (b) Demonstration of a cubic voxel with the side length of 100 μm made of PDMS mixed with Cobalt iron oxide. Source: Zhang et al. [40]/American Association for the Advancement of Science (AAAS).

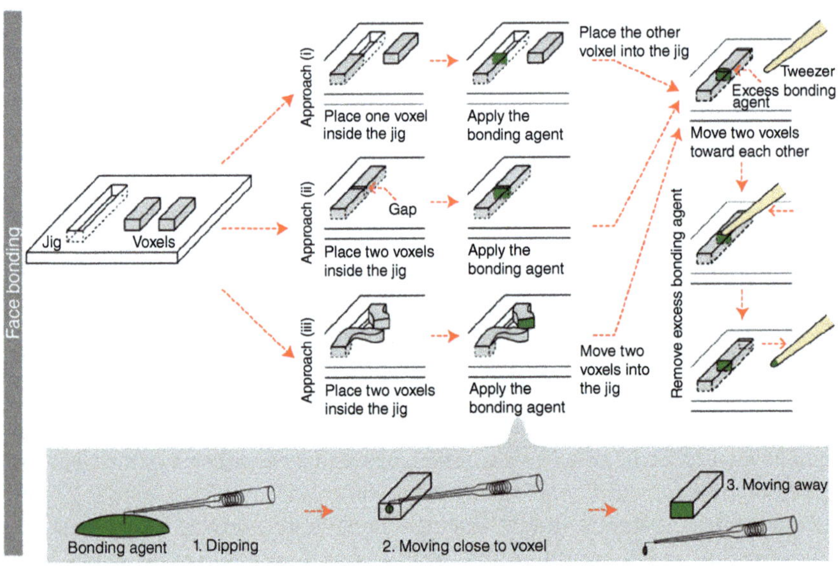


**Figure 2.12** Schematic illustration of the face-bonding approach. Source: Zhang et al. [40]/American Association for the Advancement of Science.

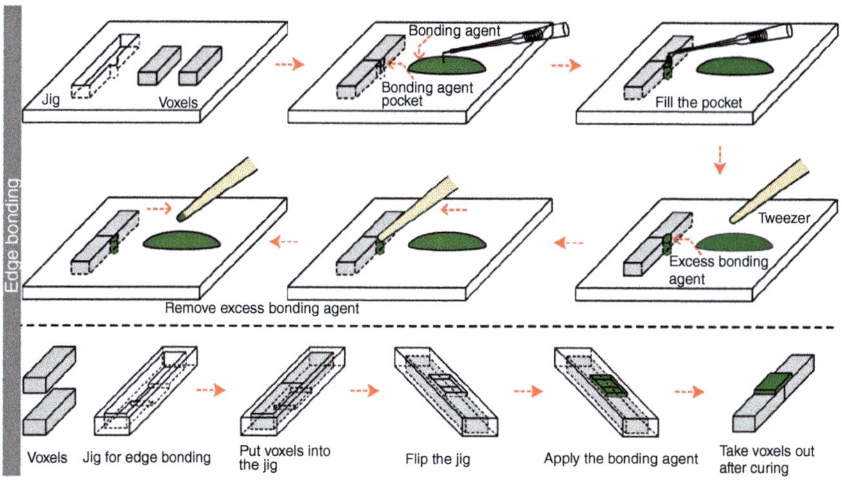


**Figure 2.13** Schematic illustration of the edge-bonding approach. Source: Zhang et al. [40]/American Association for the Advancement of Science.

knobs which can be seen on two sides can be applied to regulate the zoom and focus. Iron needles (Agani 30G needles, Terumo) and tweezers (5-SA Outils Rubis SA, Switzerland) are the primary tools for the assembly process. The bonding agent was operated by a Seque/Pro capillary containing a copper wire inside with the diameter of 0.03 mm. The magnetization direction of a voxel can be distinguished by magnetized iron needles.

As shown in Figure 2.12, by applying bonding agent in the middle, two voxels were connected in "face-to-face" manner, leading to the formation of a continuous geometry without joints. Two voxels pushed together have little bonding agents at the interfaces between the two parts. Three approaches are included for the face-bonding strategy: (i) Deploying only one voxel in the jig. (ii) Adding the other voxel into the jig and locating it beside the former one with a distance left between them. (iii) Fixing two voxels in the jig and the end portions of their bodies were placed outside. A nested copper wire dipped with some bonding agents at its tip was used as the tool. Making the wire approach the voxel, the surface was covered by the liquid-bonding agent, after which, the tool was removed. The bonding agent would spread across the voxel surface automatically. Then, we can settle the second voxel in the jig (i) or change its position in the jig closing to the first voxel. For the case (iii), we push the parts of two voxels outside in the jig. A needle or a tweezer can be used for removing excess bonding agent. Since the face bonding is symmetric, the fabricated structure does not exhibit a preference of direction in bending.

An asymmetric connection for neighboring voxels is formed with an explicit joint through an edge-bonding strategy. Particularly, we deploy two voxels in a jig which leaves enough space at the bottom or on the side. After casting the bonding agent into this pocket, a needle or a tweezer was applied to clean extra bonding agent. The shape of the jig can help to control the amount of bonding agent during the assembly process. Due to no bonding agent between the contacting faces, the separation of two voxels would easily occur during deformation. As shown in Figure 2.8, the voxels at the top side were immobilized by coating the bonding agent at the top edge. The arrangement allows the joint can be easily opened from the bottom side. The joint with this design can open and the voxels can disengage when $\boldsymbol{B}_c$ pointed up. A large deformation can be generated by the open joint with edge bonding under a small torque. The joint would become close and the voxels would compress into each other when $\boldsymbol{B}_c$ pointed down. The closed joint would drive the soft machine to bend and results in a relatively smaller deformations under the same magnetic torque. Traditionally, achieving a large deformation contains two factors: an actuating field that generates strong magnetic force or torque and soft materials which are easy to deform under external stimuli, while a large deformation can be generated by using the asymmetric edge-bonding method under a relatively weak magnetic field, which could give more freedom in design space [69].

After the assembly of these voxels, the curing of the bonding agents was performed at a designated temperature (using an oven or nonmagnetic hot plates), which was determined by the curing requirements of the bonding agents.

#### 2.2.4.3 Demonstration of Miniature Soft Machines Fabricated by the Jig-Assisted Assembly Approach

Firstly, two interconnected circular rings were applied to illustrate the proposed assembly approach, as shown in Figure 2.14. The outer diameter of each ring was 1.68 mm. Three kinds of elastomers were used to fabricate the voxels with different material properties, including PDMS (Sylgard 184, Dow Corning), Ecoflex 0010 silicone rubber (Smooth-On), and Dragon Skin (Smooth-On). Soft lithography techniques were applied to the batch fabrication of diverse nonmagnetic and magnetic voxels with various geometries. Voxels made by magnetic soft materials consist of soft elastomers such as silicone rubber doped with hard MMPs such as NdFeB microparticles, encoded with individual magnetic moments $\boldsymbol{m}_v$ with various orientation and magnitude. The heterogeneous arrangement of these voxels causes the formation of programmable 3D magnetization profile. With a uniform external magnetic field ($\boldsymbol{B}_c = 50$ mT), the soft machine exhibited the property of a negative Poisson's ratio as a 3D mechanical metamaterial with the shrinking of each ring in its radial direction.

Figure 2.14 shows the detailed assembly process of the interconnected circular rings, as well as the design of each voxel. The whole soft machine consists of 18 nonmagnetic voxels and 32 magnetic voxels. PDMS (Sylgard 184, Dow Corning) mixed with NdFeB microparticles (MQP-15-7, Magnequench) at a mass ratio of 1:3 was used as the magnetic soft materials. The densities of the polymer matrix and MMPs are $1.03\,\mathrm{g\,ml^{-1}}$ and $7.61\,\mathrm{g\,cm^{-3}}$, respectively. The resultant magnetization of the magnetic voxels was $188\,\mathrm{kA\,m^{-1}}$. Silicone elastomer (Dragon Skin 30, Smooth-On) by mixing parts A and B at 1 : 1 mass ratio was used as non-magnetic soft materials. The 3D ring in Figure 2.14 consists of two half rings and one complete ring which are assembled on a 2D plane. Firstly, an assistant jig was adopted to fix the orientations of the batch-fabricated voxels and a 1.8 T magnetic field was generated by a VSM to magnetize all voxels at once. Then, uncured silicone rubber (Ecoflex 0010, Smooth-On) was used as the bonding agent for connecting these voxels via both

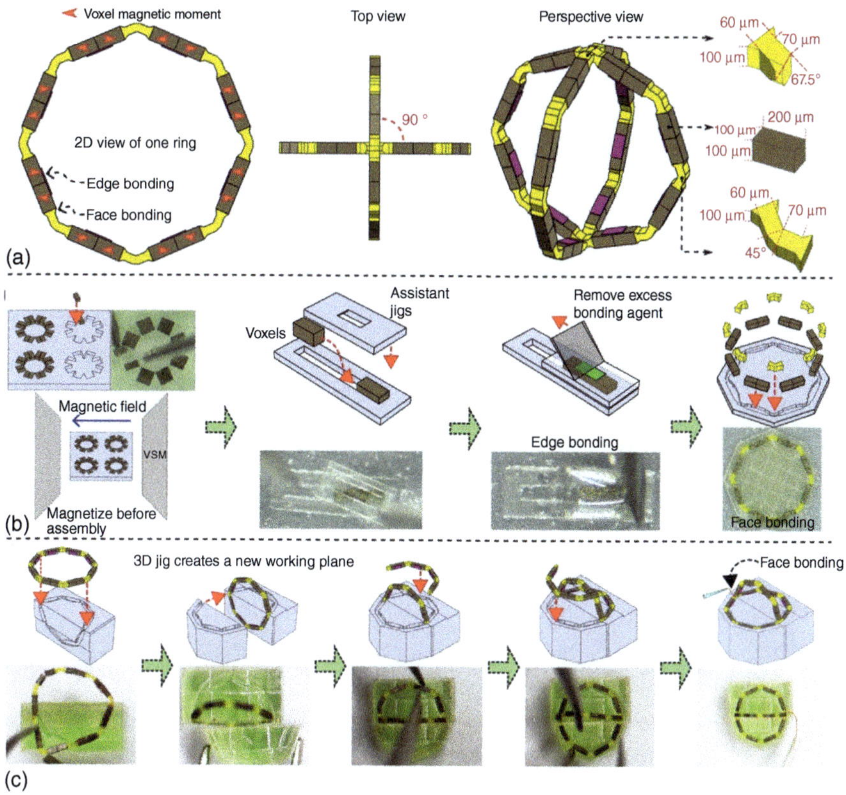


**Figure 2.14** Demonstration of the fabrication of a 3D ring structure. (a) Schematic illustration of the ring structure. (b) Schematic illustration and experimental demonstration of the fabrication process. (c) A 3D soft ring structure was fabricated by face bonding. Source: Zhang et al. [40]/American Association for the Advancement of Science (AAAS).

face-bonding and edge-bonding strategies. A desired magnetization profile of the soft machine was formed by the 3D assembly of these voxels. In addition, the assembly sequence can also be changed. For instance, the non-magnetized voxels can be assembled first followed by the magnetization of the whole 2D structure under the deformed state to encode a non-uniform magnetization across the machine. The 2D voxel-assembled parts were transferred to a 3D jig and assembled by face-bonding technique. Since the user does not demand the consideration of the magnetic interaction during this assembly sequence, the fabrication efficiency is augmented.

## 2.3 Reprogrammable Soft Machines

Soft elastomers doped with hard magnetic particles have been widely adopted due to their high coercive force, while led to a weakened structural reprogramming ability. Therefore, it is needed to develop facile strategies to realize the reprogramming ability of hard magnetic soft materials. Currently, two kinds of strategies have been proposed, one is to adopt materials with phase transition capability, and the other is to use hard magnetic particles with low Curie temperature or low coercive fields [44, 45, 70, 71].

Song et al. proposed a reprogrammable magneto-elastomer consisting of a silicon elastomer matrix and magnetic particles encapsulated in oligomeric polyethylene glycol (PEG) [70]. As shown in Figure 2.15, PEG 4000 (in the molten state) and NdFeB particles are thoroughly mixed at a mass ratio of 7 : 3, and then ground

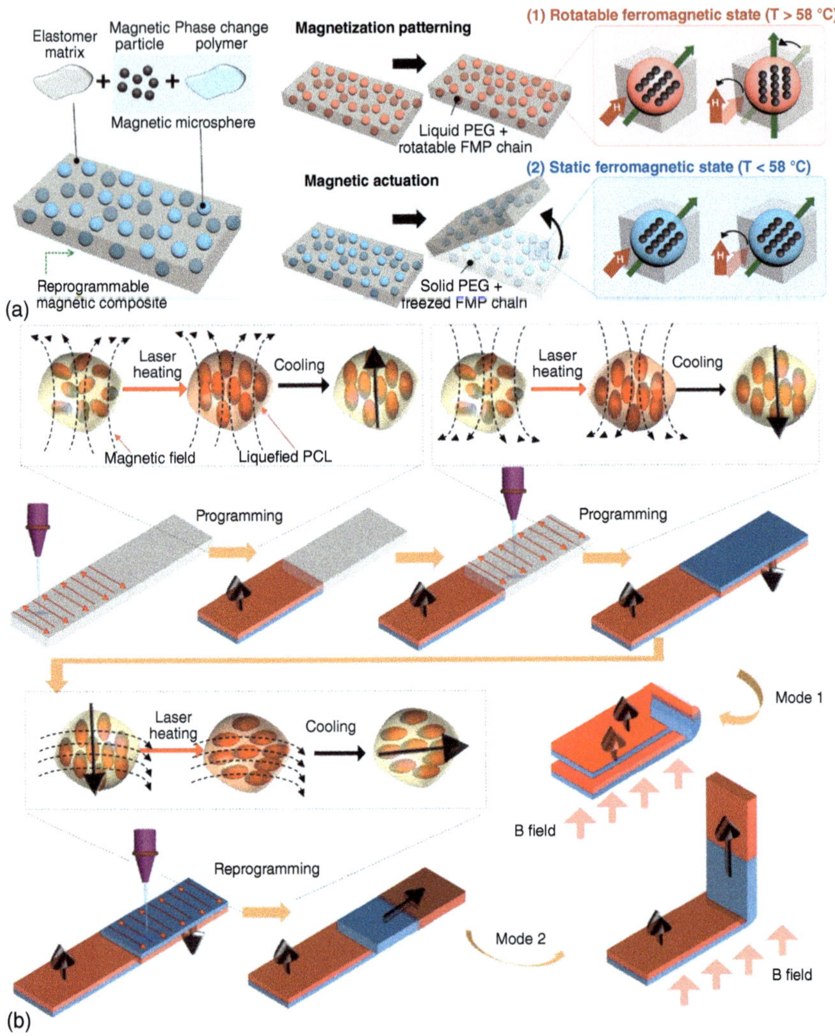


**Figure 2.15** Schematic diagram of a reconfigurable magnetoelastomer. (a) Reconfigurable composite film composed of hard magnetic particles, PEG oligomer, and elastomer matrix. At high temperature, with the melting of PEG oligomers, the hard magnetic particles can freely rotate with the change of the external magnetic field, thereby realizing the reprogramming of the magnetic domains. Source: Song et al. [70]/American Chemical Society. (b) Reconfigurable composite film composed of hard magnetic particles, PCL, and elastomer matrix. Localized heating using a laser scanning system improves the freedom and precision of reprogramming. Source: Deng et al. [71]/Reproduced from John Wiley & Sons/CC BY 4.0.

into composite magnetic particles (diameter: 53–150 μm) in a low-temperature environment. This oligomer has a low melting point, which enables the hard magnetic particles encapsulated in it to freely rotate when heated. Further, the composite magnetic particles (mass ratio: 1 : 1) were doped in the silicon elastomer matrix, and cured in a 100° environment. To achieve reprogramming of the magnetic domains inside the elastomer, it was reheated to 100 °C to enable the PEG 4000 to melt. Using a new template or manually deforming the elastomer into the target shape, a magnetic field of 300 mT was applied and maintained for three minutes to achieve the rearrangement of the orientation of the ferromagnetic particles. As the ambient temperature decreases, the orientation of ferromagnetic particles is gradually fixed, which in turn leads to the formation of new magnetic domain distribution.

The above direct heating approach has the disadvantage of limited freedom of reprogramming. In contrast, the laser heating method can achieve high precision and high-degree-of-freedom reprogramming for magnetic domain distribution, as shown in Figure 2.15. Deng et al. doped ferromagnetic particles encapsulated by phase-changeable polycaprolactone (PCL) in silicon elastomer [71]. The presence of magnetic particles can improve the photothermal conversion efficiency, so that the fabricated composite film can absorb laser energy to heat and melt the PCL. Deng et al. used a custom-made direct laser-writing system to reprogram the magneto-elastomers, and used laser-induced graphene as a marker to achieve the alignment of laser and elastomer. The direct laser-writing system is able to generate transient heat on the elastomer surface, limiting the degradation of resolution caused by thermal diffusion, and at the same time guaranteeing fast immobilization of magnetic domains after reprogramming. As shown in Figure 2.15, the premagnetized ferromagnetic particles are reoriented by melting the PCL shell and adjusting the external magnetic field (generated by a rotatable magnet).

Alapan et al. proposed a general strategy for reprogramming using magnetic particles with a low Curie temperature [44]. The PDMS matrix is mixed with chromium dioxide ($CrO_2$) ferromagnetic particles (Curie temperature: 118 °C), and a collimated near-infrared laser is used to achieve local heating of the elastomer. During the laser heating process, the magneto-elastomer can realize the magnetization of the structure and the reorientation of the magnetic domain under the action of a 15 mT magnetic field. When the ambient temperature reaches the Curie temperature again, the magnetoelastomer will be completely demagnetized. Meanwhile, this heating process has few impact on the mechanical properties of PDMS elastomers. Experimental results show that the resolution of this reprogramming method can reach 38 μm, which shows powerful functions in reconfigurable metamaterial structures and soft robots.

## 2.4 Multi-Stimuli Responsive Transformation of Magneto-Elastomer Based Soft Structures

The transformation of morphable structure in a quasi-static state and the exotic property of the structure (e.g. unusual thermal, acoustic, and mechanical properties of metamaterials) have been widely studied. Nevertheless, proper actuation approaches and structural design are still needed to explore complex anisotropic transformation modes and unfold underlying mechanisms of dynamic behaviors. The multi-stimulus responsiveness (solvent and magnetic field-responsive properties) of magneto-elastomers provides an enabling technology for dynamic regulation and multimodal transformation of soft structures.

### 2.4.1 Solvent and Magnetic-Responsive Behavior of Magneto-Elastomers

Template-assisted method was used for the fabrication of the surface-attached elastomer structures. For instance, the preparation process for surface-attached strip structures is as follows: a laser processing machine (Universal Laser Systems PLS6.75) was used to process various groove patterns such as rectangular holes on acrylic sheets. The following hydrophobic treatment was applied to acrylic sheets: Plasma treatment was performed to acrylic molds for 10 minutes followed by a silane steam hydrophobic treatment with triethoxy(1H,1H,2H,2H-perfluoro-1-octyl)silane (POTS) for three hours. A glass substrate was treated by plasma for 10 minutes to obtain a hydrophilic surface. Clips were used to hold the glass substrate and the acrylic mold together. We poured the uncured magneto-elastomer onto the top of the acrylic mold. Then, the cavity inside the acrylic mold was thoroughly filled with the uncured elastomer by performing vacuum degassing for five minutes. After the vacuum degassing, the polymer was cured at a specific temperature (60 °C for 20 min) provided by an oven. By removing the acrylic mold, the elastomer structure could attach on the glass substrate with a strong interface bonding resulting in a stable non-freestanding structure. The top of the elastomer structure was covered by a layer of silicone elastomer with embedded fluorescence powder at a mass ratio of 1 : 0.08 to achieve the fluorescent effect of the structure.

The impacts of the magnetic stimulation and organic solvent on the morphological transformation of the surface-attached strip structures were systematically studied. As shown in Figure 2.16, by immersing the structure in the toluene solution, a wave shape transformation would occur at the top of the strip structure. According to Figure 2.17, the deformation amplitude of the magneto-elastomers gradually becomes stable after about 20 minutes. A strong pulse magnetic field was applied to the structure in the deformed state. A magneto-optical sensor (MagViewS, Matesy) was used to measure the magnetization profiles of the magneto-elastomer before and after magnetization. The magnetic flux density of the strip structure presents a wave-shape distribution. By applying a rotating magnetic field, the strip structure deformed into a 3D wave shape, and the position of the troughs and peaks varied with the direction of external magnetic field. Figures 2.18 shows the impact of geometrical parameters, e.g. the height, length, and width of the strip structures, on the shape-morphing result in toluene. The transformation of the structures under the same magnetic input but different geometrical parameters was studied (Figure 2.18a–c). The result in Figure 2.18a demonstrates that the length of the structure needs to fulfill a buckling condition to form the wavy pattern: $L > L_c$. $L$ is the length of the structure and $L_c$ represents a critical length. When $L$ is larger than

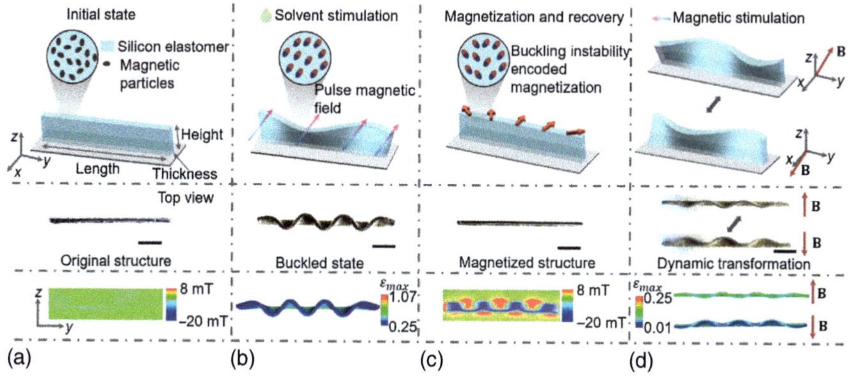


**Figure 2.16** Demonstration of the original shape (a), buckling transformation (b), magnetization (c), and magnetic responsive (d) of a strip magneto-elastomer. A magneto-optical sensor was used to measure the magnetization profile of the magneto-elastomer before and after the magnetization process. Source: Xia et al. [41]/Reproduced from Springer Nature/CC BY 4.0.

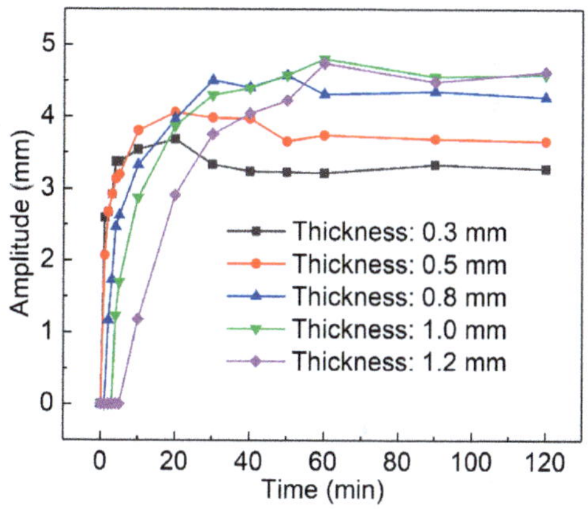


**Figure 2.17** The changing of deformation amplitude of magneto-elastomers (five different thicknesses) with swelling time. Source: Xia et al. [41]/Springer Nature/CC BY 4.0/Public domain.

$L_c$, the impact of the length on the amplitude and wavelength of the wavy shape was negligible. In addition, with the raising of the thickness and the height, the amplitude and wavelength of the wavy shape also increase.

Programmable magnetic inputs including magnetic field direction and magnetic field strength could actively control the shape-morphing results. Figure 2.18d shows that there is a linear relationship between the magnetic field strength and the amplitude of the formed wavy pattern in the structure with different heights. The direction of the external magnetic field causes a complex impact on the shape-morphing results. The direction angle in Figure 2.18e was the angle between the pulse magnetic field direction used for magnetization and the applied actuation magnetic field. The strip structures would exhibit a twist shape (e.g. the inset of Figure 2.18e with label "1") with a 90° direction angle. When the height of the structure was 4.3 or 5.0 mm and a 180° direction angle was applied, no wavy pattern was generated as shown in the inset of Figure 2.18e with label "2," while when the height of the structure was

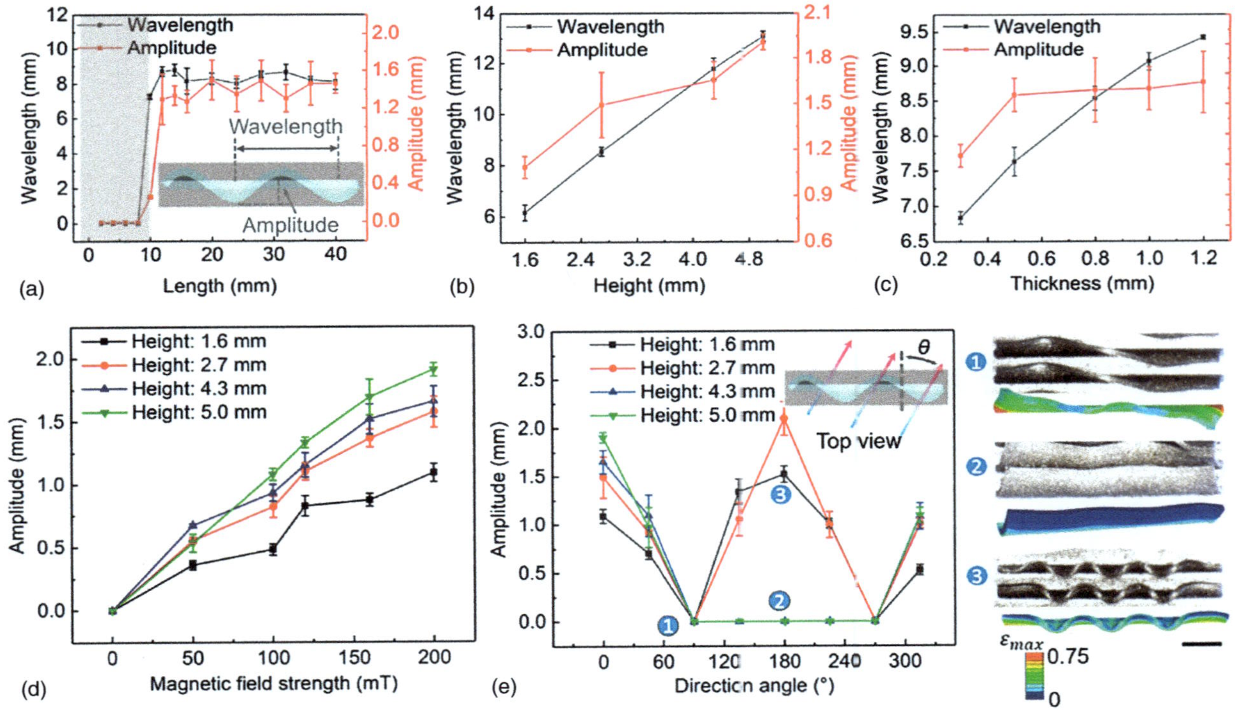


**Figure 2.18** Characterization of the impacts of geometry parameters and magnetic field parameters on the morphological transformations upon magnetic stimulation. (a–c) The quantitative relationship between length (a), height (b), thickness (c), and the deformation (wavelength and amplitude) of the strip. (d–e) The impacts of the magnetic field strength (d) and the direction angle (e). The simulation and experimental results were shown in the inset images. With a direction angle of 90°, twisted strip structures were generated (image with label "1"). With a direction angle of 180°, collapsed strip structures were generated (image with label "2," height: 5.0 mm). With a direction angle of 180°, strip structures with wavy patterns were generated (image with label "3," height: 1.6 mm). Source: Xia et al. [41]/Reproduced from Springer Nature/CC BY 4.0.

1.6 or 2.7 mm, wavy patterns can still be generated (e.g. the inset of Figure 2.18e with label "3"). The experimental and simulation results demonstrate that for a too high structure, the collapse is easy to happen under the magnetic actuation.

Apart from the stimulation (including organic solvent and external magnetic field) applied separately on these elastomer structures, the impacts of the coupled stimulation on the morphological transformation were studied and factors including stimulation sequences, magnetization patterns, and liquid environments were taken into consideration.

There are two types of magnetization profiles encoded in the magneto-elastomers including UM and buckling-encoded magnetization (BM). Toluene solution and deionized (DI) water were used for the liquid environment. Thus, four kinds of coupled stimulations are studied (Figure 2.19). Waveforms can be generated by magneto-elastomers with BM and varied with the change of the direction of external magnetic field. When immersed in organic solvent, the elastomer structure always exhibits a swelling-induced buckling state. We define a phase shift parameter: $\emptyset = \frac{\Delta x}{\lambda} \times 360$ for the quantitative description of the geometrical transformation (Figure 2.19b). $\lambda$ represents the wavelength and $\Delta x$ is the displacement of the location of the peak. The phase shift of the BM structure changes with the magnetic field direction. A form of traveling wave was presented in the dynamic transformation process. Only a decrease of the deformation amplitude occurred when changing the liquid environment to DI water, yet the deformation form was not influenced. In contrast, the UM structure presented relatively few deformation modes upon the coupled stimulation. Only bending deformation was observed when using DI water. With the varied magnetic field direction and toluene stimulation, the peak location would move back and forth along the length direction.

Finite element analysis (FEA) was performed by the structural mechanics module in COMSOL to confirm the morphological transformation upon the coupled stimulation, as shown in Figure 2.19c. The results of the coupled stimulation illustrate that the magnetic control strategy enables us to explore controllable buckling deformation of 3D morphable structures, and the buckling instability-encoded magnetization endows the soft structures with more degrees of deformation freedom, which provides great potentials in the areas like droplet manipulation, microfluidic devices, and soft robotics.

### 2.4.2 Investigation of the Dynamic Transformations of Cellular Structures

Figure 2.20 illustrates the impact of different connectivity types on the morphological transformation of the elastomer structure. The deformation of the plate S1 would be influenced by the existence and arrangement of plates B1–B4. There are four kinds of connectivity types based on the distribution of B1–B4. The length of S1 in Figure 2.20a was fixed at 8 mm allowing the generation of a full waveform during the buckling transformation. For connectivity type I, the ends of S1 presented two left-handed shapes. The two ends of S1 with connectivity types I, II, or IV coincided with the node positions of wave-shape structures, while for connectivity type III, the

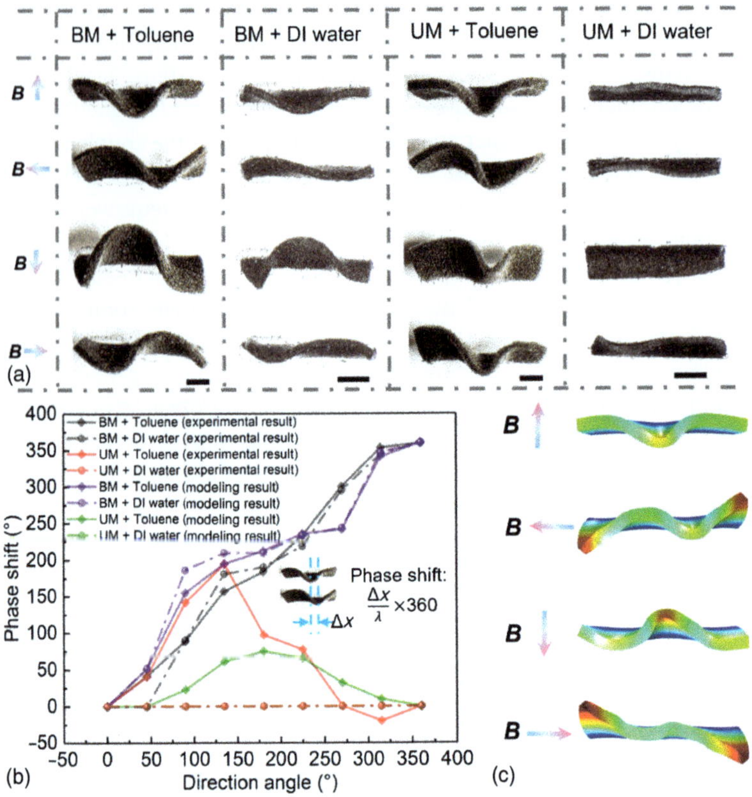


**Figure 2.19** The morphological transformation of a magneto-elastomer upon the coupling stimulation including the solvent and magnetic field. (a) Investigation of four kinds of coupled stimulations. UM and BM represent uniform magnetization and buckling instability-encoded magnetization, respectively. (b) The quantitative relationship between the phase shift, magnetic field directions, as well as stimuli types. (c) Simulation results of the BM magneto-elastomer under the coupling stimulation. Source: Xia et al. [41]/Reproduced from Springer Nature/CC BY 4.0.

ends of S1 occupied the troughs of the formed wave-shape, which was consistent with the simulation results. With a different length of S1 (e.g. 4 mm), the locations of nodes or peaks formed during the buckling transformation will also change. For example, the ends of S1 coincide with the peak locations of the deformed structure with connectivity type IV. The morphological transformations of these structures are not deterministic, because these waveforms can be either down-up or up-down.

The transformations of various lattice structures were systematically studied (Figures 2.21 and 2.22). Firstly, we focus on a square lattice. Tunable magnetic inputs, e.g. magnetic field strength, direction, and gradient, were generated by a cuboid permanent magnet with the size of 40 mm × 40 mm × 25 mm. An ordered sinusoidal pattern was generated by the square lattice submerged in toluene solution, and each plate transformed into the structure with half-wavelength shape. When an in-plane magnetic field coincides with the direction of the pulse magnetic field used for magnetization, a pattern similar to the buckled shape upon the

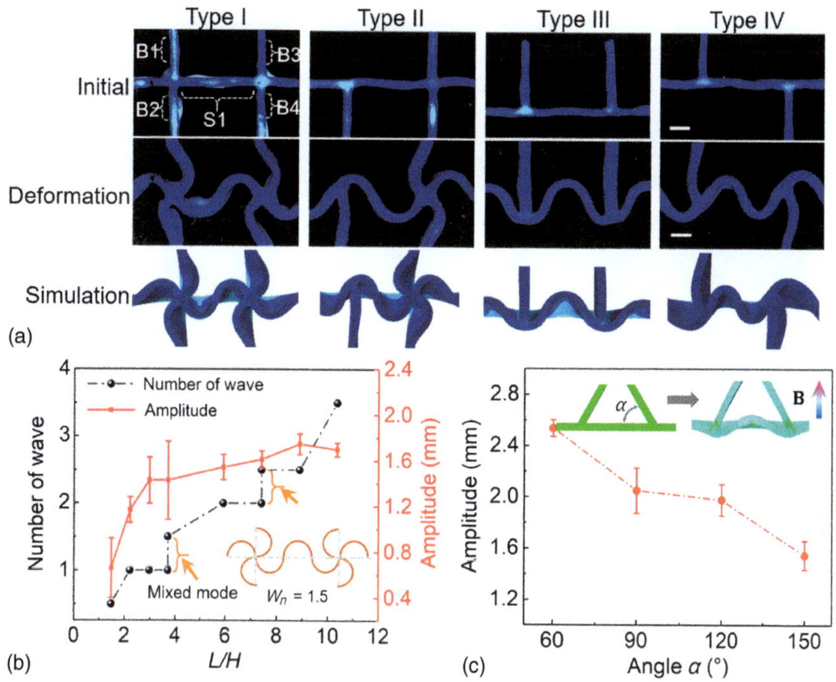


**Figure 2.20** The geometrical transformation of soft structures with different connectivity types. (a) The buckled configuration of structures with four different kinds of boundary types in toluene solution. (b) The impact of the geometrical parameter (*L/H*) on $W_n$ and amplitude upon magnetic stimulation. (c) The impact of the boundary angle on the deformation amplitude. Source: Xia et al. [41]/Reproduced from Springer Nature/CC BY 4.0.

solvent stimulation was presented by the square lattice. By reversing the magnetic field direction, the lattice structure exhibited a symmetrical wavy pattern. The experimental results demonstrate that the external magnetic field could regulate the node handedness in the lattice structure. Anisotropic patterns were induced by the application of out-of-plane magnetic field (Figure 2.21d,e). The vertical plates in the lattice structure exhibit a weak deformation amplitude, and the horizontal plates in the lattice structure deform toward the same direction but with inhomogeneous deformation amplitudes. In addition, a more complex pattern with a gradual wavy shape can be generated by placing the square lattice in the area with a strong magnetic field gradient, as shown in Figure 2.21f.

The transformations of elastomer structures with different cut designs and geometries are also investigated. As shown in Figure 2.22, different achiral patterns with controllable rotation of each vertices can be generated by a hexagon lattice under magnetic stimulation. A negative Poisson's ratio property of the staggered brick-wall elastomer structure has been widely reported [72, 73]. A diamond-like pattern or a re-entrant hexagonal pattern can be presented by the structure upon magnetic stimulation. Predesigned artificial defects can be exploited to regulate the transformation of these lattice structures. Figure 2.22c,d shows the deformation of the structure with two indents arranged on plates along the *x*-axis. The experimental results

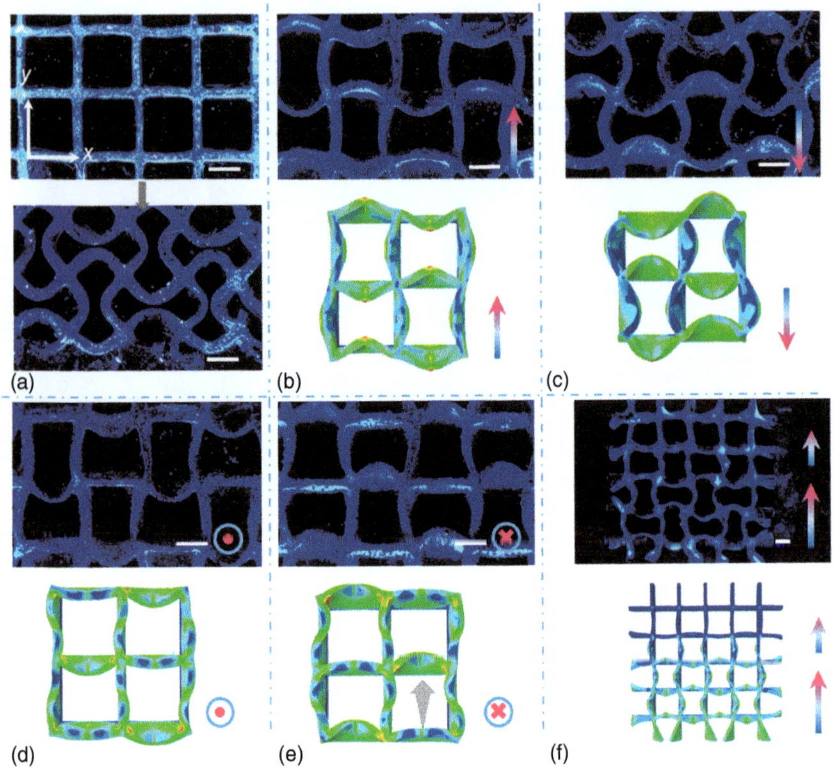


**Figure 2.21** The geometric transformation of a square lattice. (a) Images of the fabricated square lattice and the buckling shape in toluene. (b–e) The transformations of the square lattice by applying magnetic field along different directions. (f) The deformation of the square lattice by applying magnetic field with high field gradient. Source: Xia et al. [41]/Reproduced from Springer Nature/CC BY 4.0.

demonstrated that the node location and the number of waves of the deformed shape were affected by the arrangement of indentation.

## 2.5 Multimodal and Bioinspired Locomotion Adopted for Elastomer-Based Robot

Utilizing the controllable deformation capability of magneto-elastomers and programmable time-varying magnetic fields, soft robots can achieve multimodal locomotion according to different environments. As shown in Figures 2.23 and 2.24, Hu et al. presented eight motion modes of soft robots, which are based on the rigid rotation of elastomeric bodies and flexible deformation [10]. The fabrication and magnetization patterns of this magnetic soft robot are introduced in Section 2.2.1. As shown in Figure 2.23, the robot can swim against the direction of gravity by imitating the locomotion mode of jellyfish, and its shape alternates between the "V" shape and "C" shape. When the robot reaches the liquid surface, due to the effect

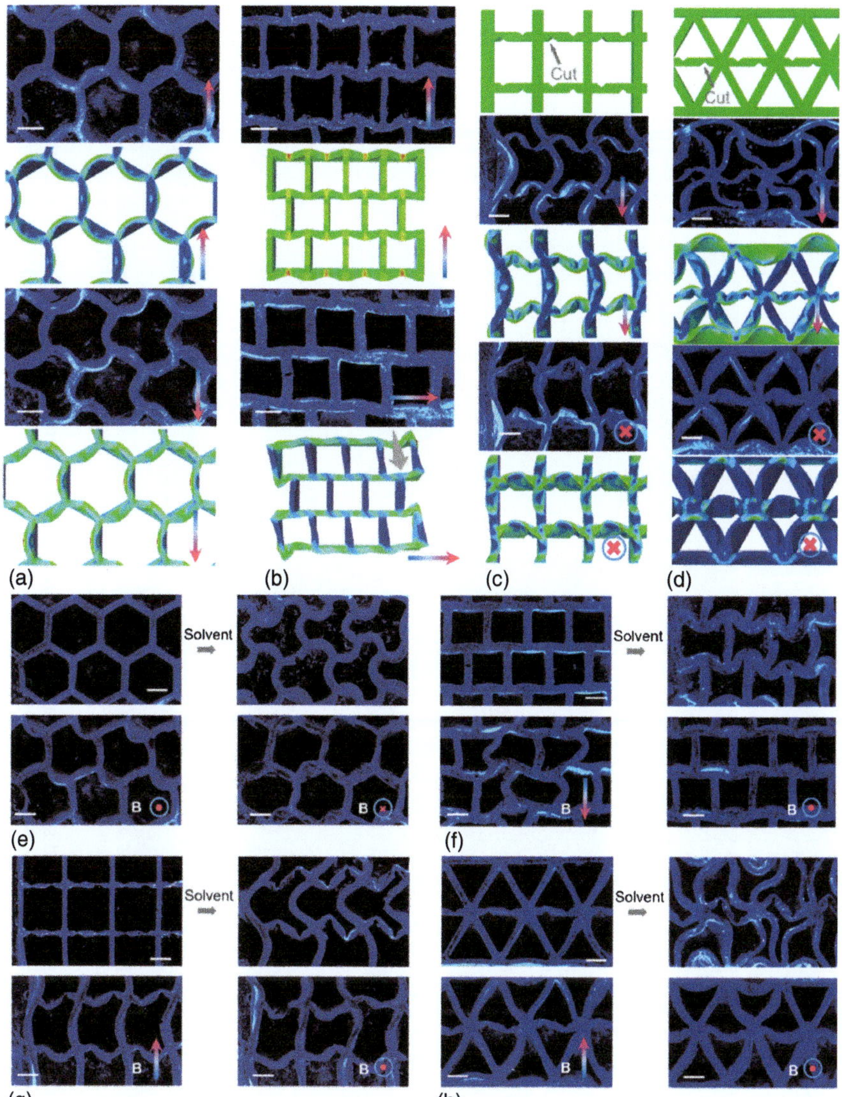


**Figure 2.22** The morphological transformation of cellular structures with different shapes and artificial defects. (a–b) The transformations of lattice structures in hexagonal and staggered square shapes upon magnetic stimulation. (c–d) The transformations of lattice structures with artificial defects upon magnetic stimulation. (e–h) Solvent and magnetic stimulation for hexagonal lattice (e), staggered square lattice (f), square lattices with artificial defects (g), and triangular lattices with artificial defects (h). Source: Xia et al. [41]/Reproduced from Springer Nature/CC BY 4.0.

of hydrophobic material and surface tension, the robot can stay on the liquid surface stably. Subsequently, the robot could roll with a C-shape to climb along the water meniscus, and then get out of the water to reach the adjacent solid surface (air environment).

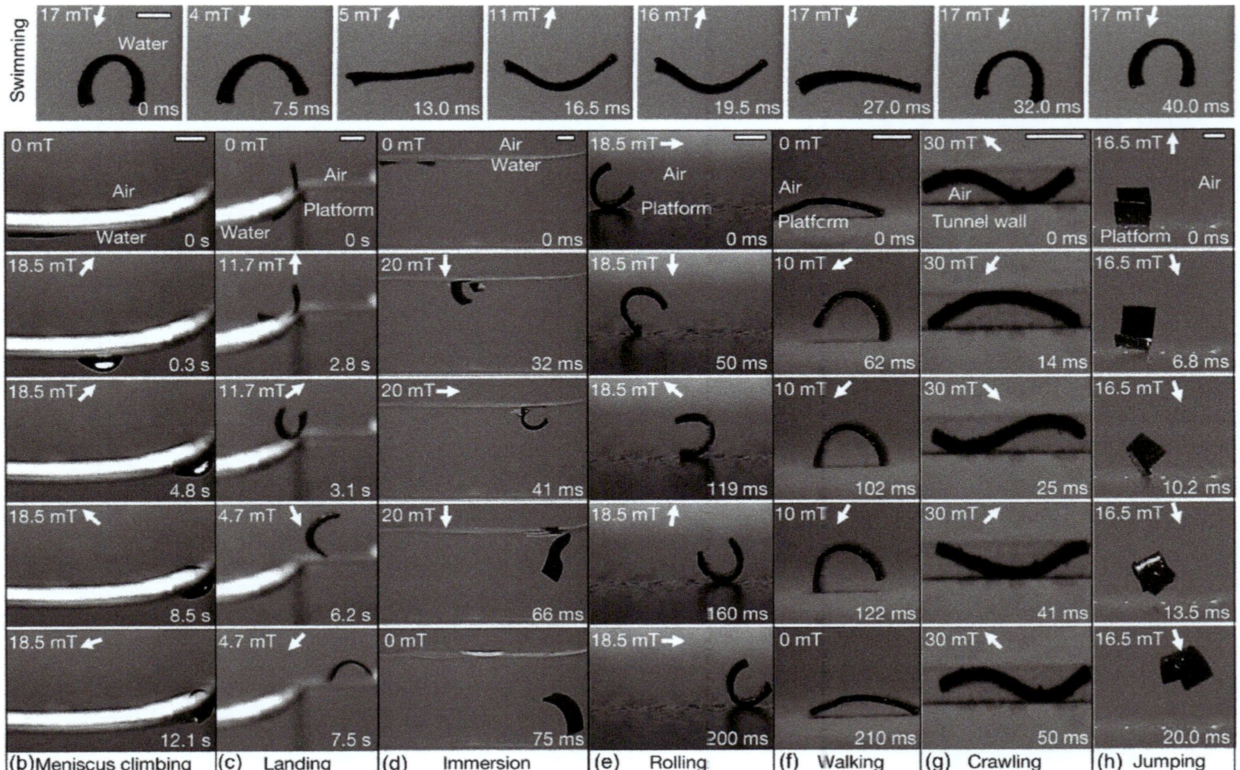


**Figure 2.23** Multimodal locomotion of magneto-elastomer based soft robots. (a–h) swimming, meniscus climbing, landing, immersion, rolling, walking, crawling, and jumping. White arrows indicate changes in magnetic field direction according to the time series. Source: Hu et al. [10]/Reproduced from Springer Nature.

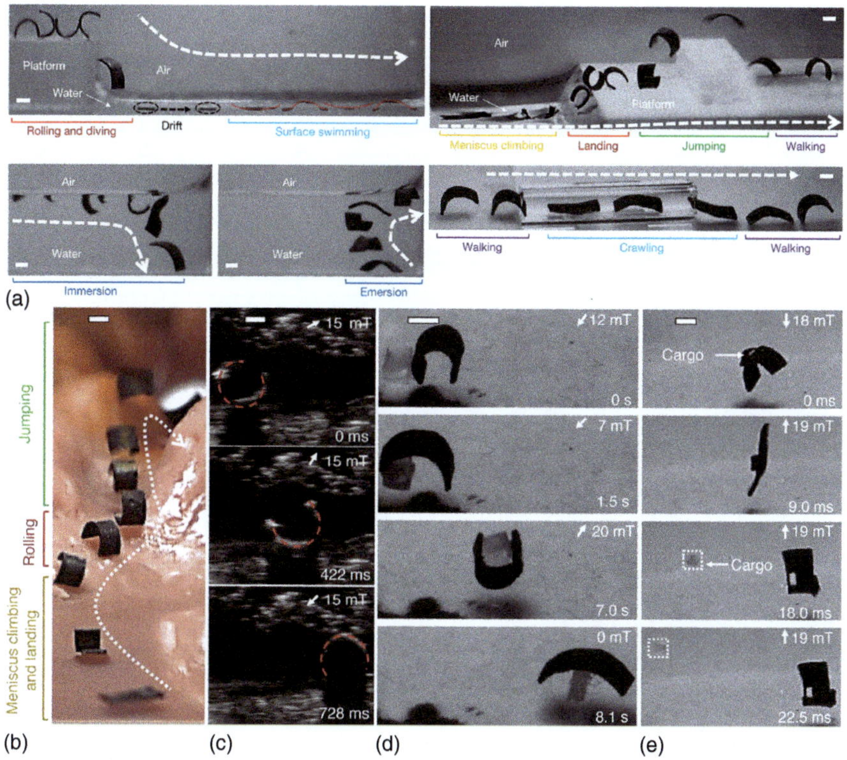


**Figure 2.24** Continuous locomotion of soft robots in mixed solid–liquid and unstructured environments. (a) The soft robot moved from the platform to the liquid surface and returned to the platform, then walked through the tunnel. (b) The robot moved in the stomach model. (c) Locomotion of the robot in chicken muscle tissue under ultrasound guidance. (d–e) Picking, delivery, and release of goods by robot. Source: Hu et al. [10]/Reproduced from Springer Nature.

In addition, the robot can break away from the liquid surface and re-enter the liquid environment. The robot could also move in a rolling manner after curling, or crawl by imitating the locomotion pattern of an inchworm, produce undulating motion gaits in narrow spaces, or move in a jumping manner. Figure 2.24 shows the multi-mode movement of the soft robot in a solid–liquid mixed environment and an unstructured environment (stomach model, chicken muscle tissue). The locomotion state of the robot can also be observed in real time through ultrasonic imaging, and the tasks of grasping, delivering and releasing can be completed.

Many organisms in nature can adapt to complex living environments with agile locomotion gaits. For example, multi-legged organisms could exhibit their ability of overcoming obstacles under the synergy of leg structures, reduce friction with different terrain, and enhance the freedom of movement. Based on the multi-legged structure in nature, Lu et al. and Gu et al. proposed two different types of crawling robots, as shown in Figure 2.25 [74, 75]. Lu et al. used PDMS doped with magnetic particles to build soft robots. A large number of tapered feet are distributed on the surface of a PDMS film to tune the movement of the robot. Under the magnetic field

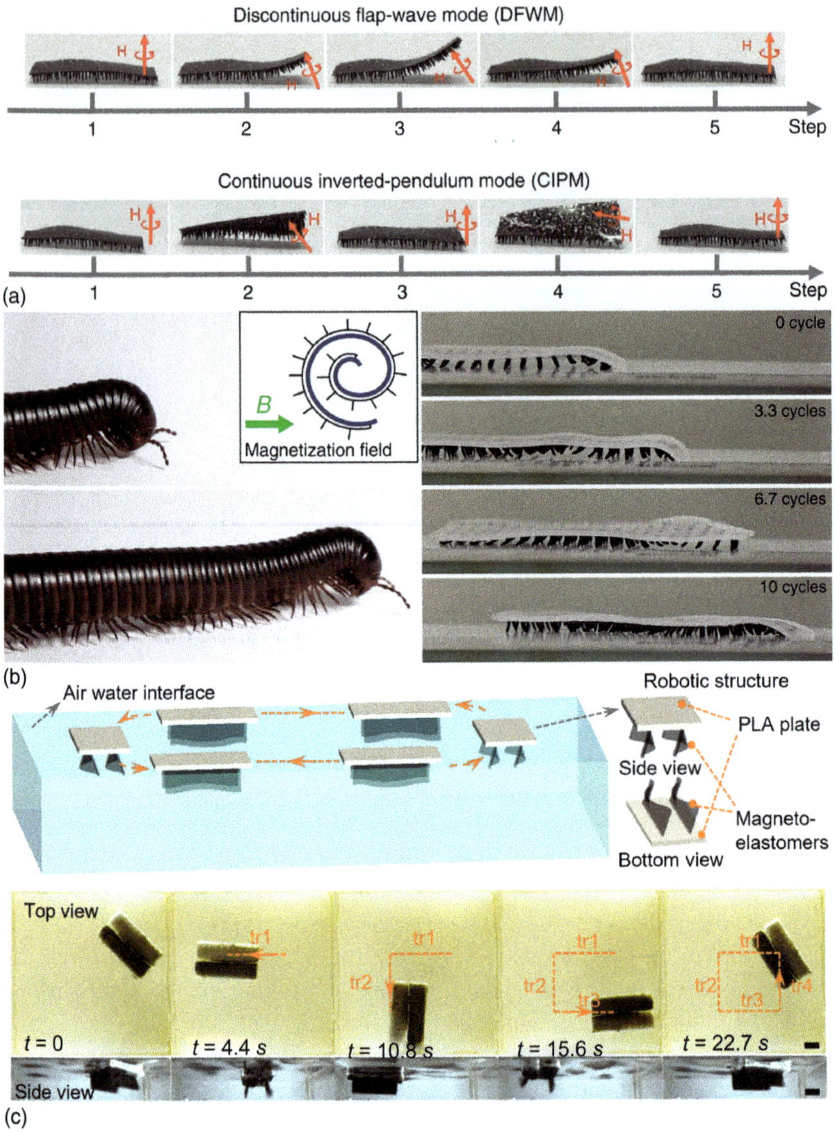


**Figure 2.25** Demonstration of magneto-elastomer robots with bioinspired locomotion modes. (a) Two locomotion modes of multilegged robot including discontinuous flap-wave (DFW) and continuous inverted-pendulum (CIP) locomotion. Source: Lu et al. [74]/Reproduced from Springer Nature/CC BY 4.0. (b) Giant African millipede-inspired multilegged robot with traveling-wave locomotion. Source: Gu et al. [75]/Reproduced from Springer Nature/CC BY 4.0. (c) Schematic illustration of the swimming robot. The actuation of the swimming robot along a rectangular trajectory, and the deformation of the magneto-elastomers in the swimming robot. Source: Xia et al. [41]/Reproduced from Springer Nature/CC BY 4.0.

generated by a permanent magnet, the robot will generate a motion pattern of flap wave or inverted pendulum, which is similar to the continuous walking of human. Gu et al. proposed a soft robot inspired by giant millipedes, in which complex magnetization distributions are encoded by template-assisted magnetization, enabling the structure to exhibit a metachronal wave under a dynamic magnetic field. Under the action of the metachronal wave, the legs of the soft robot will push the robot structure forward continuously.

Underwater creatures could control their wave-like fins to improve the stability of movement or generate effective propulsion. The morphological transformation behaviors have offered fruitful inspiration for the design of biomimetic swimming robots. Electromechanical systems have been developed to control the ribbon fin-based propulsions at the centimeter scale [76], while building small-scale untethered robots with such propulsion mechanism still remains challenging because of the limited onboard space. Magnetic control strategy has been widely adopted for the actuation of untethered miniature soft robots due to its high programmability and controllability. In this chapter, an untethered swimming robot was developed with dynamic ribbon fins controlled by programmable magnetic field inputs and buckling instability-encoded magnetization patterns. As shown in Figure 2.25, a flat plate with the length of 20 mm made of polylactic acid was used as the substrate, coated by a thin layer of silicone rubber which could help in the formation of the robust bonding between the magnetic strips and the substrate. The length and height of the strip are 16 and 5 mm, respectively. The morphological transformation of the strip was actuated by an oscillating magnetic field (frequency: 12 Hz, and magnetic field strength: 9 mT) using three-axis Helmholtz coils. Figure 2.25 demonstrates that the swimming robot could achieve efficient propulsion at the air–water interface. Omnidirectional locomotion or movement with a rectangular trajectory can be performed by continuously tuning the magnetic field direction. The further miniaturization of the proposed biomimetic swimming robot could be achieved by using a stronger external magnetic field to improve the control capability.

## 2.6 Potential Biomedical Applications of Miniature Magnetic Soft Machines

New wireless miniature soft machines were presented using the above bottom-up assembly approach to demonstrate potential biomedical functionalities. Firstly, a functional miniature capsule with the shape of hollow prolate spheroid was constructed, as shown in Figure 2.26. The height and largest diameter of the capsule were 450 and 500 μm, respectively. Two plate-shaped and seven ring-shaped voxels were used to construct the capsule. The surface rolling of the miniature capsule enables it to get access to a targeted hard-to-reach position. Meanwhile, the capsule can also execute tasks including transporting and on demand releasing of a liquid cargo, as well as be retrieved by rolling away. The developed capsule promises broad applications in in situ liquid biopsy, e.g. microbiota sampling and tumor screening.

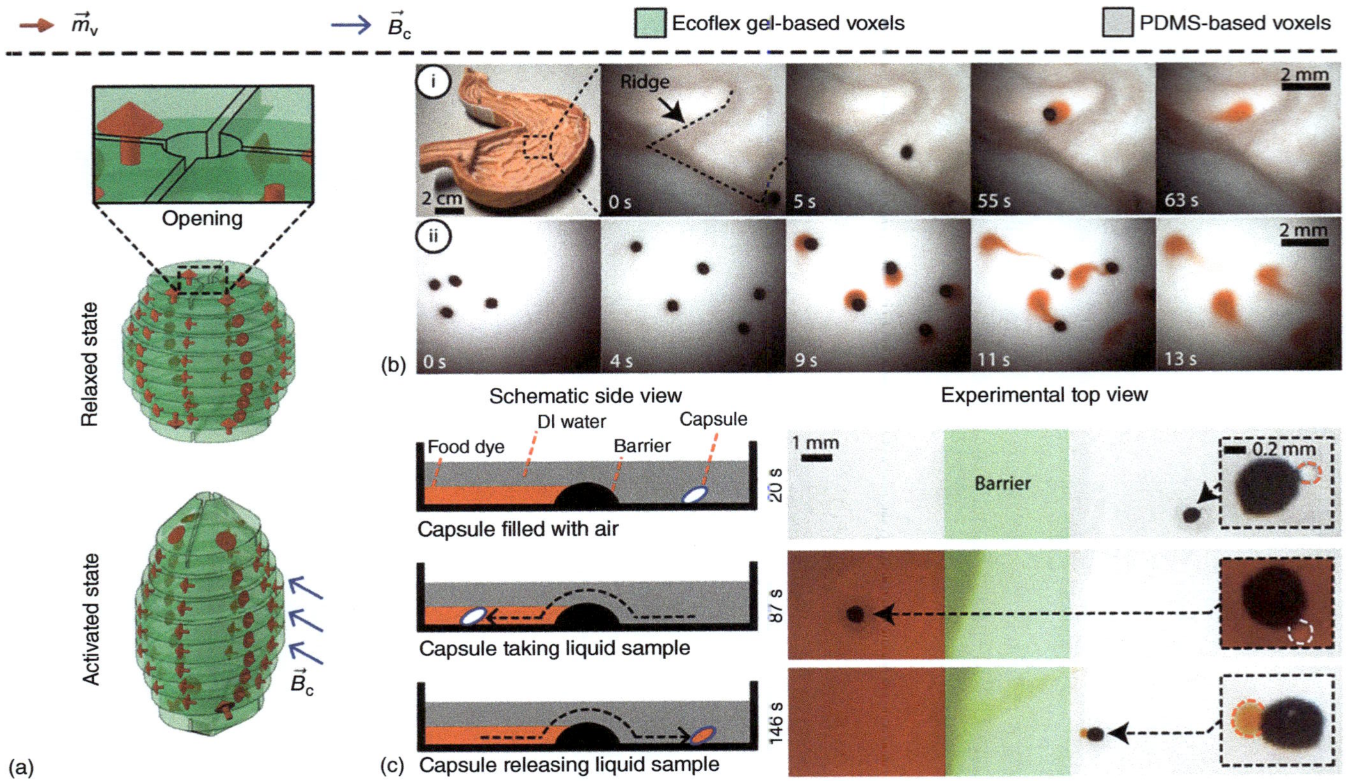


**Figure 2.26** A miniature soft capsule and its potential biomedical application. (a) Schematic illustration of the soft capsule and its function of cargo release. (b) The locomotion and on-demand release of five capsules. (c) Fluid samples were carried by the soft capsule for potential biopsy applications. (d) The rolling and activation of the soft capsule carried with red food dye in clear water. (e) Air and red food dye were carried by the soft capsule and were ejected in clear water. Source: Zhang et al. [40]/American Association for the Advancement of Science (AAAS).

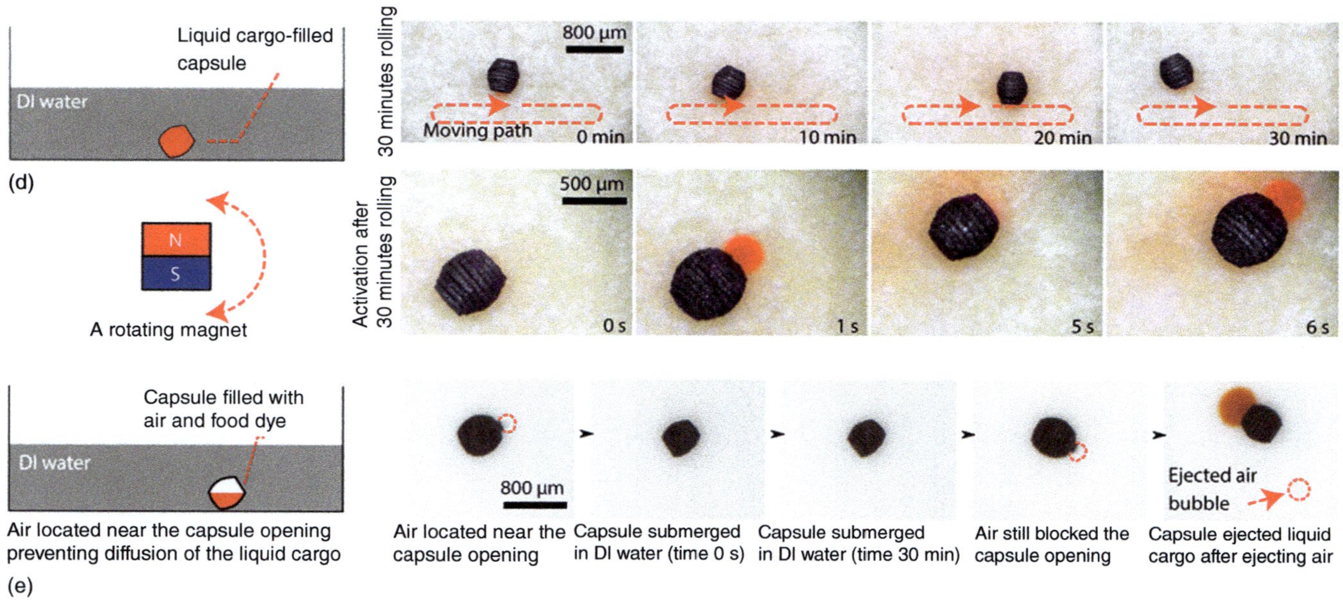


**Figure 2.26** (*Continued*)

According to the design, the capsule has a volume of ~37 nl. Each magnetic voxel in the capsule was made of Ecoflex gel mixed with NdFeB magnetic particles at a 1 : 1 mass ratio. The voxels were assembled by face-bonding technique applying the bonding agent like Ecoflex gel. The assembled machine was placed in a jig for the following magnetization. The magnetization of the magnetic voxel at last is 74 kA $m^{-1}$.

When an external magnetic field was applied, the pressure within its chamber would rise with the shrinking of the capsule, leading to the ejection of the inner content such as liquid drug, air, or food dye. Due to the existence of net magnetic moment, the miniature capsule could be moved over obstacles and rolled on uneven surfaces by applying external magnetic field. A stomach phantom was adopted to demonstrate the functionalities of the capsule. It was moved on the uneven surface of the phantom. When arriving at a targeted location, the enclosed liquid cargo (food dye) can be released on demand, after which, the capsule rolled out of the working environment. Figure 2.26 shows the simultaneous manipulation and delivery of liquid cargoes. The capsule team provides a higher supply of cargo and could be possibly commanded separately someday.

Finally, we also investigated the diffusion of the liquid cargo. A red food dye with the viscosity of ~0.89 centipoise at 25 °C was filled in the capsule. After rolling on a surface for 30 minutes, the food dye in the chamber of the capsule was ejected. The experimental result illustrates that a large proportion of liquid cargo can still be preserved in the capsule after 30 minutes. The diffusion of liquid cargo can be prevented when the opening of the capsule is blocked. Figure 2.26e shows a capsule filled with both air and liquid food dye. To achieve this, a capsule fully occupied with liquid dye was squeezed to let a little amount of the food dye out. As the recovery of the capsule, some air would flow into the chamber of the capsule. No diffusion of the air in the capsule was scrutinized after immersing it in clear water for 30 minutes. A two-step activation would be needed to eject the liquid cargo in the capsule. The air bubble that blocks the opening of the capsule needs to be ejected first followed by the ejection of the liquid cargo.

An anchoring machine with the outer diameter and length of 1.68 and 2.10 mm, respectively, is presented in Figure 2.27 with an on-demand anchoring function under a dynamic environment provided by an internal fluidic flow. The machine can also be released and moved back and forth inside a tube. Because no steady exterior energy input is demanded for anchoring capability and wireless position control, the anchoring machine could potentially be applied inside blood vessels to execute minimally invasive interventions when arriving at targeted positions in the human body which are deep and hard to enter. Under nonactuated state, the diameter of the anchoring device is larger than the inner diameter of the tube, leading to the formation of a compressive loading-based contact and high friction with the surface of the tube and robust anchoring at the targeted working space when it is deployed. The soft elastic property of the machine enables it to adapt to the variations of the tube diameter and simultaneously maintain the anchoring position. The advantage makes it possible to be used in the deploying in blood vessels with complex vessel diameter changes and strong pulsatile blood flow. Meanwhile, with a hollow internal structure, the fluid can also flow through the machine with minimum hindrance to reduce or avoid the risk of clogging.

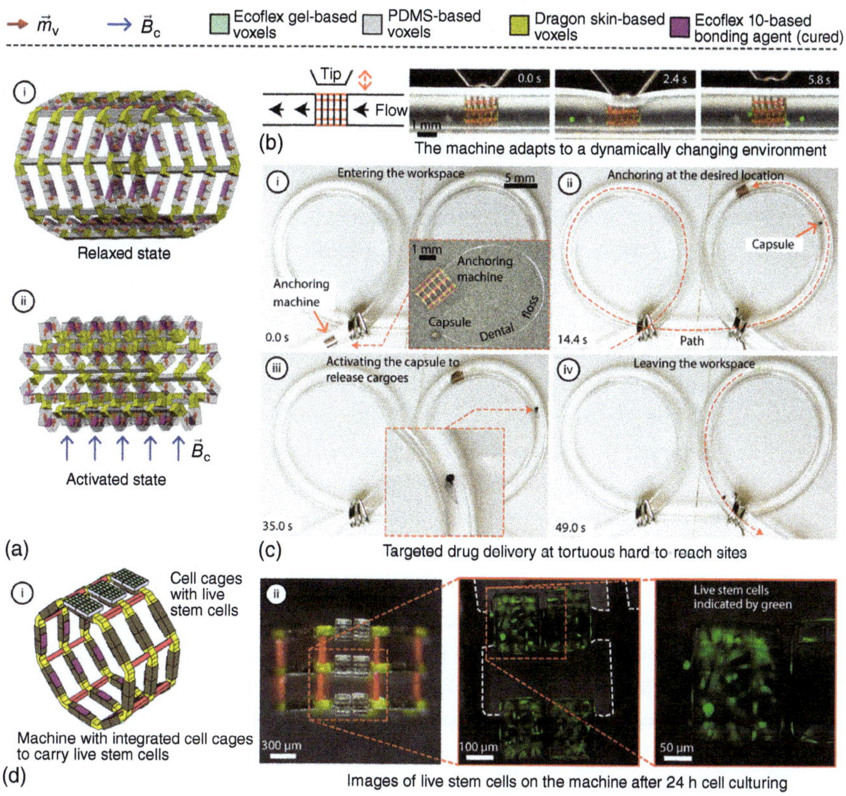


**Figure 2.27** A miniature soft anchoring machine and its potential biomedical application. (a) Schematic illustration of the soft anchoring machine with relaxed and activated states. (b) Demonstration of the adaption of the anchoring machine in a tube with changing diameter. (c) Demonstration of the functions of cargo delivery and anchoring in a synthetic tube. (d) Demonstration of the anchoring machine integrated with live stem cell for potential vascular regeneration applications. Source: Zhang et al. [40]/American Association for the Advancement of Science (AAAS).

96 magnetic voxels and 88 nonmagnetic voxels were used to construct the machine. The ring structure was adopted as a repeating functional unit. These units and the intermediate voxels were connected by face-bonding technique with uncured silicone rubber (Ecoflex 10, Smooth-On) as the bonding agent.

A flexible microfluidic tube (Tygon formula 2375 laboratory tubing, Merck KGaA) was adopted to verify the adaption capability to the dynamical change of tube diameter. The inner diameter and outer diameter of the microfluidic tube are 1.6 and 3.2 mm, respectively. An external tip compressed the tube to reduce the inner diameter. As shown in Figure 2.27, the anchoring machine could resist an internal fluidic flow, adapt to a dynamically changing working space, and maintain an anchoring position. The mean speed of the liquid flow in Figure 2.27 was $10\,\mathrm{cm\,s^{-1}}$. The material, the inner diameter, and the effective radial compliance of the tube determine the maximal compression and the flow speed that can apply to the machine. Future changes such as a micropatterning on the outer surface of the machine can be adopted to increase the tissue static friction and anchoring stability.

Magnetic gradient pulling strategy was used to wirelessly control the machine inside the microfluidic tube, demonstrating the capability of local and active cargo delivery in complex working space with fluid flows. The tube deformed into a curved shape and exhibited about 630° end-to-end bending, providing a tortuous tubular environment for the machine. It is risky and hard to manipulate conventional catheters to arrive at this tortuous environment. A dental floss with a 1.8 cm length connected the anchoring machine with an aforementioned capsule. By applying an external magnetic field with a spatial gradient generated by an external permanent magnet, the anchoring machine shrunk and exhibited a smaller size compared with the inner diameter of the microfluidic tube. Under the magnetic gradient force, the anchoring machine was moved against the fluidic flow in the microfluidic tube. When the external magnetic field was removed, the device recovered to its initial shape because of its stored elastic energy and stably anchored at the targeted location under the liquid flow whose mean flow speed was 5 cm s$^{-1}$. Then, liquid cargo loaded by the capsule was released by squeezing the capsule. The inner diameter of the used microfluidic tube is within the clinical measurement range of the average diameter of lumen in human coronary arteries.

Finally, therapeutic stem cells were carried by the anchoring machine to illustrate the potential application of vascular regeneration, e.g. the treatment of peripheral artery diseases and ischemia. As for this process, biocompatible microcages printed by two-photon lithography was used as a cell scaffold to entrap mesenchymal stem cells (MSCs) and integrated to the anchoring machine for the cell delivery process. The cell suspension was released to the cages and the MSCs could sediment into the cages. 25 μg ml$^{-1}$ fibronectin solution was used to keep the device before cell seeding to enhance the adhesion of MSCs to microcages. Then, the spreading of MSCs on the cage surface was scrutinized and no apparent toxic effect was shown by the microcages and the device for 24 hours. Microcarriers with open structures to confine stem cells have been a widely preferred method to achieve the local delivery of the cells. During the transportation, stems cells would adhere to the microcarriers. After reaching targeted location, the stems cells would experience proliferation, migration, and differentiation. The release of stem cells depends on its natural migration. The spread of cells needs more surface with the growth and proliferation of cells. In the long term, the inner section of the cages can be integrated with stimuli-responsive materials which could provoke the release of cells. The anchoring machine can also be used to the targeted delivery of drugs. After the release of the drug, the machine can be recycled from the anchoring location.

## 2.7 Fluidic Pumping by the Magneto-Elastomers

Firstly, a tubular soft peristaltic pump inspired by the intestine was demonstrated as shown in Figure 2.28. 37 annular ring-shaped plane slices (1.0 mm inner diameter, 1.2 mm outer diameter, and 100 μm thickness) were perpendicularly heaped to fabricate the pump. Each horizontal slice includes eight arc voxels. There are specific magnetization profiles formed in the slice which enables the periodical

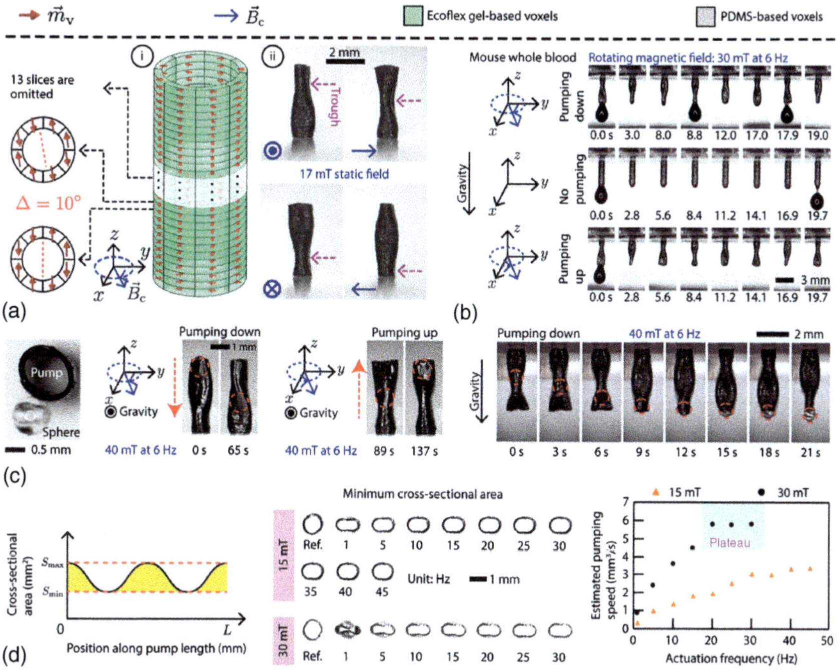


**Figure 2.28** A peristaltic soft pump and its potential application. (a) Illustration and demonstration of the peristaltic soft pump inspired by intestine. (b) The pumping of mouse whole blood by the soft machine. (c) A polystyrene sphere was transported by the peristaltic soft pump. (d) The impacts of rotating frequencies and magnetic field strengths on the pumping performance. Source: Zhang et al. [40]/American Association for the Advancement of Science (AAAS).

expanding and shrinking of the structure with a phase difference under a rotating magnetic field in the $x$–$y$ plane, leading to the formation of traveling wave along the soft peristaltic pump shown in Figure 2.28. The difference of the magnetization direction of the stacked voxels in two neighboring slices was 10° in the $x$–$y$ plane. 296 magnetic voxels with identical geometry were used to construct the pump. Each voxel was made of silicone rubber (Smooth-On) doped with NdFeB magnetic particles (MQP-15-7, Magnequench) at 1 : 1 mass ratio. After magnetization, the magnetic voxel has the resultant magnetization of $74\,\mathrm{kA\,m^{-1}}$. The neighboring voxels were connected by face-bonding technique applying the bonding agent such as uncured Ecoflex Gel (Smooth-On).

Based on the direction of the formed traveling wave, the soft machine could transport fluids such as mouse whole blood and solid objects such as polystyrene sphere forward or backward. Control experiments were conducted to analyze the impact of the gravity, as shown in Figure 2.28. The soft peristaltic pump was mounted horizontally and vertically to deliver a solid sphere, and also was controlled to transport in opposite directions. Thus, the environmental factors, e.g. gravity, can be isolated. The experimental results demonstrate that the pumping capabilities for liquid and solid objects derive from the traveling wave-like shape-morphing of the pump.

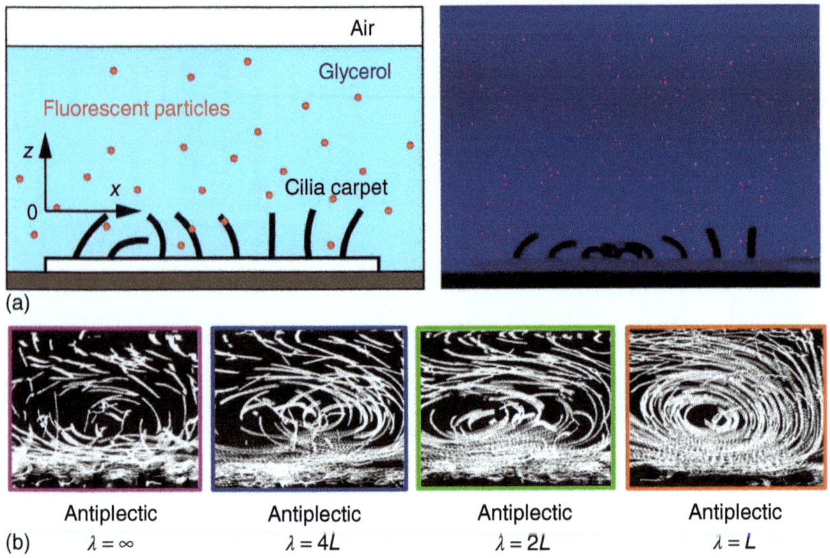


**Figure 2.29** Cilia-inspired magneto-elastomer arrays. (a) Experimental setup used to observe fluidic pumping. The liquid environment is glycerol, and the red particles are fluorescent particles. (b) Flow field induced by metachronal waves with different wavelengths. Source: Gu et al. [75]/Reproduced from Springer Nature/CC BY 4.0.

Ciliary movement is widely adopted by organisms in nature, for example, the metachronal coordination of cilia arrays can enable rapid pumping of fluids in low Reynolds number environments. Based on the ciliary array of biological systems, Gu et al. and Dong et al. developed magnetoelastomer-based artificial ciliary system and quantitatively studied the performance and parameter optimization of the pumping system [75, 77]. As shown in Figure 2.29, the columnar artificial cilia array was made of Ecoflex elastomer mixed with NdFeB particles, and its complex magnetization pattern was encoded by a curved template, so that the cilium array can exhibit metachronal coordinated locomotion under a rotating magnetic field. Fluorescent particles were added to the viscous environment (glycerol) for tracking, and the influence of the ciliary array on the fluid was characterized by calculating the average displacement of the fluid over a period. This strategy provides a quantitative physical platform to study the influence of different magnetization modes on the modulation of metachronal waves and the effect on fluid transport.

The dynamic behaviors of magneto-elastomers could help to promote fluidic transport. Many natural organisms, e.g. cuttlefish and electric ghost knifefish, could transfer momentum to liquid environment with the dynamic fluctuation of ribbon fins [56, 78]. Particle image velocimetry (PIV) technique was adopted for the investigation of the flow field caused by the dynamic transformation process. The liquid environment was provided by glycerol (dynamic viscosity: 0.876 Pa s) and fluorescent particles were adopted as tracer. The process was recorded by a digital microscope (Hirox RH-2000). An open-source package (MatPIV 1.7) was used to analyze the recorded video to characterize the induced flow field. Figure 2.30a

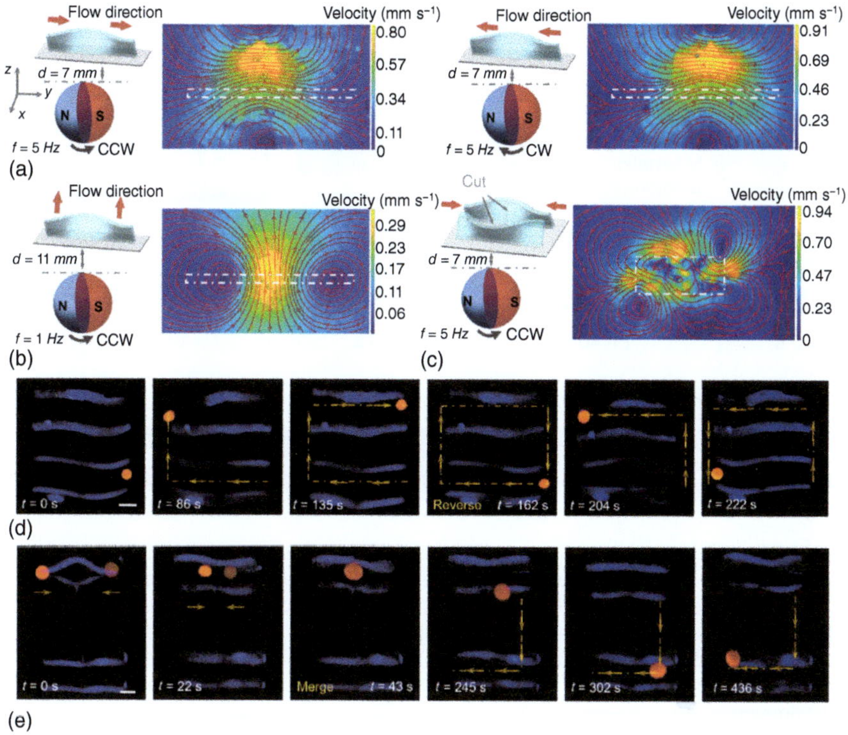


**Figure 2.30** Dynamic fluidic behaviors induced by the morphological transformation. (a–c) PIV analysis of flow field activated with different rotating directions and magnetic field strengths. (d–e) The manipulation of droplet with the induced flow field. (d) A rectangular trajectory was performed by the actuation of strip arrays. (e) Merging and transportation of droplets. Source: Xia et al. [41]/Reproduced from Springer Nature/CC BY 4.0.

shows the flow field induced by a surface-attached strip with the length of 12 mm. An indentation was added to the structure to achieve a prescribed buckling state. The rotation of a sphere permanent magnet generated an in-plane rotating field with tunable parameters including rotating frequency and magnetic field strength. As shown in Figure 2.30, with a strong magnetic field, a directional net fluid flow was generated along the $y$-axis. The direction of the flow was determined by the rotating direction of external magnetic field, while by applying a rotating magnetic field with weak strength and low frequency, a net fluid flow along the $x$-axis direction was generated. The induced maximum fluid velocity can be tuned by rotating frequencies and magnetic field strengths.

Fluidic mixing in a miniature channel holds great promise for the application of microfluidic devices. However, because of the slow diffusion effect and low Reynolds number (*Re*) environment, the efficient mixing remains challenging. Existing devices with passive fluid mixing often require specific design of microchannel, e.g. microchannel with ridge or groove microstructures, and spiral microchannels [79]. Recently, microfluidic devices with active fluid mixing were also developed by using bubble or magnetic artificial cilia, which may

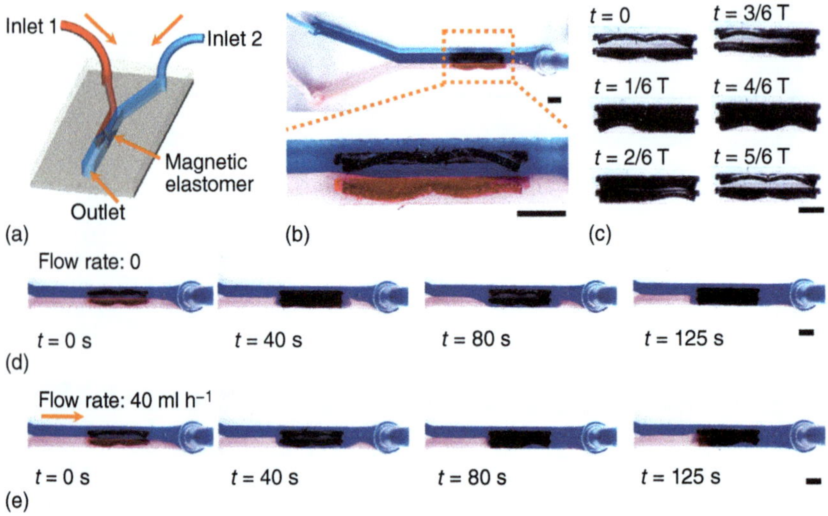


**Figure 2.31** Efficient mixing of viscous fluid using the magneto-elastomers. (a) Schematic illustration of the microfluidic device. (b) Images of the fabricated magneto-elastomers and the microfluidic device. (c) Experimental results of the actuation of the magneto-elastomers in the device using a rotating magnetic field. (d) Efficient fluid mixing in pure glycerol with no additional fluid pumped into the channel. (e) Efficient fluid mixing in pure glycerol with a flow rate of 40 ml $h^{-1}$ from the inlet. Source: Xia et al. [41]/Springer Nature/CC BY 4.0/ Public domain.

present limited mixing thoroughness or require optimized ciliary arrays. Here, the developed dynamic transformation of magneto-elastomer was used for the fluidic mixing in microfluidic devices (Figure 2.31). Glycerol colored with blue and red dyes was transported from the inlet to the microfluidic channel with a volume of 182 $mm^3$. The fluid filled in the microchannel did not exhibit apparent diffusion under static state. A rotating magnetic field with the frequency of 5 Hz was used to actuate a magnetic strip array (height: 1.6 mm) in the channel. Viscous glycerol ($Re < 0.004$) can be rapidly mixed by the microfluidic device under both static and dynamic fluid environment with a flow rate of 40 ml $h^{-1}$.

## 2.8 Other Potential Applications of the Magneto-Elastomers

In recent years, with the development of materials and control technologies, the applications of soft robots based on magneto-elastomers in other fields have also been widely explored, such as flexible electronic devices, reconfigurable surfaces, and selective capture, as shown in Figure 2.32. Kim et al. integrated electronic devices including LED lights and circuits with 3D-printed magneto-elastomers [3]. Actuated by dynamic external magnetic field, the structure will exhibit different deformation shapes, thereby realizing different electronic functions. Furthermore, the integration of magneto-elastomers with miniature magnetic sensors has been

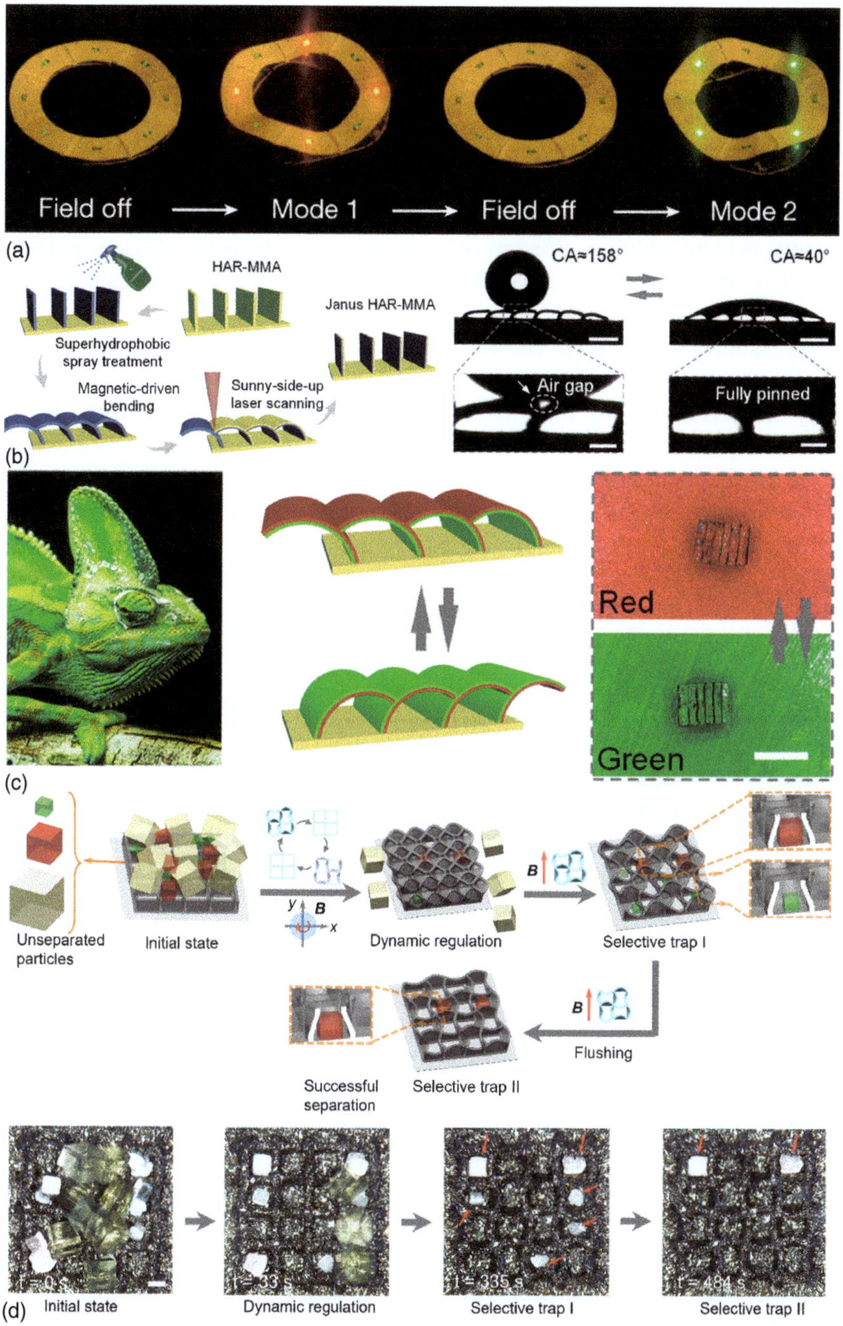


**Figure 2.32** Other functional demonstrations of magneto-elastomers. (a) Reconfigurable flexible electronic system actuated by 3D printed magneto-elastomer. Source: Kim et al. [3]/Reproduced from Springer Nature. (b) Janus magneto-elastomer with switchable wettability. Source: Jiang et al. [80]/John Wiley & Sons. (c) Chameleon-inspired Janus magnetoelastomer with color-switching ability. Source: Jiang et al. [80]/Reproduced from John Wiley & Sons, Inc. (d) Selective particle trapping via the dynamic modulation of a cellular microstructure. Demonstration and schematics for the selective trapping of 200, 400, and 600 μm particles. Source: Xia et al. [41]/Reproduced from Springer Nature/CC BY 4.0.

adopted in the development of various tactile or pressure-sensing devices [81, 82]. Jiang et al. proposed a Janus plate array based on magneto-elastomers (including PDMS matrix and NdFeB particles) [80]. For example, different treatments with hydrophobic spray and femtosecond laser scanning on both sides of the magnetic plate will make one side of the plate exhibit hydrophilic properties (contact angle: 40°), while the other side has hydrophobic properties (contact angle: 158°), which can switch the hydrophilic and hydrophobic states of the structure surface by a magnetic field. In addition, different colors were painted on both sides of the Janus plate, making it possible to dynamically tune the surface color by changing the direction of the magnetic field.

The dynamic modulation of cellular structures provides opportunities for in situ microparticles sorting and trapping. Figure 2.32 illustrates three kinds of processing methods. Activated by a rotating magnetic field, the morphology of the cellular structure could transform between open state and deformed state. The cellular microstructure can be used as a small-scale vibrating screen for sorting particles with different sizes. According to the geometry of the deformed chamber, the trapping of small particles in the closed chamber and the releasing of large particles can be realized. As shown in Figure 2.32, we fabricated a square lattice by a micromold and the length of each plate in the lattice structure was 500 μm. Cubic microparticles with the side length of 600, 400, and 200 μm were adopted to perform the selective trapping function. Firstly, actuated by a rotating magnet, 600 μm particles were extruded and 400 and 200 μm particles were reserved in the chamber. After that, the cellular structure was deformed upon a static magnetic field and the chamber remained a ~340 μm gap. By applying a flow disturbance, 200 μm particles were removed from the deformed chamber, and the boundary of the chamber could physically hold 400 μm particles. Thus, the control strategy could successfully select particles with the size of 400 μm from the group of 600, 400, and 200 μm particles.

## 2.9 Summary

Due to the wide potential applications in soft robotics, the interest in silicone elastomers is significantly growing. The additive manufacturing, bottom-to-up assembly, as well as advanced actuation strategies, especially magnetic actuation, of silicone elastomers greatly facilitate the development of reconfigurable miniature soft machines. In this chapter, biomedical functionalities of elastomer matrix-based soft machines were demonstrated. With a bottom-up fabrication approach assisted by jigs, the design freedom in geometry, magnetization profile, and material for magnetic soft machines was extended. Further optimization of structural and mechanical designs can be performed to solve specific clinical problems. With respect to materials used in this study, the platinum-catalyzed silicones have been adopted as the coating of medical catheters. Moreover, biocompatible ferromagnetic materials such as FePt nanoparticles can be applied to endow the soft machines with magnetic-responsive property so that they are more applicable to clinical applications. Future machines may be inspired by this study with advanced functionalities and improved design freedom to tackle real-world challenges. In addition, the multi-stimuli responsiveness of magneto-elastomers is explored. The swelling of magneto-elastomers by absorbing organic solvents and the arrangement of boundary constraints allow the buckling transformation of morphable soft structures, which offers a facile and template-free strategy to encode 3D heterogeneous magnetization profile inside the elastomer structures. Actuated by programmed magnetic inputs, e.g. magnetic field strength, gradient, and direction, dynamic geometric reconfiguration, multimode anisotropic transformations, as well as adaptive locomotion can be exhibited. A series of potential applications including biomedical devices, fluidic pumping, reconfigurable surfaces and electronic devices are demonstrated via the magnetic regulation.

## Abbreviations

| | |
|---|---|
| BM | buckling-encoded magnetization |
| DI | deionized |
| FEA | finite element analysis |
| MMPs | magnetic micro particles |
| MSCs | mesenchymal stem cells |
| PCL | polycaprolactone |
| PDMS | poly(dimethylsiloxane) |
| PEG | polyethylene glycol |
| PIV | particle image velocimetry |
| POTS | triethoxy(1H,1H,2H,2H-perfluoro-1-octyl)silane |
| UM | uniform magnetization |
| VSM | vibrating sample magnetometer |

## References


1 Mazurek, P., Vudayagiri, S., and Skov, A.L. (2019). How to tailor flexible silicone elastomers with mechanical integrity: a tutorial review. *Chemical Society Reviews* 48: 1448–1464.

2 Skylar-Scott, M.A., Mueller, J., Visser, C.W., and Lewis, J.A. (2019). Voxelated soft matter via multimaterial multinozzle 3D printing. *Nature* 575: 330–335.

3 Kim, Y., Yuk, H., Zhao, R. et al. (2018). Printing ferromagnetic domains for untethered fast-transforming soft materials. *Nature* 558: 274–279.

4 Zhou, L.Y., Ye, J.H., Fu, J.Z. et al. (2020). 4D printing of high-performance thermal-responsive liquid metal elastomers driven by embedded microliquid chambers. *ACS Applied Materials & Interfaces* 12: 12068–12074.

5 Kang, S.H., Shan, S., Noorduin, W.L. et al. (2013). Buckling-induced reversible symmetry breaking and amplification of chirality using supported cellular structures. *Advanced Materials* 25: 3380–3385.

6 Herzberger, J., Sirrine, J.M., Williams, C.B., and Long, T.E. (2019). Polymer design for 3D printing elastomers: recent advances in structure, properties, and printing. *Progress in Polymer Science* 97: 101144.

7 Park, K., Yuk, H., Yang, M. et al. (2022). A biomimetic elastomeric robot skin using electrical impedance and acoustic tomography for tactile sensing. *Science Robotics* 7: eabm7187.

8 Markvicka, E.J., Bartlett, M.D., Huang, X., and Majidi, C. (2018). An autonomously electrically self-healing liquid metal–elastomer composite for robust soft-matter robotics and electronics. *Nature Materials* 17: 618–624.

9 Wu, S., Hu, W., Ze, Q. et al. (2020). Multifunctional magnetic soft composites: a review. *Multifunctional Materials* 3: 042003.

10 Hu, W., Lum, G.Z., Mastrangeli, M., and Sitti, M. (2018). Small-scale soft-bodied robot with multimodal locomotion. *Nature* 554: 81–85.

11 Larson, N.M., Mueller, J., Chortos, A. et al. (2023). Rotational multimaterial printing of filaments with subvoxel control. *Nature* 1-7.

12 Zhou, L.Y., Gao, Q., Fu, J.Z. et al. (2019). Multimaterial 3D printing of highly stretchable silicone elastomers. *ACS Applied Materials & Interfaces* 11: 23573–23583.

13 Mannoor, M.S., Jiang, Z., James, T. et al. (2013). 3D printed bionic ears. *Nano Letters* 13: 2634–2639.

14 Symes, M.D., Kitson, P.J., Yan, J. et al. (2012). Integrated 3D-printed reactionware for chemical synthesis and analysis. *Nature Chemistry* 4: 349–354.

15 Wei, H., Li, K., Liu, W.G. et al. (2017). 3D printing of free-standing stretchable electrodes with tunable structure and stretchability. *Advanced Engineering Materials* 19: 1700341.

16 Guo, S.Z., Qiu, K., Meng, F. et al. (2017). 3D printed stretchable tactile sensors. *Advanced Materials* 29: 1701218.

17 Tian, K., Bae, J., Bakarich, S.E. et al. (2017). 3D printing of transparent and conductive heterogeneous hydrogel-elastomer systems. *Advanced Materials* .

18 Coenjarts, C.A. and Ober, C.K. (2004). Two-photon three-dimensional microfabrication of poly (dimethylsiloxane) elastomers. *Chemistry of Materials* 16: 5556–5558.

19 Rekštytė, S., Malinauskas, M., and Juodkazis, S. (2013). Three-dimensional laser micro-sculpturing of silicone: towards bio-compatible scaffolds. *Optics Express* 21: 17028–17041.

20 Ge, Q., Chen, Z., Cheng, J. et al. (2021). 3D printing of highly stretchable hydrogel with diverse UV curable polymers. *Science Advances* 7: eaba4261.

21 Sturgess, C., Tuck, C.J., Ashcroft, I.A., and Wildman, R.D. (2017). 3D reactive inkjet printing of polydimethylsiloxane. *Journal of Materials Chemistry C* 5: 9733–9743.

22 McCoul, D., Rosset, S., Schlatter, S., and Shea, H. (2017). Inkjet 3D printing of UV and thermal cure silicone elastomers for dielectric elastomer actuators. *Smart Materials and Structures* 26: 125022.

23 Thrasher, C.J., Schwartz, J.J., and Boydston, A.J. (2017). Modular elastomer photoresins for digital light processing additive manufacturing. *ACS Applied Materials & Interfaces* 9: 39708–39716.

24 Sirrine, J.M., Meenakshisundaram, V., Moon, N.G. et al. (2018). Functional siloxanes with photo-activated, simultaneous chain extension and crosslinking for lithography-based 3D printing. *Polymer* 152: 25–34.

25 Scott, P.J., Meenakshisundaram, V., Chartrain, N.A. et al. (2019). Additive manufacturing of hydrocarbon elastomers via simultaneous chain extension and cross-linking of hydrogenated polybutadiene. *ACS Applied Polymer Materials* 1: 684–690.

26 Wallin, T., Pikul, J.H., Bodkhe, S. et al. (2017). Click chemistry stereolithography for soft robots that self-heal. *Journal of Materials Chemistry B* 5: 6249–6255.

27 Zhao, T., Yu, R., Li, S. et al. (2019). Superstretchable and processable silicone elastomers by digital light processing 3D printing. *ACS Applied Materials & Interfaces* 11: 14391–14398.

28 Bhattacharjee, N., Parra-Cabrera, C., Kim, Y.T. et al. (2018). Desktop-stereolithography 3D-printing of a poly (dimethylsiloxane)-based material with sylgard-184 properties. *Advanced Materials* 30: 1800001.

29 Kim, D.S.D. and Tai, B.L. (2016). Hydrostatic support-free fabrication of three-dimensional soft structures. *Journal of Manufacturing Processes* 24: 391–396.

30 Li, T., Jiang, Y., Yu, K., and Wang, Q. (2017). Stretchable 3D lattice conductors. *Soft Matter* 13: 7731–7739.

31 Bruschi, A., Donati, D.M., Choong, P. et al. (2021). Dielectric elastomer actuators, neuromuscular interfaces, and foreign body response in artificial neuromuscular prostheses: a review of the literature for an in vivo application. *Advanced Healthcare Materials* 10: 2100041.

32 Liu, J., Gu, T., Shan, S. et al. (2016). Harnessing buckling to design architected materials that exhibit effective negative swelling. *Advanced Materials* 28: 6619–6624.

33 Li, Y., Qi, Z., Yang, J. et al. (2019). Origami NdFeB flexible magnetic membranes with enhanced magnetism and programmable sequences of polarities. *Advanced Functional Materials* 29: 1904977.

34 Diller, E., Zhuang, J., Zhan Lum, G. et al. (2014). Continuously distributed magnetization profile for millimeter-scale elastomeric undulatory swimming. *Applied Physics Letters* 104: 174101.

35 Venkiteswaran, V.K., Samaniego, L.F.P., Sikorski, J., and Misra, S. (2019). Bio-inspired terrestrial motion of magnetic soft millirobots. *IEEE Robotics and Automation Letters* 4: 1753–1759.

36 Yirmibesoglu, O. D., Morrow, J., Walker, S., et al. (2018). Direct 3D printing of silicone elastomer soft robots and their performance comparison with molded counterparts. *2018 IEEE International Conference on Soft Robotics (RoboSoft)*. pp. 295–302, Livorno, Italy (24–28 April 2018). (IEEE).

37 Shintake, J., Cacucciolo, V., Shea, H., and Floreano, D. (2018). Soft biomimetic fish robot made of dielectric elastomer actuators. *Soft Robotics* 5: 466–474.

38 Zhalmuratova, D. and Chung, H.-J. (2020). Reinforced gels and elastomers for biomedical and soft robotics applications. *ACS Applied Polymer Materials* 2: 1073–1091.

39 TolleyMichael, T., ShepherdRobert, F., GallowayKevin, C. et al. (2014). A resilient, untethered soft robot. *Soft Robotics* 1: 213–223.

40 Zhang, J., Ren, Z., Hu, W. et al. (2021). Voxelated three-dimensional miniature magnetic soft machines via multimaterial heterogeneous assembly. *Science robotics* 6: eabf0112.

41 Xia, N., Jin, D., Pan, C. et al. (2022). Dynamic morphological transformations in soft architected materials via buckling instability encoded heterogeneous magnetization. *Nature Communications* 13: 7514.

42 Xu, T., Zhang, J., Salehizadeh, M. et al. (2019). Millimeter-scale flexible robots with programmable three-dimensional magnetization and motions. *Science Robotics* 4: eaav4494.

43 Kim, J., Chung, S.E., Choi, S.E. et al. (2011). Programming magnetic anisotropy in polymeric microactuators. *Nature Materials* 10: 747–752.

44 Alapan, Y., Karacakol, A.C., Guzelhan, S.N. et al. (2020). Reprogrammable shape morphing of magnetic soft machines. *Science Advances* 6: eabc6414.

45 Cui, J., Huang, T.Y., Luo, Z. et al. (2019). Nanomagnetic encoding of shape-morphing micromachines. *Nature* 575: 164–168.

46 Sydney Gladman, A., Matsumoto, E.A., Nuzzo, R.G. et al. (2016). Biomimetic 4D printing. *Nature Materials* 15: 413–418.

47 Li, S., Deng, B., Grinthal, A. et al. (2021). Liquid-induced topological transformations of cellular microstructures. *Nature* 592: 386–391.

48 Mao, Y., He, Q., and Zhao, X. (2020). Designing complex architectured materials with generative adversarial networks. *Science Advances* 6: eaaz4169.

49 Pal, A., Restrepo, V., Goswami, D., and Martinez, R.V. (2021). Exploiting mechanical instabilities in soft robotics: control, sensing, and actuation. *Advanced Materials* 33: 2006939.

50 Cheng, X., Fan, Z., Yao, S. et al. (2023). Programming 3D curved mesosurfaces using microlattice designs. *Science* 379: 1225–1232.

51 Han, M., Guo, X., Chen, X. et al. (2022). Submillimeter-scale multimaterial terrestrial robots. *Science Robotics* 7: eabn0602.

52 Xue, Z., Jin, T., Xu, S. et al. (2022). Assembly of complex 3D structures and electronics on curved surfaces. *Science Advances* 8: eabm6922.

53 Ni, X., Luan, H., Kim, J.T. et al. (2022). Soft shape-programmable surfaces by fast electromagnetic actuation of liquid metal networks. *Nature Communications* 13: 5576.

54 Bai, Y., Wang, H., Xue, Y. et al. (2022). A dynamically reprogrammable surface with self-evolving shape morphing. *Nature* 609: 701–708.

55 Chen, X. and Yin, J. (2010). Buckling patterns of thin films on curved compliant substrates with applications to morphogenesis and three-dimensional micro-fabrication. *Soft Matter* 6: 5667–5680.

56 Sefati, S., Neveln, I.D., Roth, E. et al. (2013). Mutually opposing forces during locomotion can eliminate the tradeoff between maneuverability and stability. *Proceedings of the National Academy of Sciences* 110: 18798–18803.

57 Yan, D., Pezzulla, M., Cruveiller, L. et al. (2021). Magneto-active elastic shells with tunable buckling strength. *Nature Communications* 12: 2831.

58 Chen, T., Pauly, M., and Reis, P.M. (2021). A reprogrammable mechanical metamaterial with stable memory. *Nature* 589: 386–390.

59 Hu, Y., Chen, X., Whitesides, G.M. et al. (2011). Indentation of polydimethylsiloxane submerged in organic solvents. *Journal of Materials Research* 26: 785–795.

60 Kim, Y., van den Berg, J., and Crosby, A.J. (2021). Autonomous snapping and jumping polymer gels. *Nature Materials* 20: 1695–1701.

61 Dong, Y., Wang, L., Zhang, Z. et al. (2022). Endoscope-assisted magnetic helical micromachine delivery for biofilm eradication in tympanostomy tube. *Science Advances* 8: eabq8573.

62 Yang, S., Wang, Q., Jin, D. et al. (2022). Probing fast transformation of magnetic colloidal microswarms in complex fluids. *ACS Nano* 16: 19025–19037.

63 Yang, L., Jiang, J., Gao, X. et al. (2022). Autonomous environment-adaptive microrobot swarm navigation enabled by deep learning-based real-time distribution planning. *Nature Machine Intelligence* 4: 480–493.

64 Sun, M., Tian, C., Mao, L. et al. (2022). Reconfigurable magnetic slime robot: deformation, adaptability, and multifunction. *Advanced Functional Materials* 32: 2112508.

65 Wang, Q., Du, X., Jin, D., and Zhang, L. (2022). Real-time ultrasound doppler tracking and autonomous navigation of a miniature helical robot for accelerating thrombolysis in dynamic blood flow. *ACS Nano* 16: 604–616.

66 Cai, M., Wang, Q., Qi, Z. et al. (2022). Deep reinforcement learning framework-based flow rate rejection control of soft magnetic miniature robots. *IEEE Transactions on Cybernetics* 1–13. https://doi.org/10.1109/TCYB.2022.3199213.

67 Yang, Z., Yang, L., Zhang, M. et al. (2022). Ultrasound-guided wired magnetic microrobot with active steering and ejectable tip. *IEEE Transactions on Industrial Electronics* 70: 614–623.

68 Yousefi, A.-M., Smucker, B., Naber, A. et al. (2018). Controlling the extrudate swell in melt extrusion additive manufacturing of 3D scaffolds: a designed experiment. *Journal of Biomaterials Science, Polymer Edition* 29: 195–216.

69 Wu, S., Ze, Q., Zhang, R. et al. (2019). Symmetry-breaking actuation mechanism for soft robotics and active metamaterials. *ACS Applied Materials & Interfaces* 11: 41649–41658.

70 Song, H., Lee, H., Lee, J. et al. (2020). Reprogrammable ferromagnetic domains for reconfigurable soft magnetic actuators. *Nano Letters* 20: 5185–5192.

71 Deng, H., Sattari, K., Xie, Y. et al. (2020). Laser reprogramming magnetic anisotropy in soft composites for reconfigurable 3D shaping. *Nature Communications* 11: 6325.

72 Lakes, R. (2001). A broader view of membranes. *Nature* 414: 503–504.

73 Lakes, R.S. (2017). Negative-Poisson's-ratio materials: auxetic solids. *Annual Review of Materials Research* 47: 63–81.

74 Lu, H., Zhang, M., Yang, Y. et al. (2018). A bioinspired multilegged soft millirobot that functions in both dry and wet conditions. *Nature Communications* 9: 3944.

75 Gu, H., Boehler, Q., Cui, H. et al. (2020). Magnetic cilia carpets with programmable metachronal waves. *Nature Communications* 11: 2637.

76 Liu, H., Taylor, B., and Curet, O.M. (2017). Fin ray stiffness and fin morphology control ribbon-fin-based propulsion. *Soft Robotics* 4: 103–116.

77 Dong, X., Lum, G.Z., Hu, W. et al. (2020). Bioinspired cilia arrays with programmable nonreciprocal motion and metachronal coordination. *Science Advances* 6: eabc9323.

**78** Ruiz-Torres, R., Curet, O.M., Lauder, G.V., and MacIver, M.A. (2013). Kinematics of the ribbon fin in hovering and swimming of the electric ghost knifefish. *Journal of Experimental Biology* 216: 823–834.

**79** Therriault, D., White, S.R., and Lewis, J.A. (2003). Chaotic mixing in three-dimensional microvascular networks fabricated by direct-write assembly. *Nature Materials* 2: 265–271.

**80** Jiang, S., Hu, Y., Wu, H. et al. (2019). Multifunctional Janus microplates arrays actuated by magnetic fields for water/light switches and bio-inspired assimilatory coloration. *Advanced Materials* 31: 1807507.

**81** Yan, Y., Hu, Z., Yang, Z. et al. (2021). Soft magnetic skin for super-resolution tactile sensing with force self-decoupling. *Science Robotics* 6: eabc8801.

**82** Hu, H., Zhang, C., Pan, C. et al. (2022). Wireless flexible magnetic tactile sensor with super-resolution in large-areas. *ACS Nano* 16: 19271–19280.

# 3

# Carbon-Based Miniature Soft Robots with Rolled-up Concept

## 3.1 Introduction

Soft actuators, as an important research branch of miniature soft robots [1, 2], have acquired great attention and enabled promising applications in multiple fields such as biomimetic devices [3–6], soft robots [4, 5], artificial muscles [7], intelligent sensors [8], and electric generators [9]. Inspired by biological behaviors changing with surrounding environments (the opening and closing of pinecones in the dry and wet conditions [10], the closing of *Mimosa pudica* when touching [11, 12], the color changes of chameleons upon being dangerous [13], directional growth of sunflowers toward sunshine [14], etc.), soft actuators are developed to convert external stimuli (humidity [4, 5, 15], heating [16], light [17, 18], electricity [19], chemicals [20, 21], and magnetism [22, 23]) into mechanical motions, displayed in the form of structural changes. Multi-layer structures with dissimilar materials, gradient structures in the same materials, and orientational material structures in/on another material are widely accepted as the deformation strategies to induce the programmable structural changes of soft actuators [4, 5, 15, 24–26]. The basic materials for building soft actuators consist of hydrogels-based materials [3], shape memory polymers [27], liquid crystal polymers [28], and carbon-based materials [18]. Among these materials, carbon-based materials are superior candidates for constructing high-performance soft actuators due to their high mechanical strength, excellent chemical and physical stability, good thermal and electrical conductivity, and outstanding light absorption capability [18, 29, 30]. In addition, carbon-based materials can be easily functionalized and oxidized for constructing smart actuators.

In this chapter, we will focus on carbon-based soft actuators by discussing their categories, working mechanisms, responsive sources, programming strategies for structural changes, and advanced applications. We expect this chapter will help readers obtain a comprehensive and adequate understanding of this field and also promote the development of real applications of carbon-based soft actuators. Additionally, the deformation mechanisms are anticipated to be applied in other related fields.

*Untethered Miniature Soft Robots: Materials, Fabrications, and Applications*, First Edition.
Li Zhang, Jiachen Zhang, Neng Xia, and Yue Dong.

## 3.2 The Choices of Carbon-Based Materials

Carbon-based materials with excellent properties and easy reactions for introducing functional groups benefiting for processing have been widely researched for constructing soft actuators. The carbon-based materials family contain fullerenes, graphite, amorphous carbon, diamond, carbon nanotubes (CNTs), graphene, and graphene-based derivatives (Figure 3.1a) [18, 29, 30]. Among these materials, CNTs, graphene, and graphene-based derivatives are widely used in constructing soft actuators, whereas graphite and amorphous carbon are only used in several actuators.

CNTs are reported to have excellent physical properties including strong mechanical strength (1 TPa), superior light adsorption (600–1000 nm), high electrical

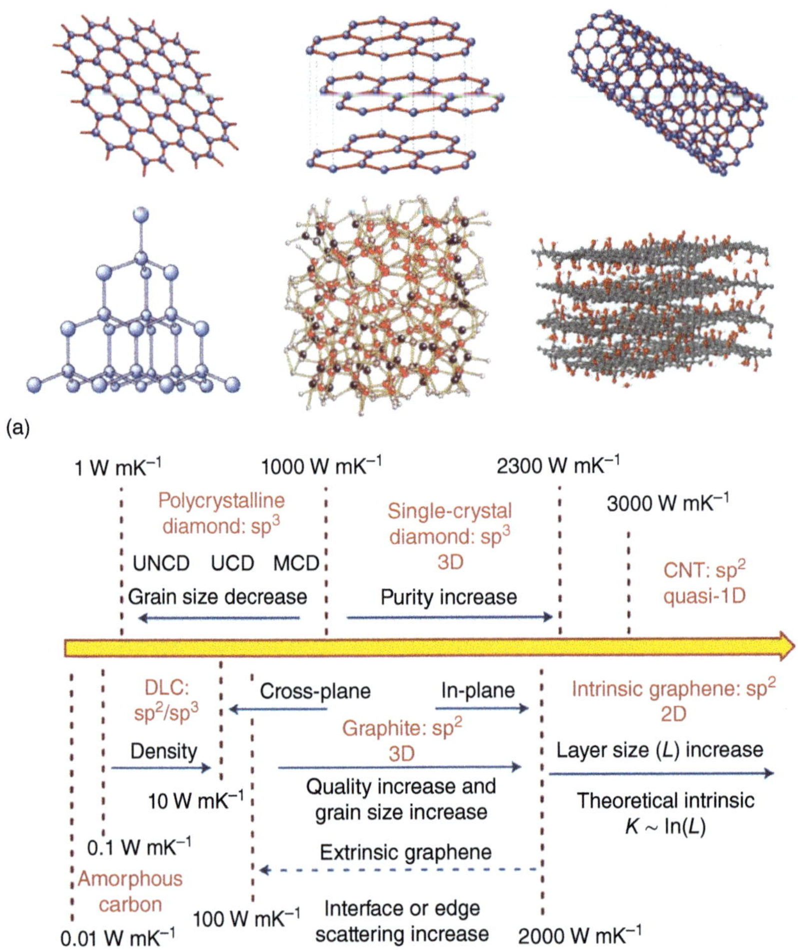


**Figure 3.1** (a) The structures of graphene, graphite, CNT, diamond, amorphous carbon, and GO. Source: Han et al., Medhekar et al., Wang et al. [18, 31, 32]. (b) Thermal conductivities of different carbon materials. Source: Balandin et al. [33]/Springer Nature.

conductivity (electron mobility of 10 000 $cm^2$ $V^{-1}$ $s^{-1}$ at room temperature), outstanding thermal conductivity (maximum: 6000 W $m^{-1}$ $k^{-1}$) (Figure 3.1b), and less coefficient of thermal expansion (CTE) [18]. Certainly, the geometrical structures of CNTs (large length-to-diameter ratio) tend to be employed to achieve directional arrangement for anisotropic structural changes of soft actuators [34, 35]. CNTs are generally classified into two types of single-walled CNTs (SWCNTs) and multi-walled CNTs (MWCNTs) based on the number of carbon layers. The optical absorptions of SWCNTs are dependent on their specific chirality, which can be adjusted for photothermal actuators with desirable wavelength-selective properties [36]. Currently developed technology for the chirality purification of CNTs has enabled the purpose. CNTs are also processed to form films and CNTs fibers with single orientation and films with scattered orientations, laying the foundation for constructing film- and fiber-based soft actuators. Additionally, CNTs are incorporated into polymers as fillers to fabricate soft actuators [37]. CNTs are also easy to be oxidized for the introduction of oxygen-containing functional groups, providing the ability of humidity response for CNTs-based actuators.

Graphene and its derivatives are the star materials in the field of soft actuators. Since the 2010 Nobel Prize in Physics was awarded for the discovery of graphene, studies on graphene have shown an explosive increase. The charming properties of graphene are highly pursued, such as the thinnest material in the universe, large specific surface area (~2630 $m^2$ $g^{-1}$), high chemical stability, strong mechanical strength (breaking strength: 42 N $m^{-1}$; Young's modulus: 1 TPa), excellent electrical conductivity (electron mobility of 20 000 $cm^2$ $V^{-1}$ $s^{-1}$ at low temperature), high light adsorption ability, distinctively negative CTE, and outstanding thermal conductivity (~5000 W $m^{-1}$ $K^{-1}$ at room temperature) (Figure 3.1b) [38]. These properties are beneficial to soft actuators. Graphene films are also successfully developed by chemical vapor deposition (CVD) and layer-by-layer assembly for film-based actuators. Graphene sheets are often filled into polymers for constructing soft actuators [39]. For the graphene-based derivatives of graphene oxide (GO) and reduced GO (RGO), although some properties are lower than graphene due to the destruction of graphene's conjugated structure and the introduction of oxygen-containing groups, the hydrophilicity and processibility of graphene-based derivatives are greatly enhanced. GO, existing in two forms of GO solutions and GO liquid crystals based on its concentration, can be easily processed into fibers, films, and foams for constructing corresponding GO-based soft actuators [4, 40]. Additionally, GO is easily reduced into RGO, which can be processed and assembled into film- and foam-based actuators. GO and RGO sheets also are good fillers for constructing soft actuators. Considering these advantages mentioned above, more and more graphene-based soft actuators are being developed.

Graphite is an optically opaque, black, and thermodynamically stable carbon materials, which has the properties of wide-band adsorption (ultraviolet light [UV] to near-infrared [NIR]) and high photothermal conversion ability [41]. Graphite can be obtained by stacking multilayers of graphene that are combined by weak van der Waals forces. This special structure of graphite grants it soft and brittle mechanical properties, so the graphite is easily slid and separated and cannot be used as a standalone structure. Graphite tends to be attached onto the surface of soft materials in a drawing style to construct soft actuators.

Amorphous carbon has unique isotropic nature in contrast to anisotropic carbon materials of CNTs, graphene, and graphite. Amorphous carbon materials are often incorporated into polymers (SU-8 or poly(dimethylsiloxane) [PDMS]) to form a uniform thermal conductive network for photothermal actuators [42]. Though the current actuators can move quickly on the water surface relying on the Marangoni effect, they do not exhibit structural changes. So, the amorphous carbon-based actuators are not in the list discussed below.

We will mainly focus on CNTs-, graphene-, and graphite-based soft actuators.

## 3.3 The Working Mechanism of Carbon-Based Materials

The working mechanism of carbon-based materials depends on the power source.

Light, heating, humidity, electricity, and magnetic field have been chosen as the power source. Heat will be produced by light irradiation by virtue of the excellent photothermal effect of carbon-based materials. Additionally, untethered soft robots are reviewed and discussed in this book, while the electricity-driven soft actuators are wire-connected types owing to the required input of current. Also, the magnetic field-driven soft actuators mainly rely on magnetic elements and have less relationship with carbon-based materials. Thus, we will focus on the working mechanisms of light- and humidity-driven soft actuators, which have been classified into many types in some review papers. We will summarize and discuss more mechanisms that reflect the important and dominant roles of carbon materials in this chapter.

### 3.3.1 Light-Induced Expansion

Light-induced expansion can also be defined as photothermal expansion due to the thermal expansion induced by the photothermal effect of carbon-based materials. The shape deformation of materials induced by photothermal expansion is a widely accepted phenomenon, in which the light irradiation lifts the surrounding temperature of materials due to the increasing molecular kinetic energy. Photothermal expansion tends to be found in the asymmetric materials system, particularly in actuators with layered structures. Individual and uniform materials with the same CTE have exhibited isotropic shape changes. CTE, an intrinsic parameter of materials, is often used to determine the degree of thermal expansion. The combination of materials with different CTEs is a prerequisite for designing responsive soft actuators. As mentioned above, carbon-based materials have relatively small and even negative CTE compared with some common polymers such as poly(vinylidene fluoride) (PVDF), PDMS, biaxially oriented polypropylene (BOPP), wax, polypropylene (PP), polyethylene (PE), polyurethane (PU), and polycarbonate (PC) [43]. This indicates that carbon/polymer actuators have more possibilities for high-performance soft actuators. Carbon materials, serving as a light-absorbing and heat-conduction layer, will convert light energy into heat that increases the local temperature of actuators. Due to the distinct CTE of carbon and polymer materials, temperature-induced expansion differences result in the large mismatch at the materials interface that makes actuators roll up.

Graphene nanoplatelets (GNPs) are uniformly dispersed into the PDMS and further combine with pure PDMS to form a GNPs-PDMS/PDMS actuator [44]. The actuator displays different bending modes upon IR light stimulus from its two sides (Figure 3.2a,b). Tubular GNPs-PDMS/Au bilayer actuator is also developed utilizing

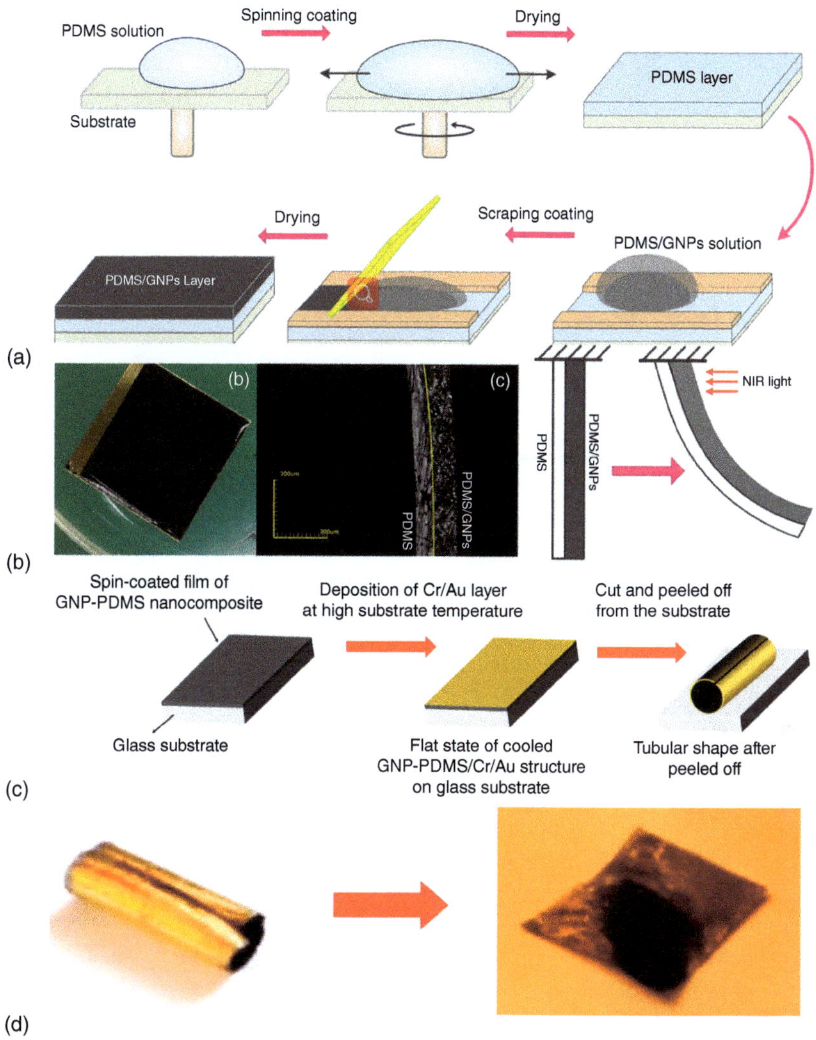


**Figure 3.2** (a) The schematic diagram of fabricating GNPs-PDMS/PDMS actuator. Source: Niu et al. [44]/Springer Nature/CC BY 4.0/Public domain. (b) The optical image, cross-sectional image, and the expansion mechanism under IR light of actuator. Source: Niu et al. [44]/Reproduced from Springer Nature/CC BY 4.0. (c) The fabrication process and (d) the structural changes under IR light of tubular GNPs-PDMS/Au bilayer actuator. Source: Leeladhar et al. [45]/Reproduced from American Chemical Society.

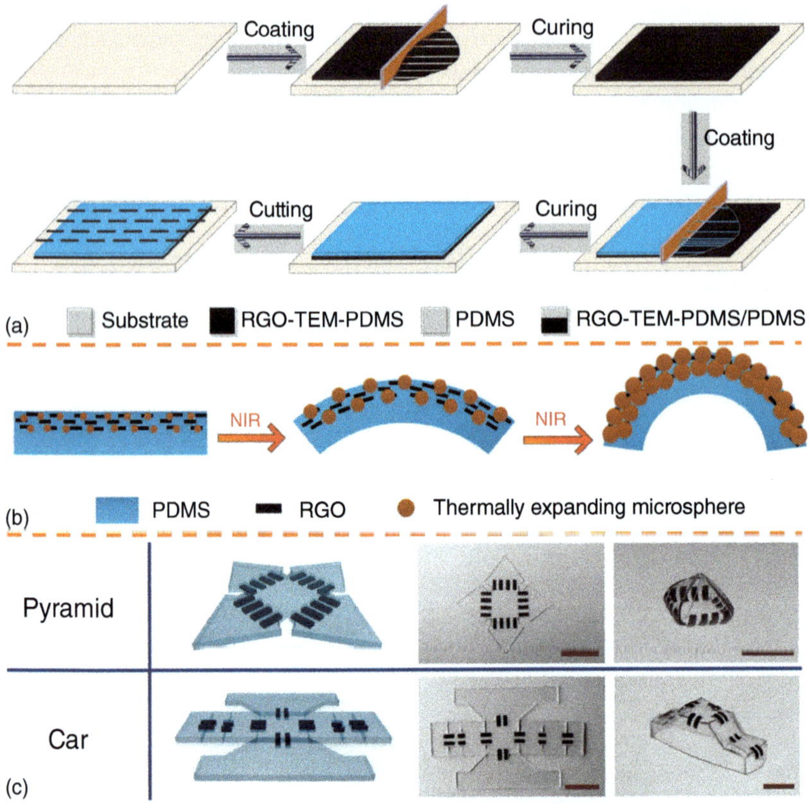


**Figure 3.3** (a) The fabrication process of RGO-TEMs-PDMS/PDMS bilayer actuator. (b) The schematic diagram of the expansion mechanism of the bilayer actuator irradiated by IR light. (c) The formed 3D structures of pyramid and car after IR light exposure. Source: Tang et al. [46]/Reproduced from John Wiley & Sons, Inc./CC BY 4.0.

the residual stress when the actuator is peeled from the glass substrate, showing a large, fast, and reversible response when irradiated by IR light (Figure 3.2c,d) [45].

Tang et al. also incorporated RGO and thermally expanding microspheres (TEMs) into the PDMS for fabricating RGO-TEMs-PDMS/PDMS bilayer actuator (Figure 3.3a,b) [46]. RGO adsorbs the IR light and converts it into heat to increase the temperature of the actuator. The actuator exhibits enhanced deformation behavior under IR light due to the large volume expansion of TEMs ($\approx$4000%) (Figure 3.3b) [46]. The actuation behavior of the actuator is irreversible, which results in permanent 3D deformation structures without the continuous supply of light energy. It also has been employed as hinges with tunable sizes to form various 3D structures (Figure 3.3c) [46].

Chen and coworkers achieved the transfer of stacked graphene assembly (SGA) onto the PE film (Figure 3.4a), and the resulting SGA/PE actuators display swift morphing behaviors under variable temperature and IR light illumination (Figure 3.4b) [47]. By employing pre-heating of SGA/PE and subsequent cooling, rolled actuators with PE layer wrapped are constructed. Lateral IR light illumination causes the

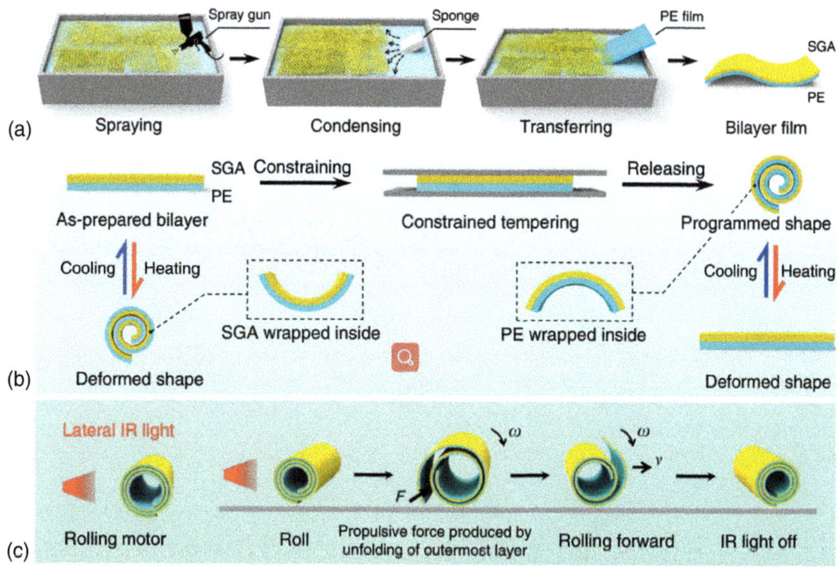


**Figure 3.4** (a) Schematic illustration of the fabrication process of SGA/PE bilayer actuators using the Langmuir–Blodgett method. (b) Schematics showing the process of constrained tempering on SGA/PE actuators, and the thermal-induced shape-morphing of an SGA/PE film and a tempered SGA/PE film stimulated by temperature and IR light. (c) SGA/PE bilayer roll as a rolling motor under the stimulus of lateral IR light. Source: Wang et al. [47]/Springer Nature/CC BY 4.0/Public domain.

rolled actuator unrolling and rolling locomotion (Figure 3.4c). These rolled actuators also are assembled into multiple-wheel systems to achieve controlling locomotion.

Dong et al. developed GO/PPy actuators by local chemical reaction with agarose stamp, the expansion of PPy under IR light and high temperature makes the actuator programmable structural changes [4]. They also proposed GO/wax actuators by printing wax patterns onto the GO film, the expansion of wax under IR light induces the programmable rolling up of GO/wax actuators [5]. Additionally, the light-driven expansion mechanism is also used to construct these graphene-based actuators including small GO (SGO)/PVDF [48], GO/PVDF [49], GO/BOPP [50], RGO/BOPP [51], and GO/PC actuators [52].

Apart from graphene-based materials, the CNTs layer acquired by natural-drying of CNTs suspension is combined with a PDMS layer to form the rolled CNT/PDMS bilayer actuator (Figure 3.5a) [37]. Due to the CTEs difference between CNT and PDMS layer, the thermal stress generated in the PDMS curing process is released to form the rolled actuator, which exhibits sunlight-driven actuations with fast response, large deformation, and multiple biomimetic motions (jumping, wobbling, and crawling) (Figure 3.5b,c) [37]. Next, the same group develops a curled droplet-shaped CNT/PDMS bimorph actuator, realizing self-oscillating motion in a wavelike fashion with a light-mechanical negative feedback loop created by constant white light irradiation [53]. The specific chirality of SWCNTs was used to investigate SWCNT/PC bilayer actuators with wavelength-selective properties. These actuators

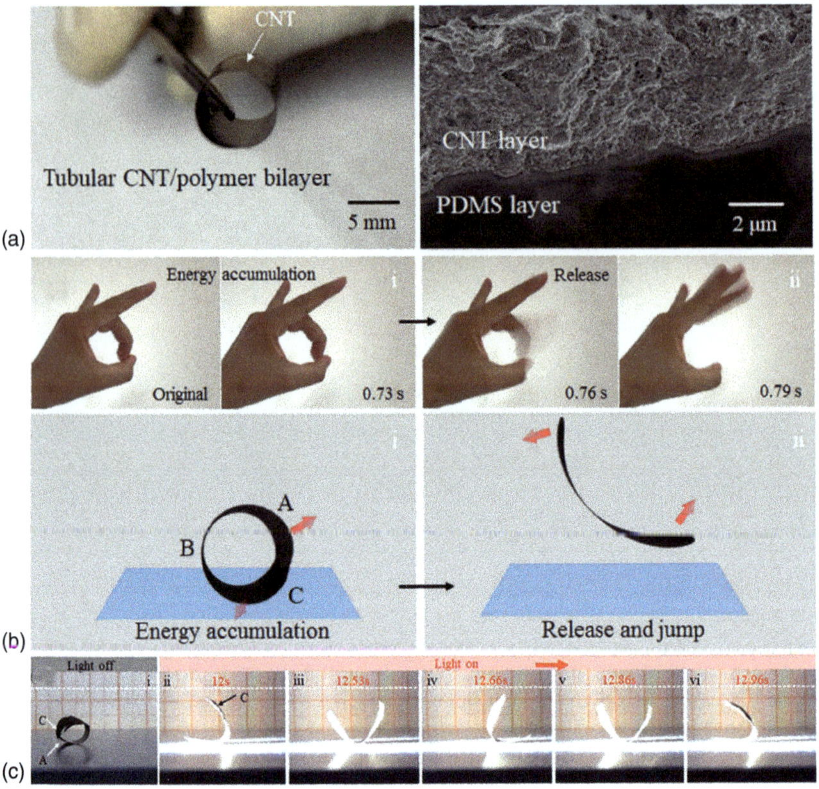


**Figure 3.5** (a) Optical image and cross-sectional SEM image of the rolled CNT/PDMS bilayer actuator. (b) Schematic illustration of the jumping of the tubular soft CNT/PDMS actuator mimicking the flicking finger behavior. (c) The optical images of light-induced wobbling motion with different incident directions. Source: Hu et al. [37]/Reproduced from John Wiley & Sons, Inc.

offered fast response, large deformation, and high sensitivity to selected wavelength ranges [36]. The orientations of CNTs play an important role in improving the actuating performance of actuators due to stronger mechanical strength along the axis. Peng et al. controlled the orientations of aligned CNTs (ACNTs) to construct ACNT-Wax/PI actuators, endowing these resulting actuators with fast response (milliseconds) to IR light and phototropic/apheliotropic bending behaviors [34]. There are also some other actuators working by light-induced expansion, such as CNT-Paper/BOPP [54], CNT-methylcellulose/BOPP [55], CNT/PVDF [56], ACNT/polyethylene terephthalate (PET)-ink [57], and CNT-PDMS/PVDF/PDMS actuator [58].

Additionally, a graphite-based actuator in Graphite/Paper/BOPP sandwiched structure was also developed, and the actuator showed a large bending actuation driven by light (Figure 3.6a,b) [59]. Carbon black particles with a graphite-like structure are dispersed onto PU film to form the programmable graphite/PU actuator, displaying complex reversible deformations under IR light irradiation (Figure 3.6c)

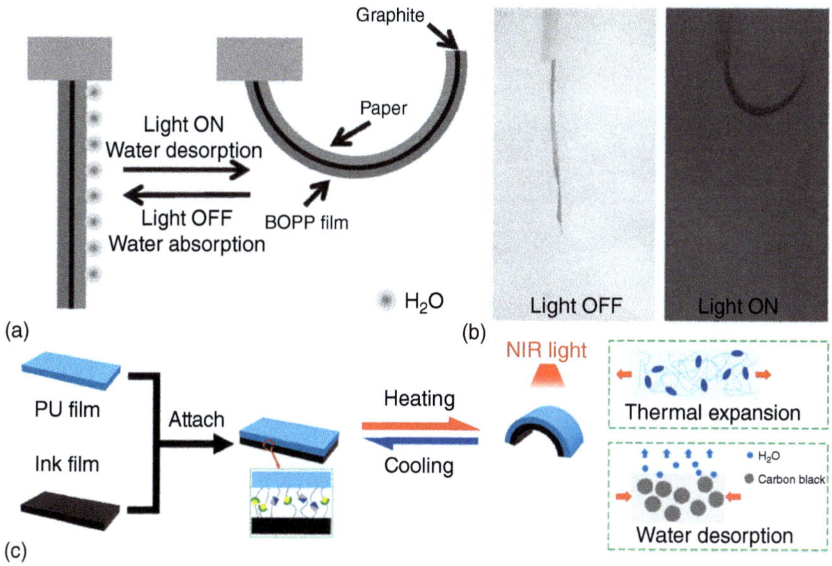


**Figure 3.6** (a) The schematic diagram and (b) optical images for the structural changes of Graphite/Paper/BOPP sandwiched actuator. Source: Weng et al. [59]/Reproduced from John Wiley & Sons, Inc. (c) Schematic illustration of reprogrammable multi-functional actuators based on carbon black/PU bilayer film. Source: Lin et al. [60]/Elsevier.

[60]. To further enhance the physical and actuating performance of the actuators, the combination of carbon materials may be a good choice. Graphite-CNT/PP [61] and CNT-PDMS/GO [62] actuators have also been reported.

### 3.3.2 Light-Induced Shrinkage

Apart from photothermal expansion, the light-induced shrinkage can also change the volume of materials due to the desorption of guest molecules from materials under light illumination. These materials tend to have reversible adsorption and desorption ability. For example, the typical guest molecule of water molecules can be easily adsorbed by some hydrophilic materials (cellulose, polydopamine [PDA], poly(*N*-isopropylacrylamide) [PNIPAAm]) in the environment with high humidity, and these water molecules are rapidly released with the increasing temperature. Meanwhile, the volume of these materials is increased and reduced with the adsorption and desorption of water molecules, respectively. Hence, the combination of hydrophilic materials (water adsorption), carbon materials (heat production), and hydrophobic materials is the basic design principle for photothermal actuators. SWCNTs have been incorporated into the network of PNIPAAm hydrogel strips, acting as hinges on the low-density PE (LDPE) film to develop programmable SWCNTs-PNIPAAm/LDPE bilayer actuator with the reversible thermal and light responsiveness (Figure 3.7a,b) [63]. The resulting actuator can fold itself into a cube when it was subjected to IR light irradiation and then recovered to its initial shape when the IR light was removed. It was believed that the water

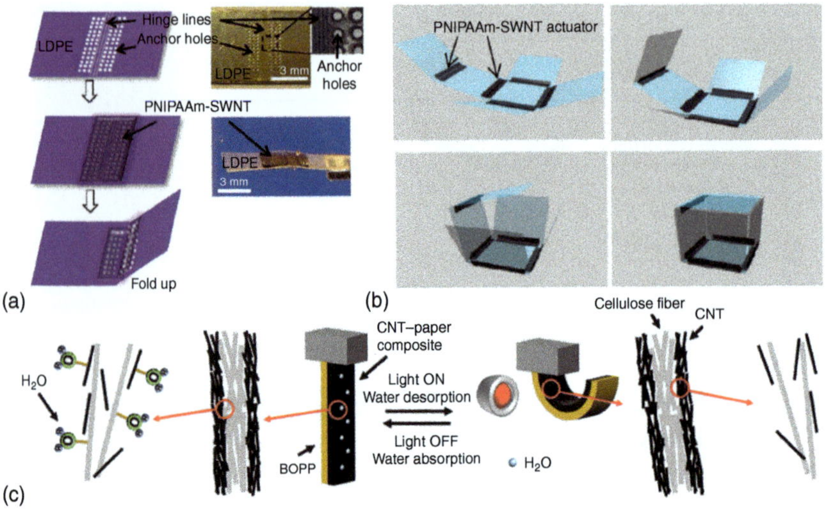


**Figure 3.7** (a) Fabrication scheme for making SWCNTs-PNIPAAm/LDPE bilayer actuators and the corresponding optical images. Source: Zhang et al. [63]/Reproduced from American Chemical Society. (b) Schematic illustration of the structural changes for this bilayer actuator. Source: Zhang et al. [63]/American Chemical Society. (c) Schematic illustration for the structural changes mechanism of CNT Paper/BOPP bilayer actuators. Source: Zhou et al. [54]/Royal Society of chemistry.

desorption from the network of PNIPAAm hydrogel caused the shape deformation of SWCNTs-PNIPAAm/LDPE actuator. Acid-treated CNTs are also bonded onto the cellulose fiber of paper by strong hydrogen bonds to construct CNT-Paper/BOPP bilayer actuators. The desorption of water molecules absorbed onto the cellulose is a reason for the shape changes of the actuators under IR stimulus (Figure 3.7c) [54].

Ji et al. employed hydrophilic PDA-modified RGO and Norland Optical Adhesive (NOA)-63 to construct PDA-RGO/NOA-63 bilayer actuator, showing fast bending performance under IR light irradiation (Figure 3.8a,b) [64]. The conversion of NIR light by RGO into thermal energy has promoted the desorption of water molecules in the PDA–RGO layer. By utilizing the hydrophilicity of PDA and GO and the hydrophobicity of RGO, PDA-modified GO is combined with RGO to form the PDA-GO/RGO bilayer actuator. Due to the excellent photothermal effect of GO and RGO, the water molecules adsorbed in the PDA-GO can be released quickly under the NIR light irradiation, increasing the deformation speed and degree of actuators [65].

Specially, GO is both a carbon material and a hydrophilic material, making it a popular material for photothermal actuators. Unmodified GO has also been directly integrated onto the hydrophobic material layers to construct bilayer actuators such as GO/PPy [4], GO/wax [5], GO/RGO [66], GO/MXene [67], SGO/PVDF [48], GO/PVDF [49], GO/BOPP [50], and GO/PC actuators [52]. Water molecules will be released from the GO layers of all these bilayer actuators after light irradiation, inducing the shrinkage of GO layers and further the deformation of actuators. Certainly, as discussed in the light-induced expansion part, these hydrophobic layers of

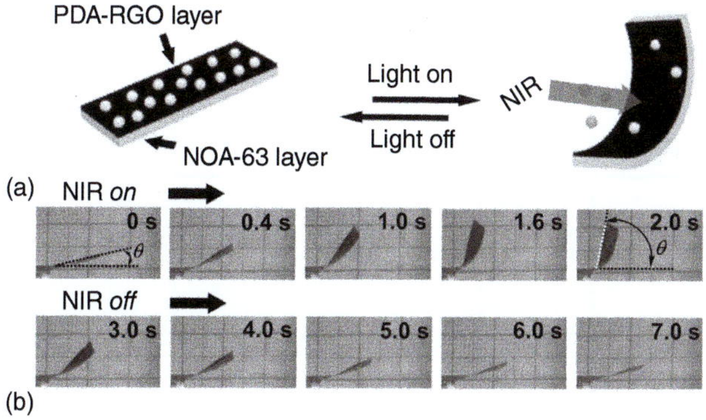


**Figure 3.8** (a) The schematic diagram for the bending/unbending actuation of a PDA-RGO/NOA-63 actuator with periodic NIR irradiation. (b) The bending and unbending processes of a PDA-RGO/NOA-63 bilayer actuator with NIR light turning on and off. Source: Adapted from Ji et al. [64]/Reproduced from John Wiley & Sons, Inc.

actuators other than GO/RGO and GO/MXene also contribute to the deformation of actuators. The expansion of the RGO and MXene layer after light irradiation can be neglected. Notably, most of the GO/RGO actuators are fabricated by light-induced reactions. Thus, a chemical reaction (thermal or optical reduction of hydrophilic oxygen groups) may occur when the GO layer is irradiated by light, which makes the GO not stable and brings serious challenges for long-term applications of actuators. Nevertheless, deformation based on light-induced shrinkage is superior to chemical reactions due to the rapid deformation of actuators in a non-contact manner.

### 3.3.3 Humidity-Induced Deformation

Humidity-responsive carbon-based actuators mainly refer to GO-based actuators. The modification of hydrophilic materials onto other carbon-based materials also has the potential in humidity-responsive actuators. Compared with graphene, CNTs, and graphite, GO has a strong interaction with water molecules due to the presence of oxygen-containing groups (epoxy, carboxyl, and hydroxyl groups). The binding energy values of these three groups are calculated to be 0.23, 0.36, and 0.20 eV per $H_2O$, respectively (Figure 3.9) [25]. Apart from the strong GO–water interaction, the ultrafast diffusion rate of water molecules within GO sheets also benefits the humidity sensitivity of GO-based actuators. Geim and coworkers demonstrated the ultrafast permeation of water through a stacked GO film [68]. The water molecules transport within GO film in three different mechanisms based on the situations of in-plane defects, sheets edges, and interlayer space. GO film also allows relatively high pure water flux, which is attributed to two major points of strong GO–water interaction and the unique nanolayer structure of GO film acting as quantum-confined superfluidic channels for water transportation (Figure 3.9) [25]. When the GO film adsorbs water molecules, its interlayer space increases, resulting in its rapid swelling. Conversely, the interlayer space of GO film decreases when it releases water molecules, leading to its rapid shrinkage. The

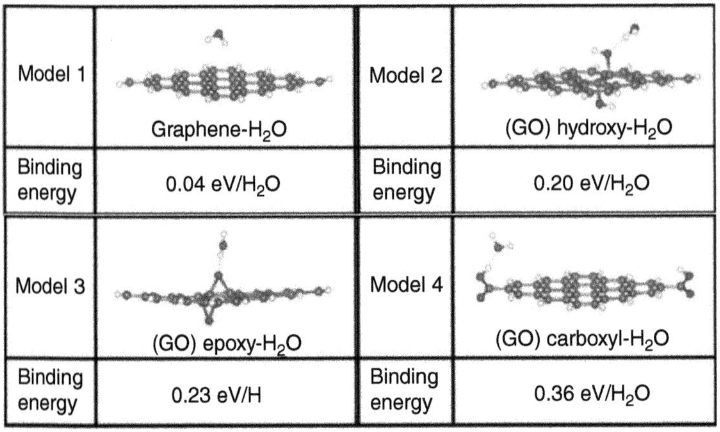


**Figure 3.9** The binding energy between a water molecule and graphene as well as GO with different oxygen groups based on first-principles study. Source: Zhang et al. [25]/John Wiley & Sons.

sensitive volume change of GO film with the changing humidity makes it ideal material for constructing actuators. However, the isotropic property of individual and uniform GO film cannot cause any apparent deformation. Several approaches have been applied to change the isotropic property of GO film for the formation of GO-based actuators. They have been classified into three types of forming layered configuration with other materials, forming structure gradient, and forming humidity gradient.

The layered configurations of GO-based actuators are achieved by combining GO film with other humidity-inert material layers. Ruoff and coworkers combined GO and MWCNT through vacuum filtration to form the GO/MWCNT bilayer actuator [69], of which the MWCNT appears to be inert to water molecules (Figure 3.10a,b). With the increase of relative humidity (RH), the GO/MWCNT bilayer actuator deforms toward the MWCNT side due to the expansion of GO layer (Figure 3.10c). The oriented MWCNT is also combined with GO film through vacuum filtration to form the all-carbon actuators with the ultrafast response (0.08 seconds), the ultralarge deformation (angle change per length 70° $mm^{-1}$), and the controllable deformation direction (Figure 3.10d,e) [70].

Dong et al. achieved the patterning of PPy layer and wax layer onto the GO film by agarose stamp and wax printing techniques [4, 5]. The developed bilayer actuators display programmable and complex deformation when stimulated by increasing RH. Following the same principle, humidity-responsive actuators of GO/PDMS [71], GO/SU-8 [24], GO/MXene [67], GO/BOPP [50], GO/PVDF [49], GO/CNT-PDMS [62], and GO/Nano-G@PVDF [72] are also fabricated and explored.

Generally, the combination of GO and other materials within these above actuators is through relatively weak van der Waals force. Thus, these actuators may delaminate between layers after frequent deformation. To solve this issue, GO films or fibers tend to be partially reduced by chemical/physical strategies to get the GO/RGO actuators. These strategies consist of laser reduction, UV reduction,

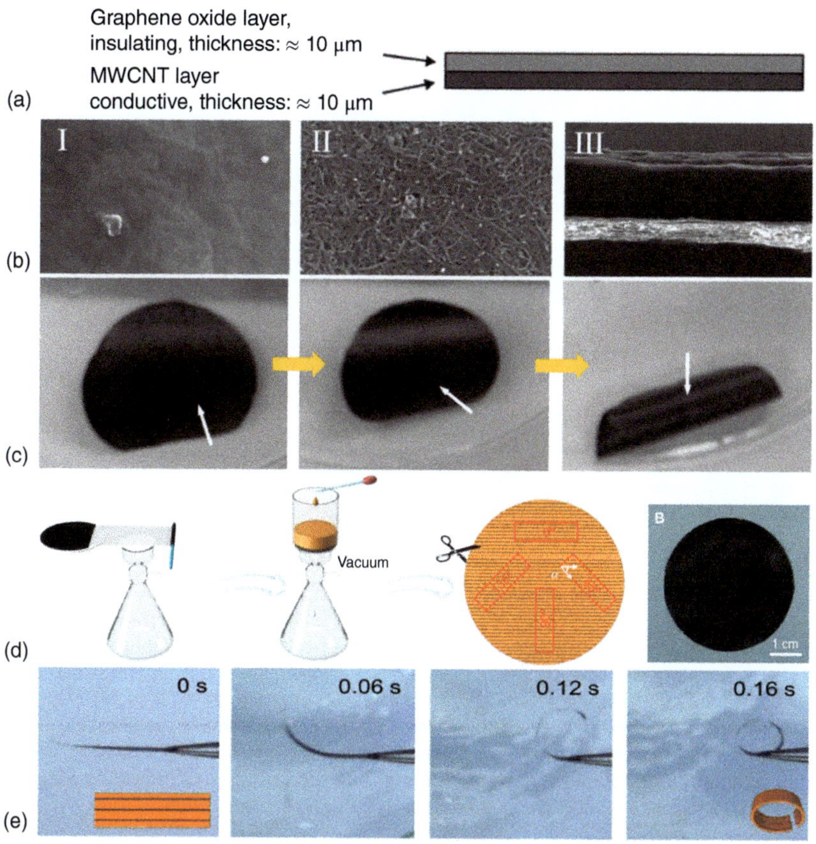


**Figure 3.10** (a) The schematic diagram of GO/MWCNT bilayer actuator. Source: Park and coworkers [69]/John Wiley & Sons. (b) The SEM images of GO side and MWCNT side as well as the cross-sectional SEM image of GO/MWCNT bilayer actuator. Source: Park and coworkers [69]. (c) The structural changes of GO/MWCNT bilayer actuator under the stimulus of humidity. Source: Park and coworkers [69]/Reproduced from John Wiley & Sons, Inc. (d) The fabrication process and (e) structural changes of oriented MWCNT/GO all-carbon actuator. Source: Li et al. [70]/Reproduced from American Chemical Society.

sunlight reduction, and metal reduction. The limited light transmittance enables the gradient of oxygen-containing groups within GO films and fibers along the direction of light. Qu and coworkers reported laser-selective reduction of GO fibers for developing GO/RGO fiber actuators with sophisticated shape changes (Figure 3.11a–d) [73]. Sun and coworkers reported the self-controlled photoreduction (UV [66] and sunlight [74]) of GO film to develop GO/RGO actuators with programmable shape changes (Figure 3.11e–g). In addition to light reduction, the gradient assembly of nanomaterials (magnetic particles [75] and polymer nanospheres [26]) within GO film is also achieved to avoid delamination during frequent deformation due to the absence of material interfaces.

The structure gradient mainly relies on creating the asymmetrical structure of GO film. These different structures on the GO film determine the arrangement

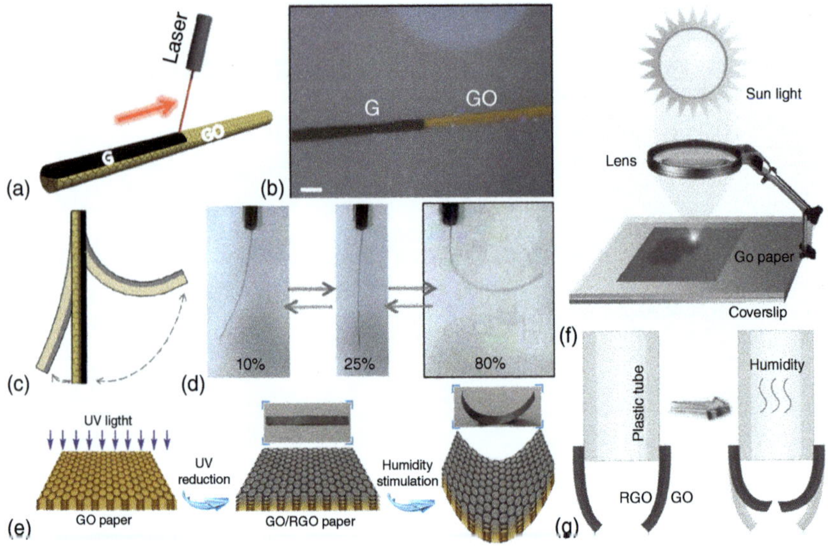


**Figure 3.11** (a) The schematic diagram for laser reduction on one side of a GO fiber. Source: Cheng et al. [73]/John Wiley & Sons. (b) The optical image of the top surface of the as-prepared asymmetric GO/RGO fiber actuator. Source: Cheng et al. [73]/Reproduced from John Wiley & Sons, Inc. (c) The schematic diagram and (d) optical images for the structural changes of GO/RGO fiber actuator. Source: Cheng et al. [73]/John Wiley & Sons. (e) The schematic diagram for UV light reduction on one side of GO film to fabricate GO/RGO actuator and their structural change. Source: Han et al. [66]/John Wiley & Sons. The schematic diagram for (f) sunlight reduction on one side of GO film to fabricate GO/RGO actuators and (g) their structural change. Source: Han et al. [74]/John Wiley & Sons.

and orientation of GO sheets, which cause apparent differences in water sensitivity between structures and further induce the deformation of GO-based actuators. The principle of forming a structure gradient is inspired by the flat-to-helical transition of the opening seed pods induced by the internal cell arrangement and orientation [76]. Cheng et al. designed an asymmetrical GO film with a rough surface on one side, exhibiting excellent ability of structural changes when altering the RH [40]. Liu and coworkers employed soft lithography to create GO film with unilateral microstructures that act as quantum-confined superfluidic channels to facilitate the adsorption, transport, and desorption of water (Figure 3.12a–c) [25]. The asymmetrical GO film displays fast bending against the grating direction toward the flat side after changing the surrounding RH (Figure 3.12d). Wang et al. proposed a geometrical design of period-gradient structures in GO films (Figure 3.12e) [77]. They exhibit excellent responding performance and could bend to 1000° in several seconds at a high RH (Figure 3.12f). 3D printing and constrained drying are combined to control the high compaction and alignment of GO sheets in GO film [78]. The developed GO film displays apparent humidity response behavior. These above GO films without material interfaces may have improved durability and stability.

The humidity gradient enables the fast and large-scale bending of uniform carbon-based films. Ge and He et al. reported uniform GO film [79] and GO/PDA

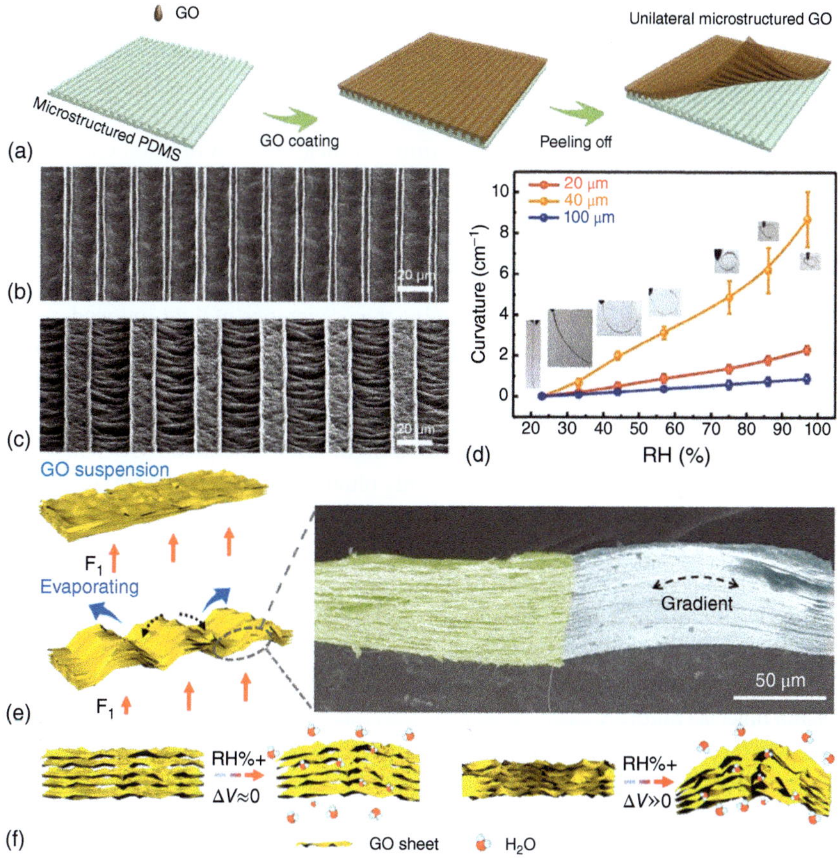


**Figure 3.12** (a) The schematic diagram of the fabrication of structured GO film; top view SEM images of structured GO film with periods of (b) 20 μm and (c) 40 μm; (d) The curvature changes of the fabricated actuator with increasing RH. Source: Zhang et al. [25]/Reproduced from John Wiley & Sons, Inc. (e) Fabrication mechanism and cross-sectional SEM image of the GO film with a sectionalized microstructure. Source: Wang et al. [77]/Reproduced from American Chemical Society. (f) Schematic diagram of the actuating mechanism in the microstructure. Source: Wang et al. [77]/American Chemical Society.

film [80]. The one side of the film directly exposed to the humidity will adsorb more water molecules than the other side, resulting in the deformation of these films. However, the humidity gradient is hard to precisely control compared with forming layered configurations and structure gradient. Thus, the humidity gradient may be a complementary strategy to induce the deformation of actuators [15].

## 3.4 The Programming Shape Changes of Actuators and Their Applications

The structural changes of actuators tend to determine their functions, which further determine their application fields. The designable and controllable structural changes of actuators should be well explored. There are many programming strategies that have been reported to achieve the aim by distributing deformation factors within the actuators. The main principle is to control the anisotropic behaviors for different parts of actuators. These strategies include light-induced local modification, non-light-induced local modification, self-healing, and oriental control of elements in actuators.

### 3.4.1 Light-Induced Local Modification

The light-induced local modification contains the occurrence of light-induced reactions and productions. Sun and coworkers reported the light (sunlight, UV light, and laser)-induced local reactions to form RGO patterns on the GO film [7, 66, 74]. Under the exposure of focused sunlight, the GO was selectively reduced into the expanded RGO along the movement path of focused sunlight. The expanded RGO will suppress the sunlight penetration, causing the only photoreduction on the surface layer of GO film to form a bilayer RGO/GO actuator. The developed actuators are used in smart claws, orientable transporter, and crawler. They also confirm that photoreduction mainly depends on the UV light in sunlight. Next, UV light was used to achieve a local reduction of GO film. A smart graphene "tendril" was programmed by making RGO diagonal stripes on the GO film under the assistance of a shadow mask, which can curl to a helical structure under high humidity for picking up and releasing objects (Figure 3.13).

The same group also applied one-step laser printing of programmable photothermal patterns on a bilayer structure of poly(methyl methacrylate) and GO embedded with gold nanorods to develop light-driven hand, Venus flytrap, and spider robot (Figure 3.14a) [7]. The light-driven hand with programmable nodes has the ability of light-addressable manipulation and every node could bend freely under laser irradiation (Figure 3.14b,c). The midrib and the cilia region of Venus flytrap were controlled by switching a filament lamp on and off to perform reversible capture and release actions (Figure 3.14d,e). The spider robot can bend its legs to deform into 3D structures under light irradiation. By irradiating its legs one by one, the legs bend one after another, resulting in the directional displacement of the gravity center of the spider robot. In this way, the robot can walk freely through the remote control to lean forward and move toward the right direction at an average speed of 2.5 mm $s^{-1}$ (Figure 3.14f,g). The group also reported laser-printing-specific RGO patterns on a bilayer structure of polymer tape and GO. The fabricated programmable actuators with bidirectional bending performance are used in the smart robotic crane and curtain (Figure 3.14h) [81].

Qu and coworkers achieved precise laser reduction of GO on the GO fiber and GO film. The construction of region-specifically asymmetric RGO/GO fiber is conducted in a well-controlled "spot reduction" upon laser irradiation. Single RGO region (0.5–1.0 mm long), multiple RGO regions, alternating RGO segments, and spiral RGO patterns are achieved in GO fibers to program GO-based actuators with complex and reversible deformation [73]. These actuators deform like a hinge, hook, S-shape, and spring structure once exposed to moisture, respectively (Figure 3.15a).

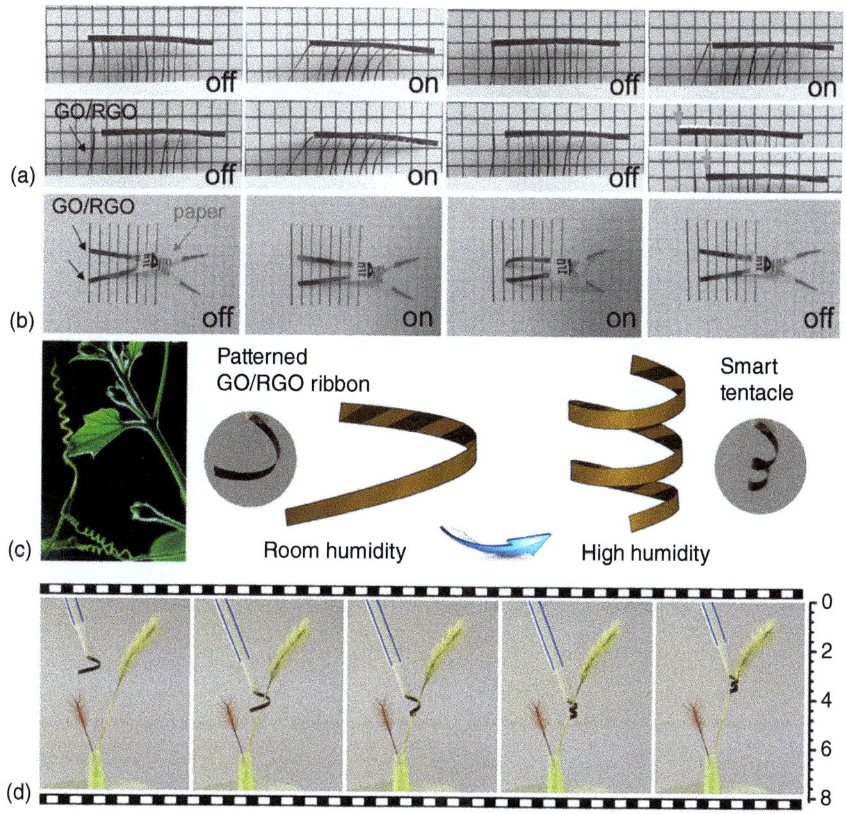


**Figure 3.13** (a) An orientable transporter fabricated by assembling 7 upright GO/RGO ribbons into a linear array, which is used to drive a paper slice toward the right. Source: Han et al. [74]/Reproduced from John Wiley & Sons, Inc. (b) A crawler paper robot fabricated by using GO/RGO bilayer paper as smart legs, which can crawl to right by turning on and turning off the humidity alternately. Source: Han et al. [74]/Reproduced from John Wiley & Sons, Inc. (c) The schematic diagram of smart graphene "tendril" inspired from tendril climber plant. Source: Han et al. [66]/John Wiley & Sons. (d) The smart "tendril" was used for picking up objects by curling to helical structure under high humidity. Source: Han et al. [66]/Reproduced from John Wiley & Sons, Inc.

Some developed actuators are applied as crawling robots, woven devices, and smart textiles (Figure 3.15b). Selective laser writing of RGO patterns on the GO film is reported by the same group to develop responsive actuators that are used in smart perceptual fingers and walking robots (Figure 3.15c) [8].

Apart from these light-induced reactions based on the GO film and fiber, light-induced production of other materials-based patterns on the GO film is also reported to develop programmable actuators. Sun and coworkers reported humidity-responsive GO-based actuators that have programmable shape-changing capability by coupling SU-8 patterns with GO film [24]. SU-8 micropattern arrays with specific geometries, distributions, and orientations were programmed on a continuous GO film to get better control over predictable and complex deformations,

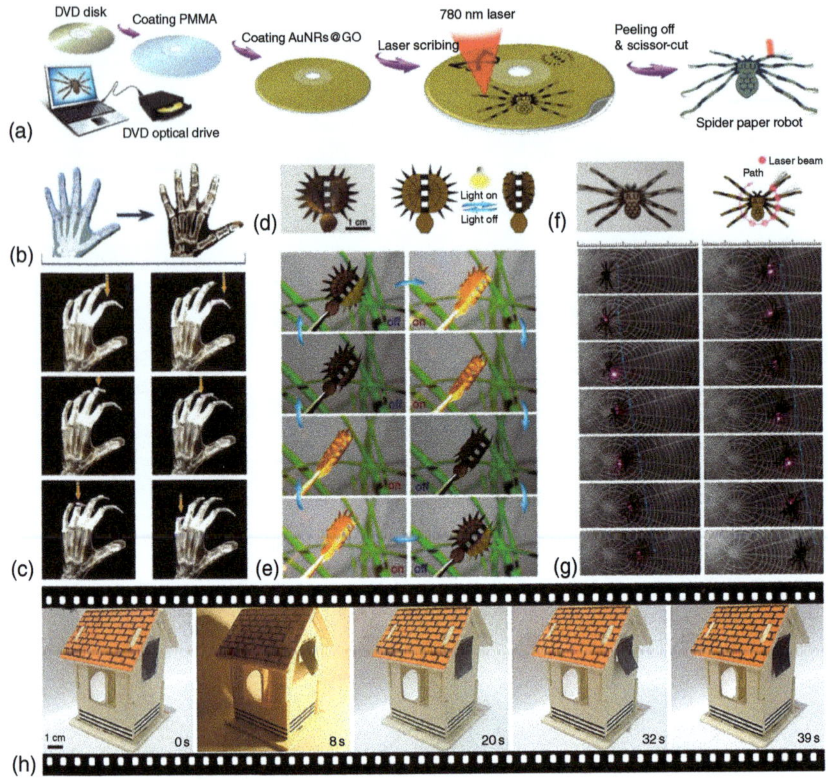


**Figure 3.14** (a) The schematic diagram for laser scribing of programmable photothermal patterns and driving mechanism of a bio-inspired robot. (b) The schematic diagram, optical image, and (c) structural changes of a light-assisted hand with a set of disabled hand skeletons. (d) The optical image and schematic diagram of a light-driven mechanism for Venus flytrap. (e) The structural changes of Venus flytrap irradiated by IR light. (f) The optical image and schematic diagram of light-driven mechanism for spider robot. (g) The structural changes of spider robot irradiated by IR light. (a–g) Source: Han et al. [7]/Reproduced from John Wiley & Sons, Inc. (h) The optical images of the smart curtain. Source: Ma et al. [81]/Reproduced from John Wiley & Sons, Inc.

including twisting, coiling, bending, 3D folding, asymmetric bending, and the combinations of these. As a proof-of-concept for practical usage, smart "dancer robots" that can dance with the humidity change and a smart "caterpillar robot" mimicking the creep of a caterpillar were developed.

### 3.4.2 Non-light-Induced Local Modification

The non-light-induced local modification strategy refers to the precise patterning of other materials on the carbon materials-based films without the action of light. The precise patterning can be achieved by chemical reactions and physical methods (printing technology and template method). Although the local modification of polypyrrole (PPy) on the GO film by self-oxidation–reduction and local reduction

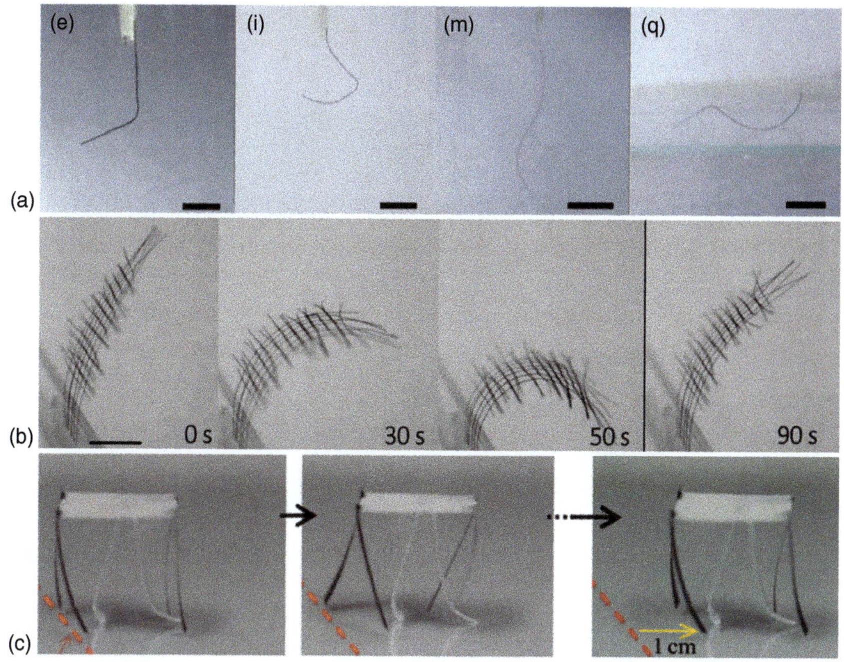


**Figure 3.15** (a) GO fibers with region-confined laser reduction. Source: Cheng et al. [73]/Reproduced from John Wiley & Sons, Inc. (b) The structural changes of these GO fibers at RH = 80%. Source: Cheng et al. [73]/Reproduced from John Wiley & Sons, Inc. (c) The optical image of a hand-woven textile-like network of RGO/GO fibers under the stimulus of humidity (RH = 80%) at different exposure times. Source: Cheng et al. [8]/Reproduced from American Chemical Society.

of GO-sodium alginate (GO-SA) film with filter paper filled with hydroiodic acid [82] have been achieved, the modification precision is poor and actuators with complex deformation are hard to construct. Dong et al. employed precise and fast local chemical reactions to fabricate programmable bilayer GO-based actuators with the assistance of a solvent-loaded agarose hydrogel stamp [4]. Compared with light-induced reactions, there are many types of chemical reactions, which are beneficial for the screening of high-performance actuators. They have achieved precise patterning of local modification of PPy, polyaniline (PANI), polyethylene dioxythiophene (PEDOT), and calcium alginate hydrogel on GO film (Figure 3.16a), and PPy on GO fiber and foam. GO/PPy bilayer actuators are chosen as representative actuators owing to their outstanding performance. The programmable structural changes of GO/PPy actuators with PPy patterns of regular triangle, square, regular pentagon, regular hexagon, and H-, U-, S-, and T-shape are explored under the stimuli of humidity and IR light (Figure 3.16b–d). Inspired by the structures of a hawk's claw and tendril climber plants, smart GO/PPy grippers with cross and helical structures at high humidity are developed to capture/release target objects dozens of times their weight (Figure 3.17a–d). Furthermore, inspired by the movement of the inchworm, a soft walking robot is programmed to move

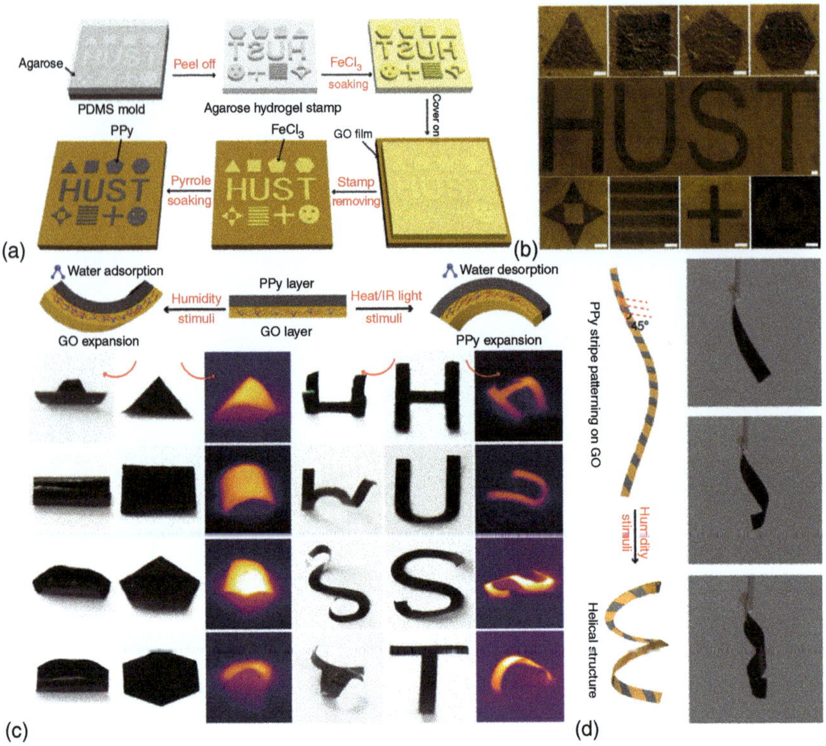


**Figure 3.16** (a) The schematic diagram of precise PPy patterning on a GO film for the fabrication of GO/PPy actuators. (b) The optical image of precise PPy patterning on a GO film. (c) The structural changes of GO/PPy bilayer actuators with different PPy patterns under the stimuli of humidity and IR light. (d) The schematic diagram of a programmable GO/PPy actuator with a helical structure under the stimulus of humidity. Source: Dong et al. [4]/Reproduced from Springer Nature/CC BY 4.0.

forward for a long distance under the interval stimulus of IR light (Figure 3.17e,f). This demonstrates that these programmed actuators can intelligently change their structures and various GO-based programmable actuators may be fabricated for some potential applications via the presented method.

To achieve the fast fabrication of high-performance actuators with superior reproductivity, Dong et al. further reported precise printing of designable wax patterns on the GO film with A4 size to construct GO/wax actuators (Figure 3.18a–c) [5]. Considering the mature and stable printing technology, the GO/wax actuators can be produced quickly and in large quantities. The hand-like pattern, the spaced rectangular pattern, the planar pattern of the seedpod, the planar pattern of the orchid, and the flower-like patterns with three, four, and five petals are printed on the GO film to fabricate programmable GO/wax actuators (Figure 3.18d–g). By employing single-sided printing, double-sided printing, and multiple printing, three smart shape-adaption grippers are developed to stably grasp and release heavy objects with different shapes under the stimulus of humidity. An autonomous soft crawling robot that can move forward by employing the self-oscillating behavior of a soft robot under alternate

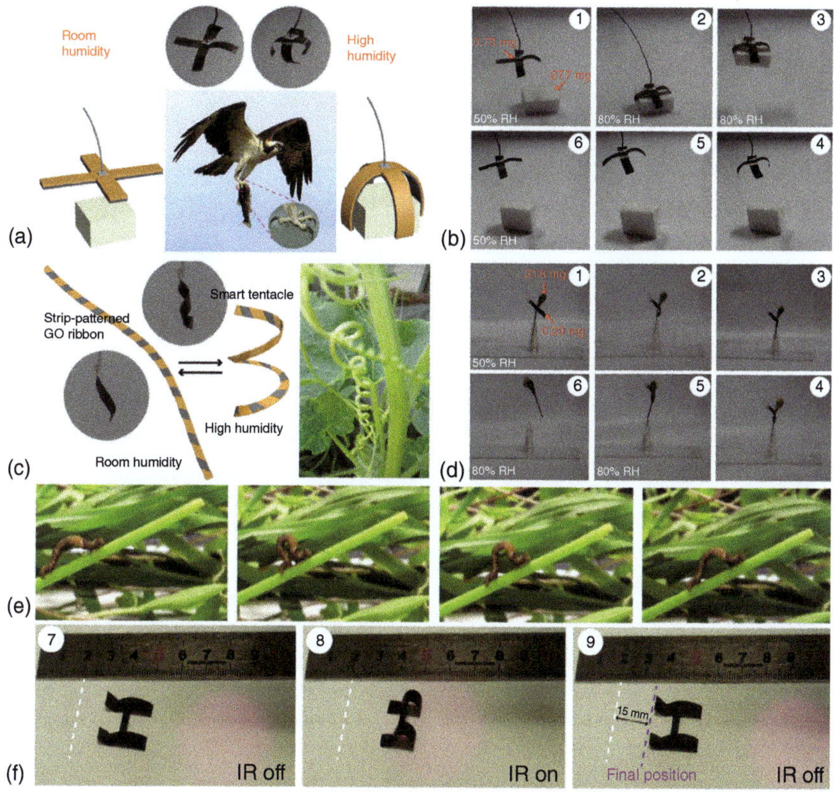


**Figure 3.17** (a) The schematic diagram of a smart GO/PPy gripper with a cross structure inspired by the claw of a hawk. (b) The picking up and releasing of a foam by the smart gripper. (c) The schematic diagram of a smart GO/PPy tendril inspired by a tendril climber plant. (d) The removal of a flower from a vase by the helical GO/PPy gripper. (e) The movement process of the inchworm. (f) The movement process of the soft inchworm-inspired walking robot under the interval stimulus of IR light. Source: Dong et al. [4]/Reproduced from Springer Nature/CC BY 4.0.

changes of surrounding temperature was also explored, which may be applied for rational utilization of ubiquitous low-grade waste heat (Figure 3.19). Wang et al. adopted a template method to precisely distribute discrete SGA patches on both sides of PE film that are pre-covered by template with specific dimensions. After transferring the SGA patterns and then peeling off the template, programmable SGA/PE bilayer actuators are generated, which have complex shapes including V-shape, spiral structure, and helical structure.

### 3.4.3 Self-Healing

The self-healing strategy mainly depends on the special property of GO film. The high content of hydrogen bond network on GO films has been demonstrated to be decisive for their self-healing ability, so they can be easily assembled for shape-programmable actuators with the assistance of water moisture. After

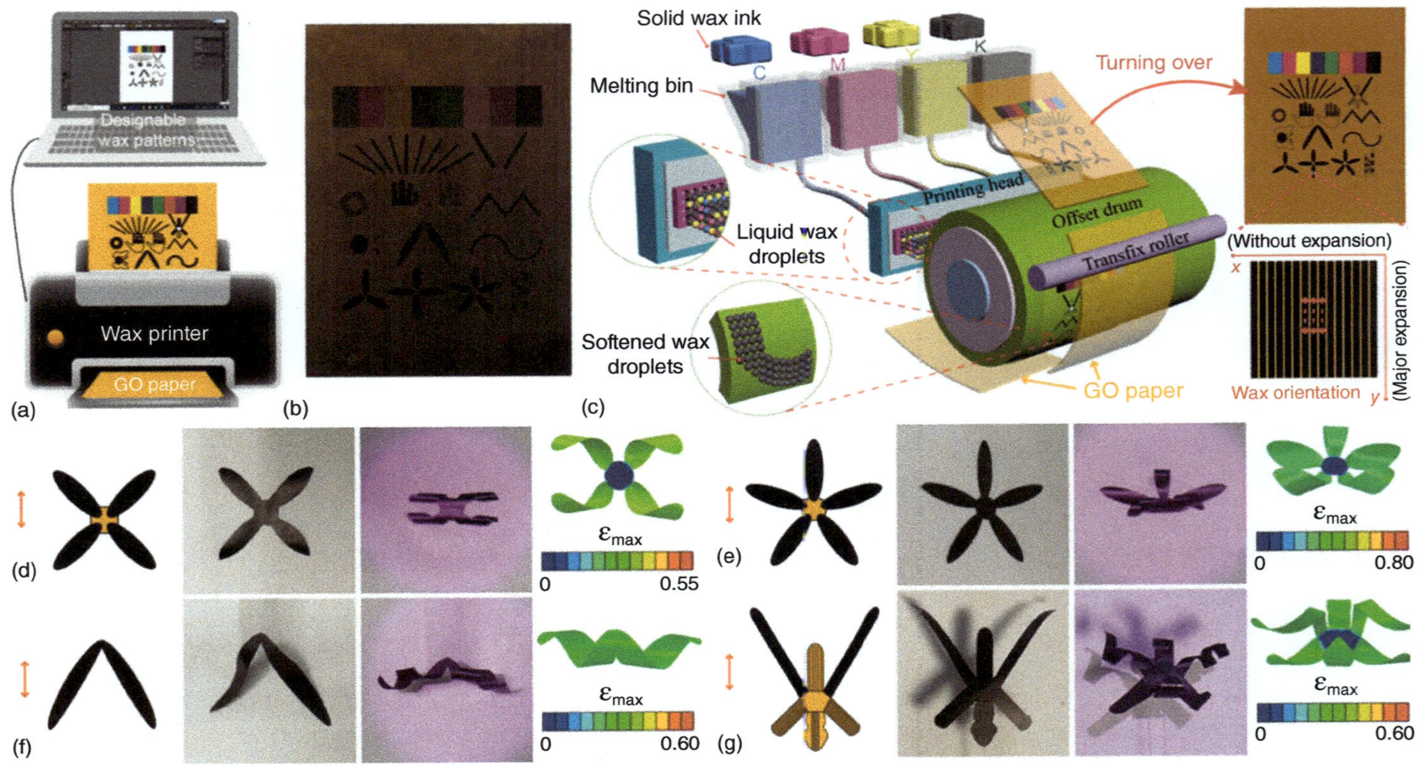


**Figure 3.18** (a, b) The total process and (c) schematic diagram for wax printing of designable wax patterns composed of orientated wax fibrils onto GO paper with A4 size for the fabrication of a programmable GO/wax actuator. (d) The diagrams, optical images, structural changes under the stimulus of IR light, and simulated deformation results of actuators with (d, e) flower-like patterns with four and five petals. (f) planar pattern of seedpod; as well as (g) planar pattern of orchid. Source: Dong et al. [5]/Reproduced from Elsevier.

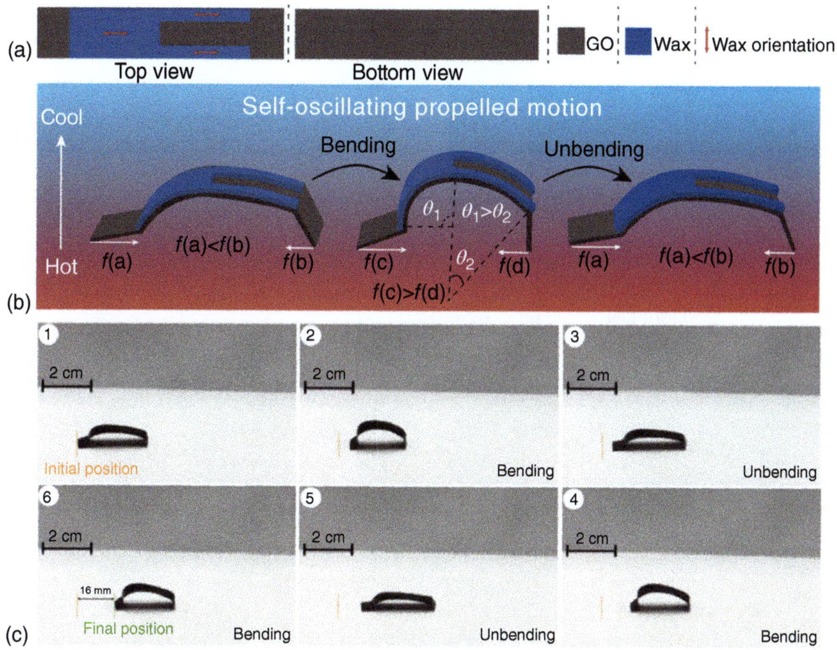


**Figure 3.19** (a) The planar structure of the soft robot and the orientation of wax fibrils. (b) The schematic diagram of the self-oscillating propelled movement of the soft crawling robot on the hot plate. (c) The actual movement process of the soft crawling robot. Source: Dong et al. [5]/Reproduced from Elsevier.

exposing the fracture surfaces of GO film to the water moisture, the GO sheets within these surfaces are wetted and swelled, which further partially breaks the hydrogen-bonding interaction between GO sheets. Next, the wetted GO films are physically fit together at their fracture surfaces to make them contact closely with each other at the interfaces owing to the interdiffusion of GO sheets and the capillary force. In this process, water has been considered a good dispersing solvent for GO sheets to endow the fluidity of GO. When the water moisture is removed, the interspacing of swelled GO sheets is gradually reduced to let individual GO sheets recombine together and reconstruct hydrogen bonds. Thus, different GO films are assembled to form a monolithic actuator. Based on the self-healing mechanism, Cheng et al. have developed a variety of smart GO-based actuators with self-healing ability. These actuators originate from an asymmetric GO film (a-GO) that has a rougher surface on one side than the other side. The ribbon-shaped a-GO actuator can bend into a semicircular arc at high RH. After cut and then healed, the a-GO actuator still maintains favorable shape changeable ability [40]. Similarly, a lizard-tail-shaped a-GO film also bends freely and works well after being cut and healed. A self-healing a-GO walker that can crawl forward on a ratcheted substrate recovers its motion capability to about 88% value of the original state. Apart from these film-based actuators, a self-healing twisted GO fiber-based actuator is demonstrated. Owing to the helical configuration of the fiber-based actuator,

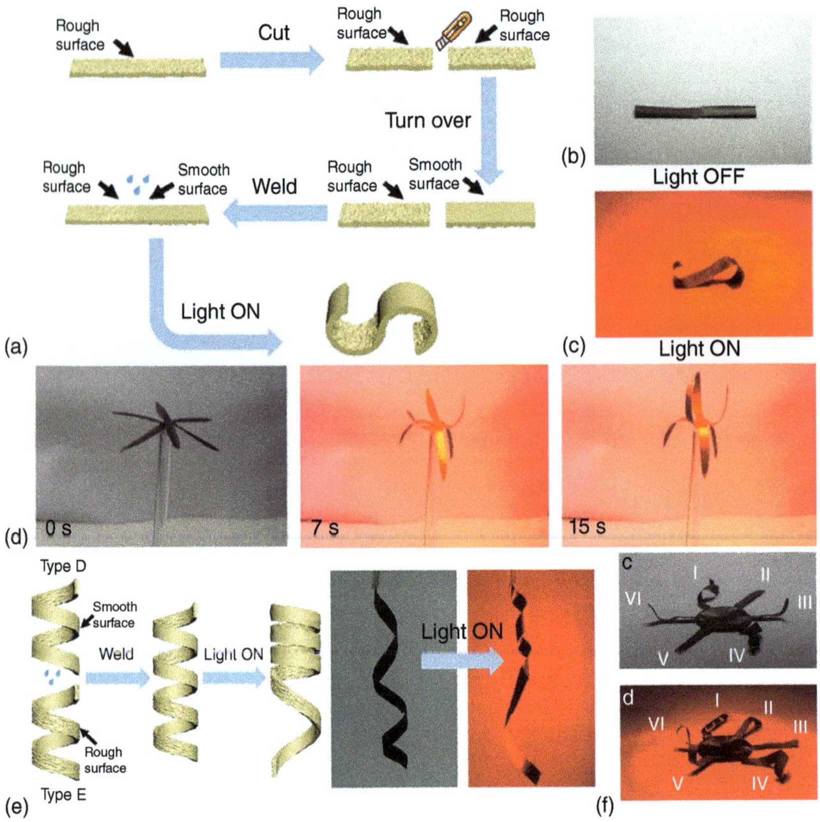


**Figure 3.20** (a) The schematic diagram of the fabrication process and actuation mechanism of the S-shape GO actuator; the optical images of the S-shape GO actuator (b) without and (c) under NIR light irradiation. (d) The optical images for the bionic flower-shape actuator irradiated by NIR light (e) The schematic diagram for the fabrication process and actuation mechanism of the tendril-shape actuator, and the optical images of this actuator irradiated by NIR light. (f) The optical images of the fabricated six-in-one octopus-shape actuator without and under NIR light irradiation. Source: Weng et al. [83]/Reproduced from American Chemical Society.

its rotating speed can reach 3700 rotations per minute. The broken actuator can recover its rotating speed to about 94% upon being healed.

Weng et al. achieve more complex a-GO shape-morphing actuators using a self-healing strategy. They widen the morphing direction of actuators by turning over certain a-GO films in the self-healing process. When the ends of two a-GO films with one turned over are welded together, an actuator with S-shaped deformation is obtained (Figure 3.20a) [83]. The left part and right parts of the actuator bend toward opposite directions upon IR light irradiation, making it deform into S-shape (Figure 3.20b,c). Six petal-like a-GO films are assembled to form a bionic flower-shaped actuator with unique blooming motion, in which six petals bend upward and downward in sequence (Figure 3.20d). In addition to the flat a-GO films, the pre-shaped a-GO films with multiple types are also assembled to form more complex actuators (Figure 3.20e,f). Additionally, only a few programmable actuators are exhibited, but this study will unlock a new way to programmable, self-healing, and light-driven actuators with tunable and complex shape-morphing.

### 3.4.4 Oriental Control of Elements in Actuators

The deformation behaviors of plants with the environment and life stage changing are controlled by aligning cellulose fibrils produced by themselves within their extracellular matrices. The expansion and shrinkage of matrices are restricted in the alignment direction of cellulose fibrils, while the deformation easily happens in the transversal direction due to less structural constraint. It was learned from that, oriental control of elements in actuators benefits the arrangement of the anisotropic behaviors in different parts of actuators. The deformation direction depends on the oriental arrangement of the physical structures of carbon materials. CNTs, with a similar shape to cellulose fibril, are a good choice to construct actuators by mimicking the deformation behaviors of plants. Peng's group presented visible-light-responsive actuators with tunable and diverse deformations by incorporating aligned CNTs into the paraffin wax layer on a polyimide film [34]. The CNTs and paraffin wax act as cellulose fibrils and extracellular matrices, respectively. These actuators hold multiple types of photomechanical deformation (from phototropic/apheliotropic bending to 3D helical buckling) controlled by tuning the CNT alignment directions. By integrating these multiple deformations, a light-manipulated mechanical arm is assembled to grasp objects intelligently. Similarly, programmable actuations are also achieved by embedding aligned CNTs into GO films [70]. The all-carbon (GO-CNT) actuators exhibit large-size, reversible deformation 2D directional bending, and 3D chiral twisting in response to humidity and NIR light. Super-aligned CNTs and carbon-based ink are modified onto both sides of PET film to develop programmable actuators [57]. The alignment direction of CNTs mainly programs the structural changes of corresponding actuators upon IR irradiation.

The aligned laser-induced graphene (LIG) microstructures developed by the direct laser writing (DLW) method are reported to serve as geometrically constraining materials to precisely program the structural changes of LIG/polyimide bilayer soft actuators (Figure 3.21a) [84]. By employing the advantages of DLW, the shape and alignment direction of actuators are easily tailored. The resulting actuators show great potential for a range of applications, including sensing, artificial muscles, and soft robotics (Figure 3.21b). GO sheets are also controlled to spontaneously align themselves in the sodium alginate (SA) matrix to form GO-based actuators with programmable deformations via a direct-ink-writing printing process [78]. The shear force during extrusion results in the alignment of GO sheets. The alignment direction is identical to the printing direction (Figure 3.21c). By designing the printing pathway of every layer, various actuators are printed, which further transform into complex 3D architectures upon stimulation by heat, light, and humidity (Figure 3.21d).

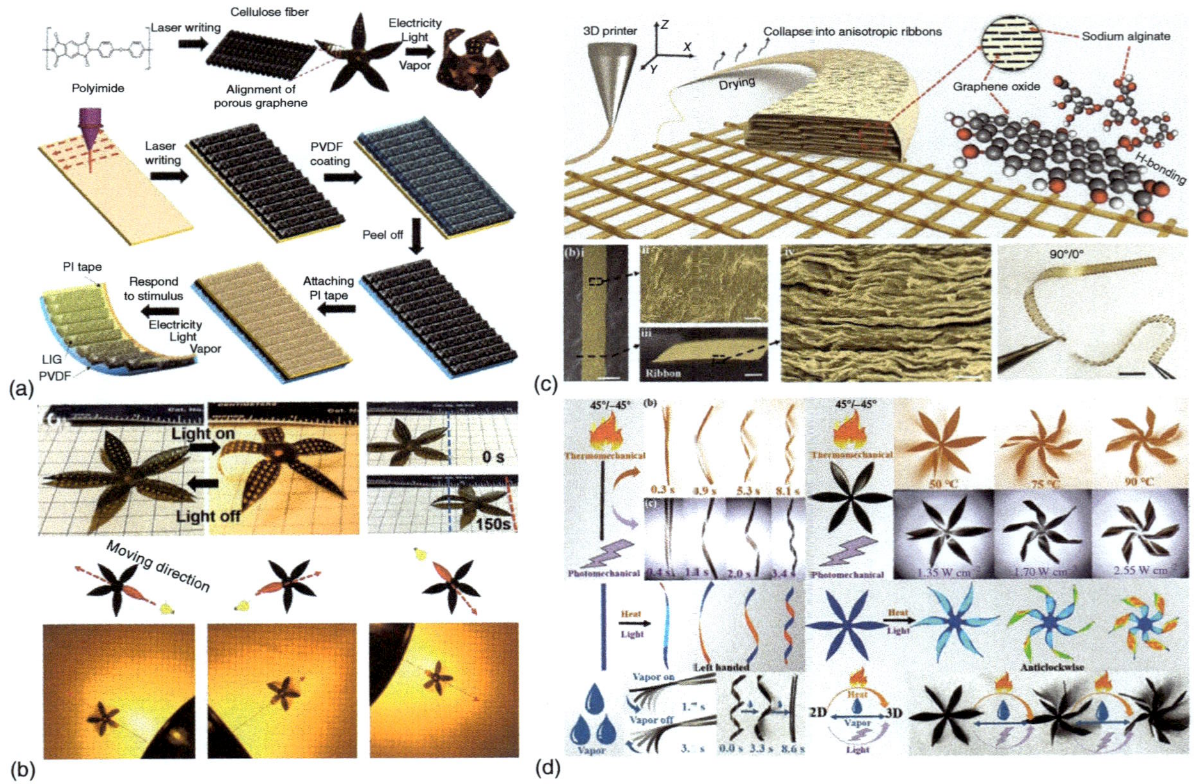


**Figure 3.21** (a) The schematic diagram of aligned LIG using the DLW method, the controllable structural changes of the resulting soft actuators, and the fabrication process for a LIG/polyimide bilayer soft actuator. Source: Deng et al. [84]/Reproduced from Royal Society of Chemistry. (b) The optical images of a PLP robot crawling by switching the light on and off (upper part), and the schematic diagram and optical images of a PLP robot with multi-way gaits under the stimulus of a light. Source: Deng et al. [84]/Reproduced from Royal Society of Chemistry. (c) The composition and the assembling structure in the printed GO ribbons, SEM images of a printed GO ribbon (i, ii), cross-sectional SEM images of a printed ribbon (iii, iv), and typical examples of printed strips with 90°/0° and 45°/−45° configurations. Source: Zhou et al. [78]/Reproduced from American Chemical Society. (d) Multi-stimuli-responsive structural changes of the printed various actuators. Source: Zhou et al. [78]/Reproduced from American Chemical Society.

## 3.5 Summary

In this chapter, we mainly introduce the development of carbon-based actuators based on the rolled-up strategy from four aspects, including the choice of carbon-based materials, the deformation mechanism of actuators, the programmable structural changes of actuators, and some special applications of actuators. By comparing the special properties of carbon-based materials, CNTs-, graphene-, and graphite-based soft actuators are discussed in detail. The deformation of soft actuators is determined by controlling the expansion and contraction of the internal materials based on mechanisms among light-induced expansion, light-induced contraction, and humidity-induced expansion. These strategies of light-induced local modification, non-light-induced local modification, self-healing, and oriental control of elements in actuators have been proposed to enable various programmable structural changes of actuators. These programmable structural changes have allowed the advanced applications of soft actuators such as smart grippers, intelligent biomimetics, and controllable soft robots. Among current progress, the programmable structural changes are critical for promoting the applications of soft actuators, and researchers have put more efforts into this aspect. However, no applications of carbon-based soft actuators have been anticipated to be applied in the real case. This is owing to the lower output force and single function of current soft actuators. Therefore, special structure design needed to be explored to increase their output force. Additionally, multiple modules needed to be integrated into one soft actuator, enabling it to respond to more stimuli, separate their functions and motions, and achieve multiple functions.

## Abbreviations

| | |
|---|---|
| ACNTs | aligned CNTs |
| BOPP | biaxially oriented polypropylene |
| CVD | chemical vapor deposition |
| CNTs | carbon nanotubes |
| CTE | coefficient of thermal expansion |
| DLW | direct laser writing |
| GO | graphene oxide |
| GNPs | graphene nanoplatelets |
| LDPE | low-density PE |
| LIG | laser-induced graphene |
| MWCNTs | multi-walled CNTs |
| NOA | Norland Optical Adhesive |
| PC | polycarbonate |
| PANI | polyaniline |
| PDA | polydopamine |
| PE | polyethylene |
| PEDOT | polyethylene dioxythiophene |
| PNIPAAm | poly(*N*-isopropylacrylamide) |

| | |
|---|---|
| PP | polypropylene |
| PU | polyurethane |
| PPy | polypyrrole |
| PVDF | poly(vinylidene fluoride) |
| RH | relative humidity |
| RGO | reduced GO |
| SGA | stacked graphene assembly |
| SWCNTs | single-walled CNTs |
| TEMs | thermally expanding microspheres |

## References


1 Hines, L., Petersen, K., Lum, G.Z., and Sitti, M. (2017). Soft actuators for small-scale robotics. *Advanced Materials* 29: 1603483.

2 Kim, H., Ahn, S.K., Mackie, D.M. et al. (2020). Shape morphing smart 3D actuator materials for micro soft robot. *Materials Today* 41: 243–269.

3 Le, X., Lu, W., Zhang, J., and Chen, T. (2019). Recent progress in biomimetic anisotropic hydrogel actuators. *Advanced Science* 6: 1801584.

4 Dong, Y., Wang, J., Guo, X. et al. (2019). Multi-stimuli-responsive programmable biomimetic actuator. *Nature Communications* 10: 4087.

5 Dong, Y., Wang, L., Xia, N. et al. (2021). Multi-stimuli-response programmable soft actuators with site-specific and anisotropic deformation behavior. *Nano Energy* 88: 106254.

6 Wang, X.Q., Chan, K.H., Cheng, Y. et al. (2020). Somatosensory, light-driven, thin-film robots capable of integrated perception and motility. *Advanced Materials* 32: 2000351.

7 Han, B., Zhang, Y.L., Zhu, L. et al. (2019). Plasmonic-assisted graphene oxide artificial muscles. *Advanced Materials* 31: 1806386.

8 Cheng, H., Zhao, F., Xue, J. et al. (2016). One single graphene oxide film for responsive actuation. *ACS Nano* 10: 9529–9535.

9 Wang, X.-Q., Tan, C.F., Chan, K.H. et al. (2018). In-built thermo-mechanical cooperative feedback mechanism for self-propelled multimodal locomotion and electricity generation. *Nature Communications* 9: 1–10.

10 Erb, R.M., Sander, J.S., Grisch, R., and Studart, A.R. (2013). Self-shaping composites with programmable bioinspired microstructures. *Nature Communications* 4: 1712.

11 Li, C., Peng, Z.-K., and He, Q. (2022). Stimuli-responsive metamaterials with information-driven elastodynamics programming. *Matter* 5: 988–1003.

12 Li, J.-F., Soldatov, I.V., Tang, X.C. et al. (2022). Metallic *Mimosa pudica*: A 3D biomimetic buckling structure made of metallic glasses. *Science Advances* 8: eabm7658.

13 Wang, Y., Cui, H., Zhao, Q., and Du, X. (2019). Chameleon-inspired structural-color actuators. *Matter* 1: 626–638.

14 Yang, H., Leow, W.R., Wang, T. et al. (2017). 3D printed photoresponsive devices based on shape memory composites. *Advanced Materials* 29: 1701627.

**15** Liu, Y.Q., Chen, Z.D., Han, D.D. et al. (2021). Bioinspired soft robots based on the moisture-responsive graphene oxide. *Advanced Science* 8: 2002464.

**16** Jiang, S., Liu, F., Lerch, A. et al. (2015). Unusual and superfast temperature-triggered actuators. *Advanced Materials* 27: 4865–4870.

**17** Jiang, Z.C., Xiao, Y.Y., Tong, X., and Zhao, Y. (2019). Selective decrosslinking in liquid crystal polymer actuators for optical reconfiguration of origami and light-fueled locomotion. *Angewandte Chemie* 131: 5386–5391.

**18** Han, B., Zhang, Y.-L., Chen, Q.-D., and Sun, H.-B. (2018). Carbon-based photothermal actuators. *Advanced Functional Materials* 28: 1802235.

**19** Xiao, Y.Y., Jiang, Z.C., Tong, X., and Zhao, Y. (2019). Biomimetic locomotion of electrically powered "Janus" soft robots using a liquid crystal polymer. *Advanced Materials* 31: 1903452.

**20** Zhao, Q., Dunlop, J.W., Qiu, X. et al. (2014). An instant multi-responsive porous polymer actuator driven by solvent molecule sorption. *Nature Communications* 5: 4293.

**21** Lin, H., Gong, J., Eder, M. et al. (2017). Programmable actuation of porous poly (ionic liquid) membranes by aligned carbon nanotubes. *Advanced Materials Interfaces* 4: 1600768.

**22** Dong, Y., Wang, L., Xia, N. et al. (2022). Untethered small-scale magnetic soft robot with programmable magnetization and integrated multifunctional modules. *Science Advances* 8: eabn8932.

**23** Kim, Y., Yuk, H., Zhao, R. et al. (2018). Printing ferromagnetic domains for untethered fast-transforming soft materials. *Nature* 558: 274–279.

**24** Ma, J.-N., Zhang, Y.L., Han, D.D. et al. (2020). Programmable deformation of patterned bimorph actuator swarm. *National Science Review* 7: 775–785.

**25** Zhang, Y.-L., Liu, Y.Q., Han, D.D. et al. (2019). Quantum-confined-superfluidics-enabled moisture actuation based on unilaterally structured graphene oxide papers. *Advanced Materials* 31: 1901585.

**26** Gao, Y.-Y., Zhang, Y.L., Han, B. et al. (2019). Gradient assembly of polymer nanospheres and graphene oxide sheets for dual-responsive soft actuators. *ACS Applied Materials & Interfaces* 11: 37130–37138.

**27** Hu, Y., Li, Z., Lan, T., and Chen, W. (2016). Photoactuators for direct optical-to-mechanical energy conversion: from nanocomponent assembly to macroscopic deformation. *Advanced Materials* 28: 10548–10556.

**28** Lahikainen, M., Zeng, H., and Priimagi, A. (2018). Reconfigurable photoactuator through synergistic use of photochemical and photothermal effects. *Nature Communications* 9: 4148.

**29** Chang, L., Wang, D., Jiang, A., and Hu, Y. (2022). Soft actuators based on carbon nanomaterials. *ChemPlusChem* 87: e202100437.

**30** Yang, M., Yuan, Z., Liu, J. et al. (2019). Photoresponsive actuators built from carbon-based soft materials. *Advanced Optical Materials* 7: 1900069.

**31** Medhekar, N.V., Ramasubramaniam, A., Ruoff, R.S., and Shenoy, V.B. (2010). Hydrogen bond networks in graphene oxide composite paper: structure and mechanical properties. *ACS Nano* 4: 2300–2306.

**32** Wang, C.Z. and Ho, K.M. (1994). Structure and electronic properties of diamond-like amorphous carbon. *MRS Online Proceedings Library* 349: 483–488.

**33** Balandin, A.A. (2011). Thermal properties of graphene and nanostructured carbon materials. *Nature Materials* 10: 569–581.

**34** Peng, H., Deng, J., Li, J. et al. (2016). Tunable photothermal actuators based on a pre-programmed aligned nanostructure. *Journal of the American Chemical Society* 138: 225–230.

**35** Okano, K., Noguchi, I., and Yamashita, T. (2010). Anisotropic carbon nanotube films fabricated from a lyotropic liquid-crystalline polymer. *Macromolecules* 43: 5496–5499.

**36** Zhang, X., Yu, Z., Wang, C. et al. (2014). Photoactuators and motors based on carbon nanotubes with selective chirality distributions. *Nature Communications* 5: 2983.

**37** Hu, Y., Liu, J., Chang, L. et al. (2017). Electrically and sunlight-driven actuator with versatile biomimetic motions based on rolled carbon nanotube bilayer composite. *Advanced Functional Materials* 27: 1704388.

**38** Sun, Z., Fang, S., and Hu, Y.H. (2020). 3D graphene materials: from understanding to design and synthesis control. *Chemical Reviews* 120: 10336–10453.

**39** Wang, E., Desai, M.S., and Lee, S.-W. (2013). Light-controlled graphene-elastin composite hydrogel actuators. *Nano Letters* 13: 2826–2830.

**40** Cheng, H., Huang, Y., Cheng, Q. et al. (2017). Self-healing graphene oxide based functional architectures triggered by moisture. *Advanced Functional Materials* 27: 1703096.

**41** Stankovich, S., Dikin, D.A., Dommett, G.H. et al. (2006). Graphene-based composite materials. *Nature* 442: 282–286.

**42** Maggi, C., Saglimbeni, F., Dipalo, M. et al. (2015). Micromotors with asymmetric shape that efficiently convert light into work by thermocapillary effects. *Nature Communications* 6: 7855.

**43** Chen, Y., Yang, J., Zhang, X. et al. (2021). Light-driven bimorph soft actuators: Design, fabrication, and properties. *Materials Horizons* 8: 728–757.

**44** Niu, D., Jiang, W., Liu, H. et al. (2016). Reversible bending behaviors of photomechanical soft actuators based on graphene nanocomposites. *Scientific Reports* 6: 27366.

**45** Leeladhar and Singh, J.P. (2018). Photomechanical and chemomechanical actuation behavior of graphene–poly(dimethylsiloxane)/gold bilayer tube for multimode soft grippers and volatile organic compounds detection applications. *ACS Applied Materials & Interfaces* 10: 33956–33965.

**46** Tang, Z., Gao, Z., Jia, S. et al. (2017). Graphene-based polymer bilayers with superior light-driven properties for remote construction of 3D structures. *Advanced Science* 4: 1600437.

**47** Wang, S., Gao, Y., Wei, A. et al. (2020). Asymmetric elastoplasticity of stacked graphene assembly actualizes programmable untethered soft robotics. *Nature Communications* 11: 4359.

**48** Xu, G., Zhang, M., Zhou, Q. et al. (2017). A small graphene oxide sheet/polyvinylidene fluoride bilayer actuator with large and rapid responses to multiple stimuli. *Nanoscale* 9: 17465–17470.

**49** Gao, T., Xu, G., Wen, Y. et al. (2020). An intelligent film actuator with multi-level deformation behaviour. *Nanoscale Horizons* 5: 1226–1232.

**50** Chen, L., Weng, M., Zhou, P. et al. (2017). Multi-responsive actuators based on a graphene oxide composite: intelligent robot and bioinspired applications. *Nanoscale* 9: 9825–9833.

**51** Chen, L., Weng, M., Zhou, P. et al. (2019). Graphene-based actuator with integrated-sensing function. *Advanced Functional Materials* 29: 1806057.

**52** Leeladhar, Raturi, P., and Singh, J.P. (2018). Sunlight-driven eco-friendly smart curtain based on infrared responsive graphene oxide-polymer photoactuators. *Scientific Reports* 8: 3687.

**53** Yang, L., Chang, L., Hu, Y. et al. (2020). An autonomous soft actuator with light-driven self-sustained wavelike oscillation for phototactic self-locomotion and power generation. *Advanced Functional Materials* 30: 1908842.

**54** Zhou, P., Chen, L., Yao, L. et al. (2018). Humidity- and light-driven actuators based on carbon nanotube-coated paper and polymer composite. *Nanoscale* 10: 8422–8427.

**55** Xiao, Y., Lin, J., Xiao, J. et al. (2021). A multi-functional light-driven actuator with an integrated temperature-sensing function based on a carbon nanotube composite. *Nanoscale* 13: 6259–6265.

**56** Tai, Y., Lubineau, G., and Yang, Z. (2016). Light-activated rapid-response polyvinylidene-fluoride-based flexible films. *Advanced Materials* 28: 4665–4670.

**57** Li, J., Mou, L., Zhang, R. et al. (2019). Multi-responsive and multi-motion bimorph actuator based on super-aligned carbon nanotube sheets. *Carbon* 148: 487–495.

**58** Wang, M., Zhou, L., Deng, W. et al. (2022). Ultrafast response and programmable locomotion of liquid/vapor/light-driven soft multifunctional actuators. *ACS Nano* 16: 2672–2681.

**59** Weng, M., Zhou, P., Chen, L. et al. (2016). Multiresponsive bidirectional bending actuators fabricated by a pencil-on-paper method. *Advanced Functional Materials* 26: 7244–7253.

**60** Lin, J., Zhou, P., Chen, Q. et al. (2022). Reprogrammable, light-driven and sensing actuators based on Chinese ink composite: a synergetic use of shape-memory and self-healing strategies. *Sensors and Actuators B: Chemical* 362: 131776.

**61** Amjadi, M. and Sitti, M. (2018). Self-sensing paper actuators based on graphite–carbon nanotube hybrid films. *Advanced Science* 5: 1800239.

**62** Wang, W., Xiang, C., Zhu, Q. et al. (2018). Multistimulus responsive actuator with GO and carbon nanotube/PDMS bilayer structure for flexible and smart devices. *ACS Applied Materials & Interfaces* 10: 27215–27223.

**63** Zhang, X., Pint, C.L., Lee, M.H. et al. (2011). Optically- and thermally-responsive programmable materials based on carbon nanotube-hydrogel polymer composites. *Nano Letters* 11: 3239–3244.

**64** Ji, M., Jiang, N., Chang, J., and Sun, J. (2014). Near-infrared light-driven, highly efficient bilayer actuators based on polydopamine-modified reduced graphene oxide. *Advanced Functional Materials* 24: 5412–5419.

**65** Mu, J., Hou, C., Wang, H. et al. (2015). Origami-inspired active graphene-based paper for programmable instant self-folding walking devices. *Science Advances* 1: e1500533.

66 Han, D.-D., Zhang, Y.L., Liu, Y. et al. (2015). Bioinspired graphene actuators prepared by unilateral UV irradiation of graphene oxide papers. *Advanced Functional Materials* 25: 4548–4557.

67 Li, L., Zhao, S., Luo, X.-J. et al. (2021). Smart MXene-based Janus films with multi-responsive actuation capability and high electromagnetic interference shielding performances. *Carbon* 175: 594–602.

68 Nair, R.R., Wu, H.A., Jayaram, P.N. et al. (2012). Unimpeded permeation of water through helium-leak–tight graphene-based membranes. *Science* 335: 442–444.

69 Park, S., An, J., Suk, J.W., and Ruoff, R.S. (2010). Graphene-based actuators. *Small* 6: 210–212.

70 Li, H. and Wang, J. (2019). Ultrafast yet controllable dual-responsive all-carbon actuators for implementing unusual mechanical movements. *ACS Applied Materials & Interfaces* 11: 10218–10225.

71 Wang, W., Zhang, Y.L., Han, B. et al. (2019). A complementary strategy for producing moisture and alkane dual-responsive actuators based on graphene oxide and PDMS bimorph. *Sensors and Actuators B: Chemical* 290: 133–139.

72 Zhang, Y.-L., Ma, J.N., Liu, S. et al. (2020). A "Yin"–"Yang" complementarity strategy for design and fabrication of dual-responsive bimorph actuators. *Nano Energy* 68: 104302.

73 Cheng, H., Liu, J., Zhao, Y. et al. (2013). Graphene fibers with predetermined deformation as moisture-triggered actuators and robots. *Angewandte Chemie International Edition* 52: 10482–10486.

74 Han, D.-D., Zhang, Y.L., Jiang, H.B. et al. (2015). Moisture-responsive graphene paper prepared by self-controlled photoreduction. *Advanced Materials* 27: 332–338.

75 Han, B., Gao, Y.Y., Zhang, Y.L. et al. (2020). Multi-field-coupling energy conversion for flexible manipulation of graphene-based soft robots. *Nano Energy* 71: 104578.

76 Borno, R.T., Steinmeyer, J.D., and Maharbiz, M.M. (2006). Transpiration actuation: the design, fabrication and characterization of biomimetic microactuators driven by the surface tension of water. *Journal of Micromechanics and Microengineering* 16: 2375.

77 Wang, M., Li, Q., Shi, J. et al. (2020). Bio-inspired high sensitivity of moisture-mechanical GO films with period-gradient structures. *ACS Applied Materials & Interfaces* 12: 33104–33112.

78 Zhou, G.-X., Yu, Y.G., Yang, Z.H. et al. (2022). 3D printing graphene oxide soft robotics. *ACS Nano* 16: 3664–3673.

79 Ge, Y., Cao, R., Ye, S. et al. (2018). A bio-inspired homogeneous graphene oxide actuator driven by moisture gradients. *Chemical Communications* 54: 3126–3129.

80 He, J., Xiao, P., Zhang, J. et al. (2016). Highly efficient actuator of graphene/polydopamine uniform composite thin film driven by moisture gradients. *Advanced Materials Interfaces* 3: 1600169.

**81** Ma, J.-N., Mao, J.W., Han, D.D. et al. (2019). Laser programmable patterning of RGO/GO Janus paper for multiresponsive actuators. *Advanced Materials Technologies* 4: 1900554.

**82** Zhang, L., Zhang, Y., Li, F. et al. (2019). Water-evaporation-powered fast actuators with multimodal motion based on robust nacre-mimetic composite film. *ACS Applied Materials & Interfaces* 11: 12890–12897.

**83** Weng, M., Xiao, Y., Yao, L. et al. (2020). Programmable and self-healing light-driven actuators through synergetic use of water-shaping and -welding methods. *ACS Applied Materials & Interfaces* 12: 55125–55133.

**84** Deng, H., Zhang, C., Su, J.W. et al. (2018). Bioinspired multi-responsive soft actuators controlled by laser tailored graphene structures. *Journal of Materials Chemistry B* 6: 5415–5423.

# 4
# Hydrogels-Based Miniature Soft Robots

## 4.1 Introduction

With the capability of sensing and responding to external stimuli, natural organisms could adapt to changing environments. Therefore, stimuli-responsive materials play an important role in the development of intelligent soft robots. In this respect, stimuli-responsive hydrogels with broadly tunable mechanical properties are one of the important intelligent materials. Hydrogels are a 3D crosslinked network of hydrophilic polymers, which can be formed by various mechanisms, e.g. physical entanglement, ionic crosslinking, as well as chemical crosslinking [1–4]. To date, synthesized hydrogel materials that could respond to a wide spectrum of environmental stimuli have been developed, including thermo-responsive, pH-responsive, electrical-responsive, chemically responsive, and photo-responsive hydrogels (Figure 4.1).

Thermo-responsive hydrogels could exhibit shrinking or swelling behaviors based on the variation in environmental temperature [5, 10–16]. Poly(*N*-isopropylacrylamide) (PNIPAAm) is one of the most commonly used thermo-responsive hydrogels. When the environmental temperature exceeds a lower critical solution temperature (LCST), the shrinkage of the PNIPAAm hydrogel would be present. The hydrogel matrix contains both the hydrophobic groups and the hydrophilic groups. With a temperature lower than the LCST, the PNIPAAm hydrogels swell due to the dominant effect of hydrogen bonds formed between the water molecules and hydrophilic groups. In contrast, under a high temperature (above the LCST), the polymer chains transform from coil state to globule state, resulting in the discharge of exceeding water inside the network and the shrinkage of the hydrogel [17–22]. It should be noted that the LCST of the PNIPAAm hydrogels can be tuned by the introduction of appropriate hydrophilic groups so that they can be more suitable for biomedical applications [23].

pH-responsive hydrogels that could execute tasks based on the variation of pH values have also drawn great attention [24–32]. The responsive behavior of the hydrogel matrix mainly derives from the protonation or deprotonation of the ionizable pendant groups [6, 33]. The swelling of the pH-responsive hydrogel was activated by electrostatic repulsions of charged polymer chains. Based on the charge polarity, pH-responsive hydrogels can be divided into two types including the cationic

*Untethered Miniature Soft Robots: Materials, Fabrications, and Applications*, First Edition.
Li Zhang, Jiachen Zhang, Neng Xia, and Yue Dong.

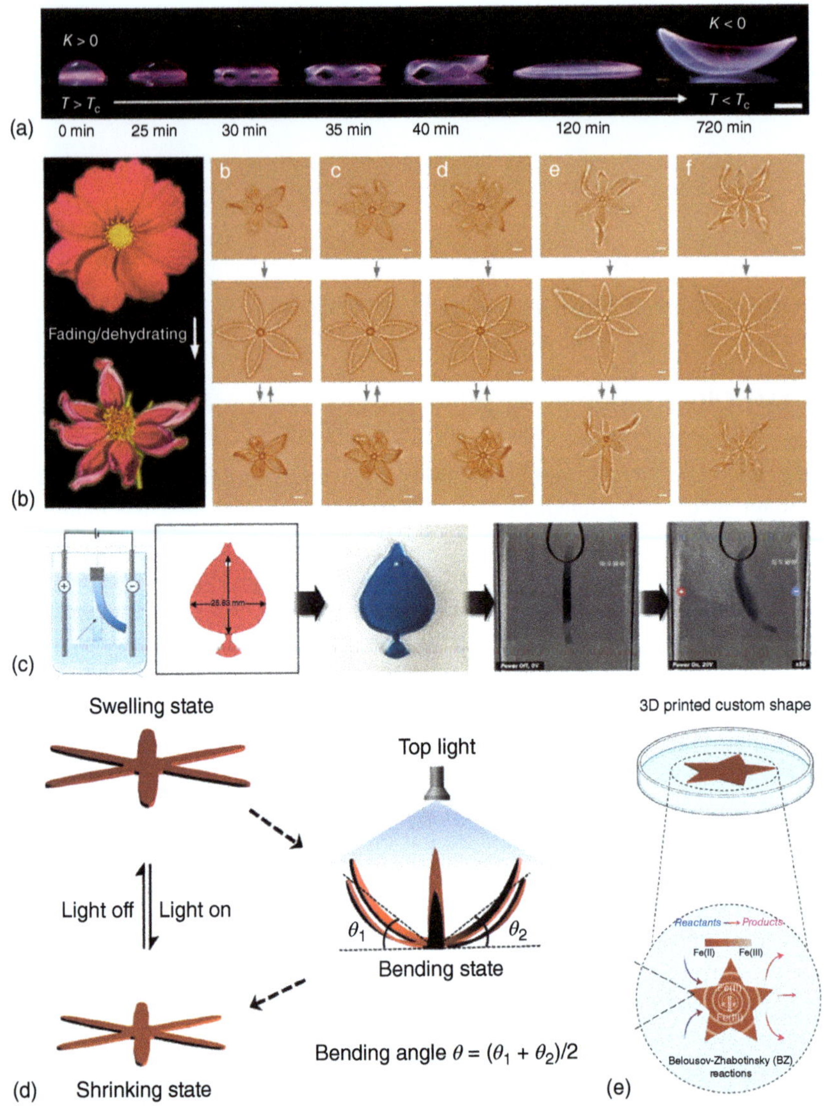


**Figure 4.1** Stimuli-responsive hydrogels. (a) Thermo-responsive hydrogels. Source: Nojoomi et al. [5]/Reproduced from Springer Nature/CC BY 4.0. (b) pH-responsive hydrogels. Source: Hu et al. [6]/Reproduced from John Wiley & Sons, Inc. (c) Electrical-responsive hydrogels. Source: Shin et al. [7]/Reproduced from American Chemical Society. (d) Photo-responsive hydrogels. Source: Li et al. [8]/Reproduced from Springer Nature. (e) Chemically responsive hydrogels. Source: Nava-Medina et al. [9]/John Wiley & Sons.

hydrogel and the anionic hydrogel [34–37]. Under an environmental pH lower than dissociation constant (p$K_a$), the cationic hydrogels with pendant groups (e.g. amide group and amine group) swell. In contrast, the anionic hydrogels with acidic pendant groups swell under an environmental pH higher than p$K_a$. pH-responsive hydrogels hold great promise for biomedical study. For instance, pH-sensitive microgrippers with various shapes (e.g. pillar array and cage) were developed by Li et al. to reversely capture cells for *in situ* manipulation or target delivery [33].

Electrical-responsive hydrogels (e.g. sulfonated polystyrene and polyvinyl alcohol), which are usually composed of polyelectrolytes with ionizable groups, experience the change of volume based on external electrical signals [7, 38–47]. By applying an electrical field, electrodes with opposite charges would attract the free ions in the environment [48, 49]. The non-uniform local gradient of the ion concentrations would be generated by the interaction between the migrating ion and the ionizable group. Subsequently, the induced osmotic pressure activates the removal or absorption of water.

Photo-responsive hydrogels usually contain photoactive moieties inside their matrix such as photosensitive groups or photoactive dopants, which allow the hydrogels to react to light [8, 12, 50–58]. A variety of mechanisms including photothermal effect, reversible isomerization, or dimerization reaction were adopted for the photo-responsive behaviors [59–63]. In addition, due to the remote manipulation capability and the high spatial and temporal resolutions, the light actuation method based on photo-responsive hydrogels has attracted growing attention.

Chemically responsive hydrogels, e.g. glucose-responsive hydrogels, utilize chemical reactions with certain substances to induce volume change [9, 64–71]. For example, through the immobilization of glucose oxidase into a pH-responsive hydrogel, the synthesized material would deswell in an environment containing glucose, because the glucose would be transformed into gluconic acid catalyzed by glucose oxidase.

## 4.2 Fabrication of Reconfigurable Hydrogel Micromachines

The development of stimuli-responsive and reconfigurable devices promises a wide spectrum of applications in biomedical engineering. To achieve complex shape transformations, the local or selective adjustment of the internal stresses, mechanical properties, or volume changes play pivotal roles. In this respect, several strategies have been proposed.

Firstly, shape transformation can be achieved by the construction of layered structures made of materials with different responsive properties [27, 72–75]. For example, a bi-layer structure was fabricated by thermo-responsive hydrogel (PNIPAAm) and non-thermo-responsive hydrogel (PAAm) [76]. With an environmental temperature higher than the LCST, the volume shrinkage of the PNIPAAm layer results in the bending of the structure, which provides a straightforward method for the shape transformation. Secondly, the introduction and arrangement of anisotropic fibers or particles inside the hydrogel matrix could locally regulate the swelling behavior [77, 78]. For instance, Gladman et al. developed thermo-responsive hydrogel structures embedded with programmable anisotropic cellulose fibrils [79]. The hydrogel structures were fabricated by the direct ink-writing technique. During the extrusion through the nozzle, the alignment of the cellulose fibrils was induced by a shear effect, leading to an anisotropic stiffness of the hydrogel filaments and a difference in the swelling behaviors in radial and axial directions. With the planning of the print path, flat structures could deform into architectures with various Gaussian curvatures. Thirdly, the shape transformation can be encoded by gradient properties, especially the crosslinking density. The volume changes of the hydrogel matrix can be regulated by the crosslinking density. Huang et al. utilized the exposure dosage during a digital printing process to regulate the crosslinking density in a single material, resulting in heterogeneous swelling or shrinking behaviors in the fabricated structure, and subsequently achieving various 3D configurations [80].

Four-dimensional (4D) printing, which refers to the printing of 3D structures presenting geometries varied with time, has been rapidly developed. A variety of mechanisms were proposed to pattern intelligent materials with heterogeneous properties inside the 3D printed structures. In this section, various reconfigurable micromachines were fabricated using direct laser writing (DLW)-based 4D printing technique and a single pH-responsive hydrogel. Printing parameters including scanning time and laser power enable the programmable modulation of mechanical properties including stiffness, crosslinking density, and the swelling ratio. These fabricated soft structures could achieve reversible 3D-to-3D morphological transformation under external stimuli.

### 4.2.1 Design and Characterization of Smart Hydrogel for 4D Printing

A polymer precursor of printable stimuli-responsive hydrogel was developed for the high-resolution 4D printing [81, 82]. The precursor solution was made by the dissolution of a variety of components, including crosslinker (DPEPA), monomer (AAc and NIPAAm), photoinitiator (EMK), and photosensitizer (TEA) into proper solvent (Figure 4.2). Typically, 1 ml EL was used to dissolve 0.8 ml AAc, 1.6 g NIPAAm, and 0.15 g polyvinylpyrrolidone (PVP), followed by vigorous stirring overnight to obtain the composite solution. 2.5 ml of the solution was then mixed with 0.5 ml TEA, 100 μl EMK/DMF solution with a weight ratio of 20%, and 0.4 ml DPEPA. A yellow polymer precursor was obtained and kept in suitable condition without unnecessary exposure. With the irradiation of laser beam, lots of radicals were produced by the excitation of the photoinitiator EMK assisted by the photosensitizer TEA. The linear copolymers between NIPAAm and AAc monomers were formed

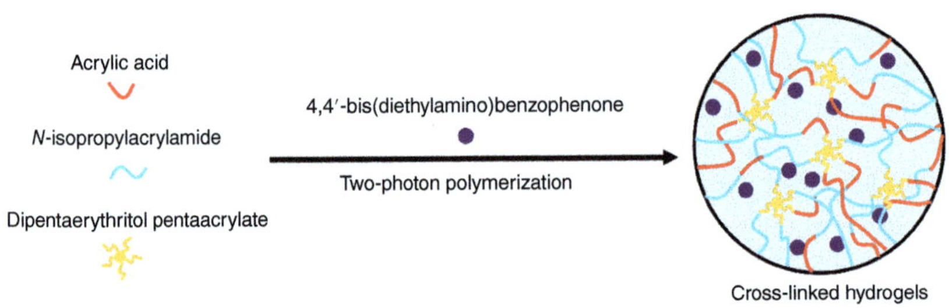


**Figure 4.2** The two-photon polymerization reaction adopted during the printing process. Source: Jin et al. [82]/Elsevier/CC BY 4.0/Public domain.

by free radical polymerization. Meanwhile, the radicals also trigger the crosslinker DPEPA for the interconnection of poly(AAc-*co*-NIPAAm) chains which in turn leads to the formation of the hydrogel framework. The viscosity of the precursor solution was tuned by PVP to prevent the drift and collapse of the printed freestanding 3D architectures.

During the printing process, a piece of cover glass was adopted as the printing substrate. Before using it, a series of treatments were performed on the glass substrate, including rinsing with deionized (DI) water and isopropanol (IPA), purging with nitrogen, as well as oxygen plasma treatment. A pipette was used to add the prepared resin to the center of the cover glass. The fabrication of hydrogel structure was performed by a commercial direct laser-writing system (Nanoscribe GmbH) which includes a 63× oil immersion objective. The bundled software of the printing system allows the predesign of parameters of laser source including the scanning speed and power and the modulation of exposure dosage during the printing process. The polymerization of the hydrogel structure was only performed in a local region with focused laser beam. In this manner, a variety of physical properties of printed structures can be locally encoded, e.g. stiffness, crosslinking density, and swelling/shrinking degree. After printing, IPA was used to remove excess resin and the developed structures were then immersed in DI water before further use. As shown in Figure 4.3, a 3D lantern-like structure was printed with a ~250 nm line width, which demonstrated excellent printing precision.

The pH-responsive property was used to evaluate the programmability of the printed hydrogel structure. Due to the usage of AAc, the obtained hydrogel matrix includes lots of pendants acidic (carboxylic) groups. When immersed in alkaline or acidic solution, the release or acceptance of protons would occur. At alkaline conditions, ionized carboxylic groups induce a strong electrostatic repulsive interaction, resulting in the hydrogel swelling, while in acidic solution, deionized carboxylic groups cause a weak electrostatic repulsion and the shrinking of the hydrogel structure. Moreover, the degrees of the swelling and shrinking can be elaborately tuned by the DLW system (Figure 4.4). Generally, the decrease in scanning speed or the increase in exposure dosage causes a larger exposure dosage, which would

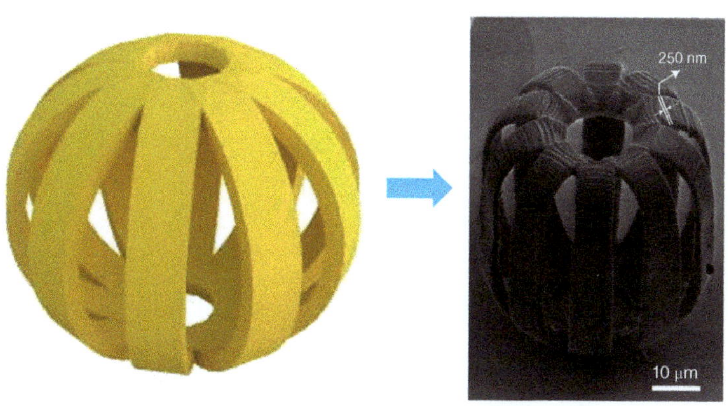


**Figure 4.3** The SEM image of the printed pH-responsive microball structure. Source: Jin et al. [82]/Reproduced from Elsevier/CC BY 4.0.

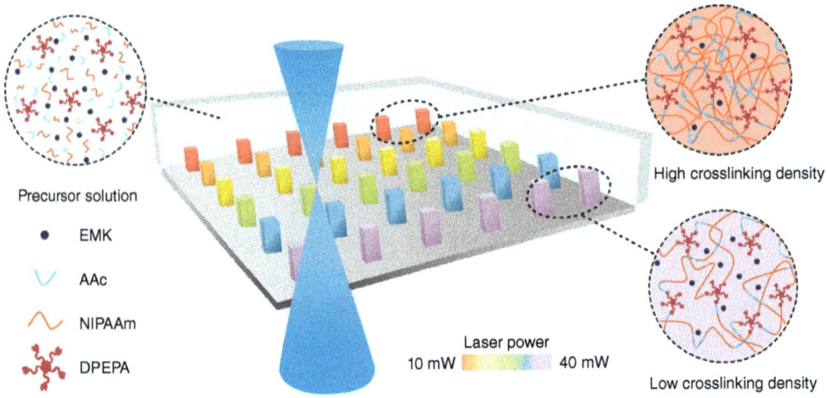


**Figure 4.4** Schematic illustration of the 4D printing process with a DLW system. Source: Jin et al. [82].

produce a highly crosslinked hydrogel matrix. Subsequently, the electrostatic effects of neighboring carboxylic groups would be weakened, resulting in a low degree of swelling/shrinking. In contrast, a loose and soft hydrogel network was formed with fewer polymer chains by a low exposure dosage, causing a dramatic increase in the swelling/shrinking degrees. Thus, the programmable photocuring parameters are vital for the fabrication of heterogeneous hydrogel structures.

As shown in Figure 4.5, the phototunable responsiveness of printed hydrogel structures was verified by cuboid samples (length: 20 μm, height: 20 μm, and width: 40 μm) with a constant scanning speed (8000 $\mu m\,s^{-1}$) and a laser power ranging from 12 to 40 mW. After fabrication, *in situ* nanomechanical testing was performed to measure the properties of printed samples by a commercial testing system (FemtoTools AG, Switzerland). The microcuboid samples were compressed by a microprobe to attain a 30% compression ratio. The impact of the laser power on the measured stiffness of the cuboid sample is shown in Figure 4.6a. The cuboid sample made by a 10 mW laser power exhibited a limited stiffness (25 $N\,m^{-1}$).

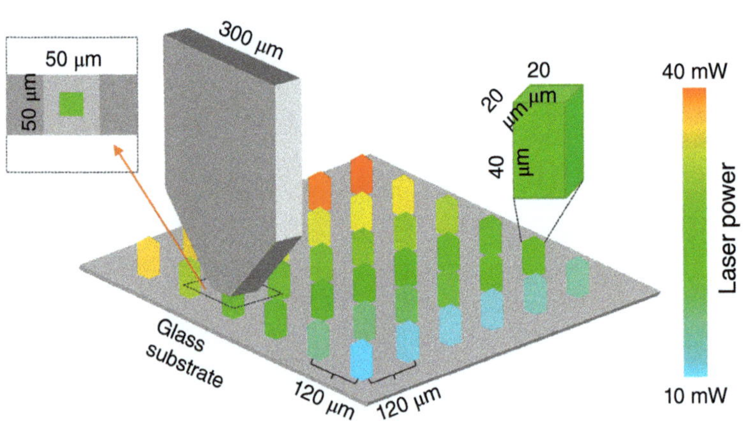


**Figure 4.5** Schematic illustration of the micromechanical measurement. Source: Jin et al. [82]/Elsevier/CC BY 4.0/Public domain.

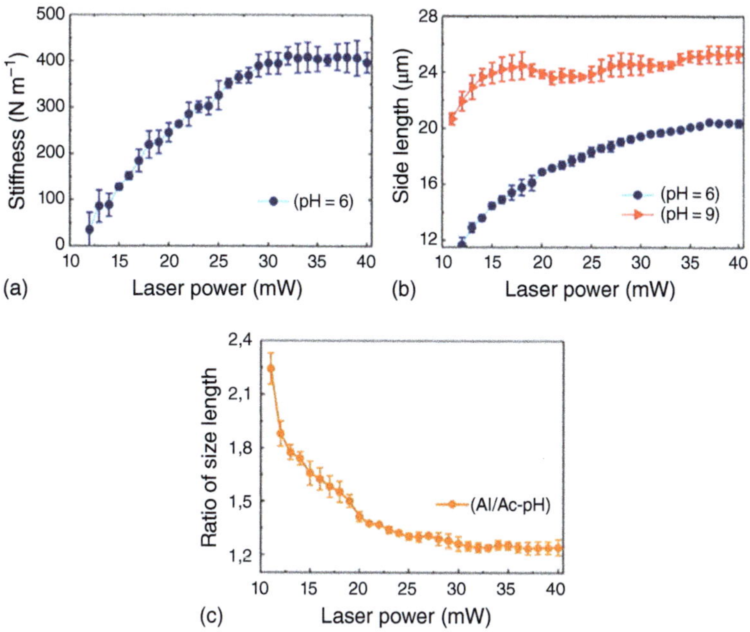


**Figure 4.6** Quantitative characterization of the mechanical property and the size change of printed material. (a) The impact of laser power on the stiffness of the printed microcuboid. (b) The impact of laser power on the side length of the printed microcuboid under different pH values. (c) The impact of laser power on the swelling ratio. Source: Jin et al. [82]/Elsevier/CC BY 4.0/Public domain.

With the rising of laser power, the stiffness grew rapidly and attained a plateau ($400\,\mathrm{N\,m^{-1}}$) after 35 mW laser power, which demonstrated the generation of denser hydrogel network upon strong laser exposure. A 3D digital microscope was used to record the swelling/shrinking behaviors of the cuboid samples. Figure 4.6b shows the relationship between the side length of the cuboid sample and the laser power under different pH environments, which could illustrate the shrinking and swelling behaviors of hydrogel structures in alkaline and acidic solutions. For instance, the side lengths of the cuboid sample made by a 12 mW laser beam attained ~20.5 and ~11.5 μm at pH 12 and 6, respectively, while for the sample made by 35 mW laser beam, the corresponding side lengths change to ~24.5 and ~20 μm. The measured result confirmed that the hydrogel structure printed by a lower exposure dosage was more sensitive to the change in pH value. Figure 4.6c shows the variation of the swelling ratio of the side length with the laser power. The experimental results demonstrate that the crosslinking density of the printed structure increases with the exposure dosage which would cause the weakening of its stimuli-responsiveness.

### 4.2.2 Reconfigurable Soft Microdevices Fabricated by Point-by-Point 4D Printing

A variety of 3D microstructures patterned with heterogeneous crosslinked hydrogels were printed by the DLW system with high resolution. Figure 4.7 presents a hollow microball consisting of 10 stripes with varying exposure dosages to demonstrate

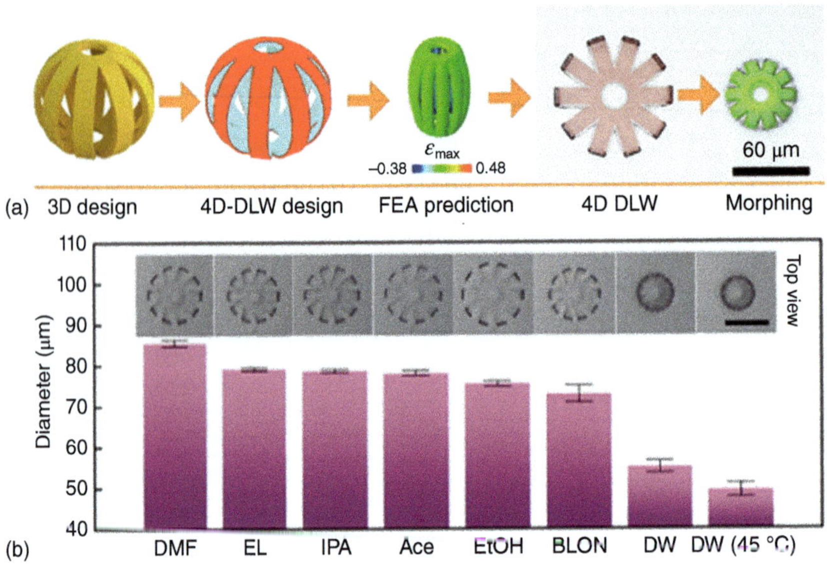


**Figure 4.7** The responsive behavior of the 4D printed microball structure. (a) Structural design, experimental results of the printed and deformed microball, as well as simulation results. (b) The responsive behaviors of the printed microball in different solvents and temperatures, including DMF (*N,N*-dimethylmethanamide), EL (ethyl lactate), IPA (isopropanol), Ace (acetone), EtOH (ethanol), BLON (1,4-butyrolactone), DW (deionized water), and DW (45 °C) (deionized water at 45 °C). Source: Jin et al. [82]/Reproduced from Elsevier/CC BY 4.0.

the 3D-to-3D shape-morphing behaviors. Each strip includes two layers of different crosslinked hydrogel matrixes. The laser powers used for the fabrication of the inner layer and the outer layer are 12 and 40 mW, respectively. By immersing in environments with different pH values, the microball exhibited a reversible morphological transformation between a slender cylinder and a spherical structure because of the heterogeneous responses of the two layers, which are consistent with the simulation results. Figure 4.7b shows the impact of various external stimuli (e.g. temperature and chemical solvent) on the deformation of the microball.

The shape-morphing behaviors of intelligent machines were determined by the programmable range of printing materials in terms of swelling/shrinking degrees, while because of the restricted forces applied to the device, most reported microdevices showed difficulty in exhibiting large deformations. In this chapter, the developed 4D printing strategy allows the construction of complex reconfigurable machines with tunable exposure dosages. As shown in Figure 4.8a, active bilayer belts connect eight rigid rods in perpendicular directions to form an active mechanical device, which exhibits a uniaxial-contraction transformation. An octagonal prismatic microtube adopted such uniaxial-contraction mechanism to activate a radial contraction which coincides with the simulation results. A microcage consisting of bilayer beams was fabricated and exhibited a biaxial contraction or spread under external stimuli. In addition, a microdevice with high aspect ratio was demonstrated (Figure 4.8c). To achieve a super-large deformation, a mechanically

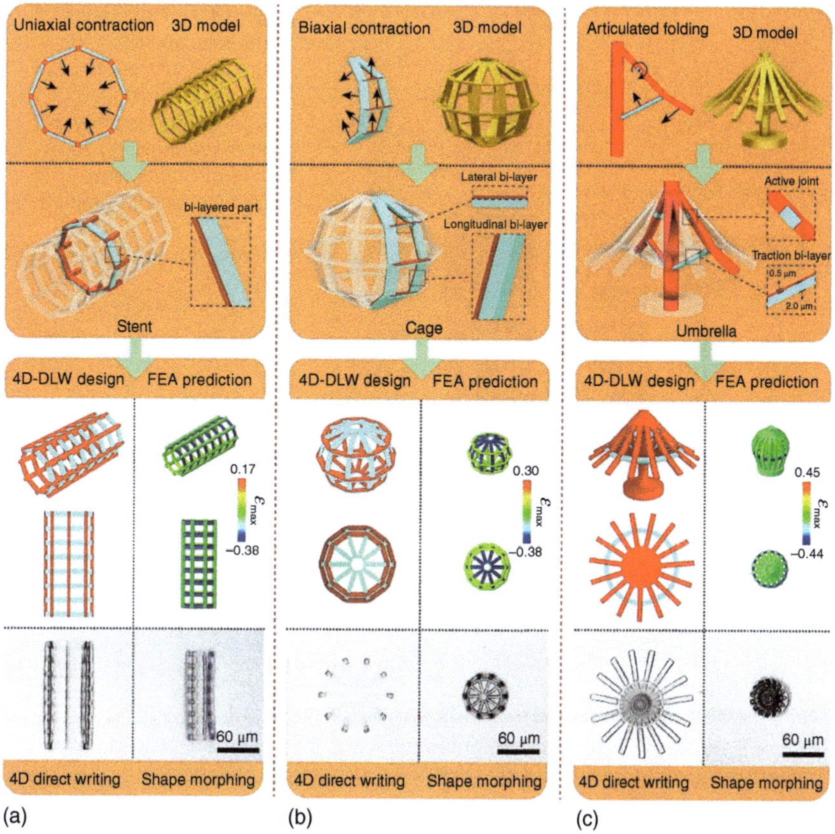


**Figure 4.8** Structural design, simulation prediction, and experimental results of 3D-to-3D reconfigurable micromachines. (a) Microstent. (b) Microcage. (c) Microumbrella. Source: Jin et al. [82]/Reproduced from Elsevier/CC BY 4.0.

amplified mechanism, called "the law of the lever," was adopted, which allows the usage of less force to deform and enhances the functionalities as well as deformation performance of the printed device. As shown in Figure 4.8c, a foldable microdevice was printed with shapes similar to the closing state and opening state of an umbrella. There are 15 deformable stretchers inside the microdevice and each of them was a bilayer structure connected to an umbrella rib. With a changed environmental pH, the rotation of the rib would be activated by the bending of the stretcher, demonstrating the folding of the articulated lever.

### 4.2.3 Reconfigurable Soft Microdevices Fabricated by Asymmetric Bessel Beam

The above point-to-point DLW technique offers a promising way for the fabrication of smart microdevices due to its high printing resolution and programmable exposure, while this method is limited by fabrication efficiency [33]. In this section, a local hologram phase modulation method was proposed by the integration of DLW

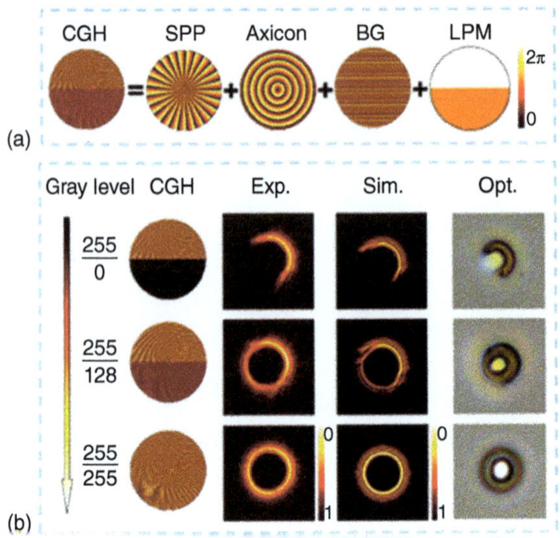


**Figure 4.9** Modulation of light field. (a) Schematic illustration of the computer-generated-hologram (CGH). SPP: spiral phase plate. BG: blazed grating. LPM: local phase modulation. (b) Experimental and simulation results of light fields with different phase depths. Source: Li et al. [33]/Reproduced from American Chemical Society.

technique and the spatial light modulation technology. An asymmetric Bessel beam can be presented by this strategy, enabling the much more efficient fabrication of smart microdevices. A ring-structured light field was generated with nonuniform distribution of energy. Therefore, various microtubes made of heterogeneous hydrogels can be obtained without the need of point-by-point scanning.

Figure 4.9 illustrates the spatial light modulator (SLM)-based printing system. As a pure phase modulator, the liquid-crystal SLM is in the range of gray level (0–255). There is a linear relationship between gray scale value (0–255) and phase delay (0–2π) on the hologram. With the computer-generated-hologram (CGH), a ring-structure Bessel beam was generated by the modulation of a Gaussian laser femtosecond beam (central wavelength: 800 nm, pulse repetition rate: 80 MHz, and pulse width: 75 fs). The CGH consists of a local phase modulation (LPM), a spiral phase plate (SPP), a blazed grating (BG), and an axicon. A zeroth Bessel beam was generated by the axicon and subsequently converted into high-order Bessel beam via the SPP. The BG facilitates the separation of the first-order light and others. Equation (4.1) shows the expression of the phase distribution of CGH without LPM.

$$\mathrm{Ph}_1(x_0, y_0) = \mathrm{mod}\left(n\theta + \frac{2\pi r}{r_0} + \frac{2\pi x_0}{d}, 2\pi\right) \tag{4.1}$$

$x_0$ represents the horizontal axis in the Cartesian coordinate system, and $y_0$ is the vertical axis. The topological charge of SPP ($n$) represents the order of Bessel beam. $\theta = \tan^{-1}(y_0/x_0)$ represents the polar angle. $r$ and $r_0$ are the polar radius and the radius of the axicon, respectively. $d$ represents the period of the BG.

The compression of the phase modulation depth was used to modulate the intensity of the light field of the Bessel beam, as shown in Eq. (4.2).

$$\mathrm{Ph}_2(x_0, y_0) = \mathrm{mod}\left(n\theta + \frac{2\pi r}{r_0} + \frac{2\pi x_0}{d}, 2\pi\right) \times \frac{\varphi}{2\pi} \tag{4.2}$$

$\varphi \in [0, 2\pi]$ represents the maximum phase modulation depth.

The diffraction efficiency $\eta$ for the light spot with all orders could be tuned by the change of $\varphi$, as shown in Eq. (4.3).

$$\eta(\varphi) = 2 \times \frac{(1 - \cos\varphi)}{(\varphi - 2n\pi)^2} \tag{4.3}$$

To regulate the local energy distribution of the ring-structure Bessel beam, a LPM was superimposed on the hologram. The phase modulation depth and the shape of the LPM could be adopted to modulate any region of the ring-structure light field. Here, the hologram is divided into two parts by using a semicircle shape of LPM. The white part represents the unchanged phase distribution, and the orange part represents a $\varphi$ phase modulation depth. Equation (4.4) illustrates the phase pattern of superimposed hologram.

$$\mathrm{Ph}_3(x_0, y_0) = \begin{cases} \mathrm{mod}\left(n\theta + \frac{2\pi r}{r_0} + \frac{2\pi x_0}{d}, 2\pi\right), & t < 0 \\ \mathrm{mod}\left(n\theta + \frac{2\pi r}{r_0} + \frac{2\pi x_0}{d}, 2\pi\right) \times \frac{\varphi}{2\pi}, & t > 0 \end{cases} \tag{4.4}$$

$t = x_0\cos\alpha + y_0\sin\alpha$. $\alpha \in [0, 2\pi]$ was defined by the angle between the horizontal axis of the hologram and boundary of the LPM.

An oil-immersion objective lens was adopted to focus the modulated Bessel beam into pH-responsive hydrogel. Figure 4.9 illustrates the simulation results and experimental results of the modulated light field with different holograms. The maximum gray scale value of the three holograms was 255, 128, and 0 from bottom to top, and the corresponding phase modulation depths were $2\pi$, $\pi$, and 0. There is an agreement between the experimental results and the simulation results.

As aforementioned, the processing energy could determine the swelling ratio of the pH-responsive hydrogel. Therefore, the phase modulation depth and the power of laser source could be adopted to control the morphological transformation of the microtubes. A ring-structure Bessel beam with nonuniform intensity distribution was modulated by the superimposition of an LPM on the hologram. Tuning the environment pH value allows the transformation from straight heteromicrotubes to bent microtubes. The local gray scale values of holograms and the laser source power can be tuned to achieve various combinations of nonuniform distribution of energy of Bessel beam. As shown in Figure 4.10, the diagram consists of three regions (I, II, III) which mean the formation of three kinds of pH-responsive microtubes. Due to the too strong laser power and the weak heterodistribution of light field, the microtube in the region I exhibits negligible pH-responsive deformation. In contrast, with a power of semicircle beam lower than the exposure threshold, the fabricated microtubes in the region III presented semicircle shape. With a moderate laser power and suitable gray level, the microtube in the region II showed pH-responsive bending under a pH value lower than 9.

The impacts of parameters (e.g. the laser power: $p$, the local gray value: $g$, and the microtube height: $h$) on the deformation (bending angle: $\theta$) of the heteromicrotubes were systematically studied. The bending angle was calculated as:

$$\theta = \arcsin(\delta/h) \tag{4.5}$$

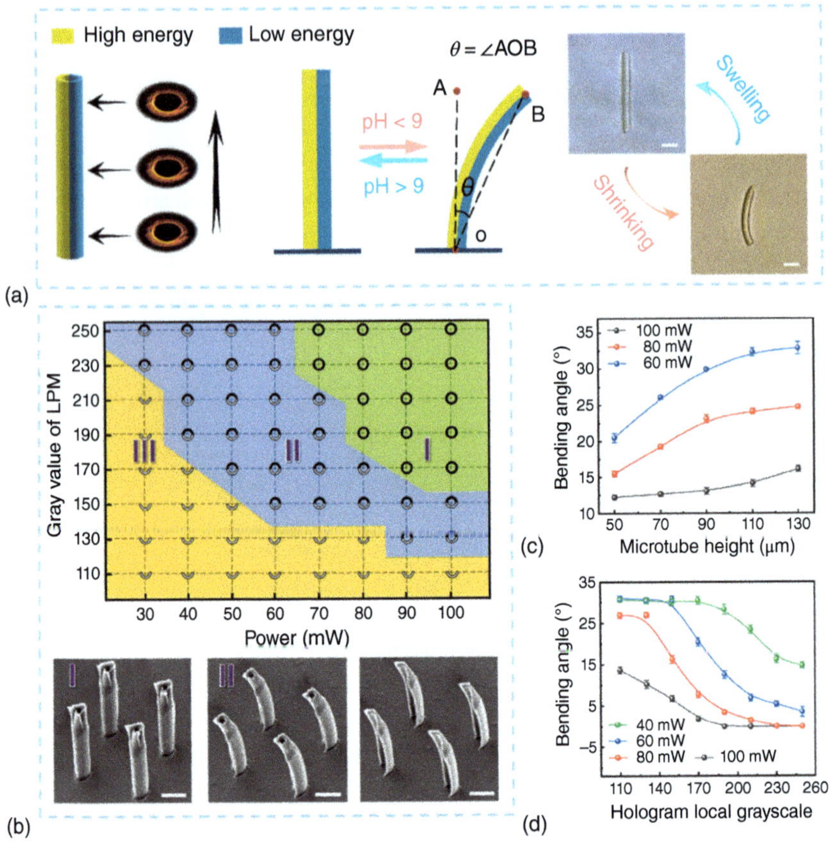


**Figure 4.10** Fabrication and characterization of pH-triggered microtubes. (a) Schematic illustration of the pH-responsive bending of the microtube. (b) Phase diagram of three kinds of microtubes fabricated by different laser powers and local gray values. (c) The impact of the microtube height on the bending angle under different laser powers. (d) The impact of the local gray level on the bending angle under different laser powers. Source: Li et al. [33]/Reproduced from American Chemical Society.

$h$ and $\delta$ represent the height of microtube, and the horizontal distance between the root of the fabricated microtube and the tipping point, respectively. It can be concluded that there is a positive correlation between the microtube height and the bending degree, and a negative correlation between laser power, local grayscale, and the bending degree. Therefore, to fabricate pH-responsive microtubes with distinct bending, a moderate laser power, local grayscale, as well as a relatively high microtube are needed.

The process shown in Figure 4.10 is only suitable for constructing microtubules that bend in a single direction. Dynamic holography technique is used to fabricate segmented microtubules, which can achieve more sophisticated deformations, as shown in Figure 4.11. When the piezoelectric platform immobilized with the hydrogel material is moved along the $z$-axis direction, continuously changing the holographic pattern can form multi-segment microtubules with complex

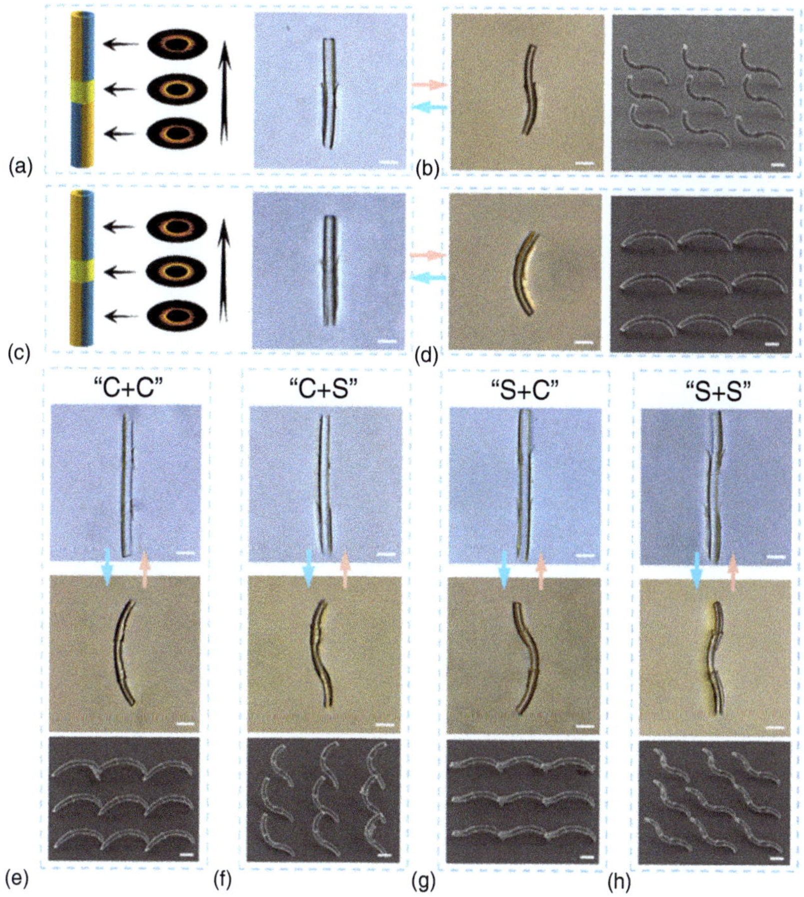


**Figure 4.11** Segmented microtubule structures fabricated by dynamic holograms. (a and b) Schematic diagram of microtubules with S-shaped deformability, fabrication results, deformation results, and SEM results under supercritical drying conditions. (c and d) Schematic diagram of microtubules with C-shaped deformability, fabrication results, deformation results, and SEM results under supercritical drying conditions. (e–h) Microtubule structures with two intermediate segments exhibit four different deformation modes. Source: Li et al. [33]/Reproduced from American Chemical Society.

deformability. Figure 4.11a,b demonstrates a microtubule that can deform to S-shape in response to environmental pH stimuli. Different colors (e.g. blue, orange, yellow) in the schematic diagram of the microtubule represent increasing exposure doses, which in turn indicate an increase in crosslink density and a decrease in swelling ratio. During the movement of the piezoelectric platform, microtubes were fabricated using three kinds of Bessel beams, two of which have opposite energy distributions. By adjusting the pH of the medium (less than 9), the upper and lower segments of the microtubules bend to the right and left, respectively. In addition, changing the energy distribution of the lower segment can make the microtubules bend into a C-shape (both upper and lower segments bend to the right). More

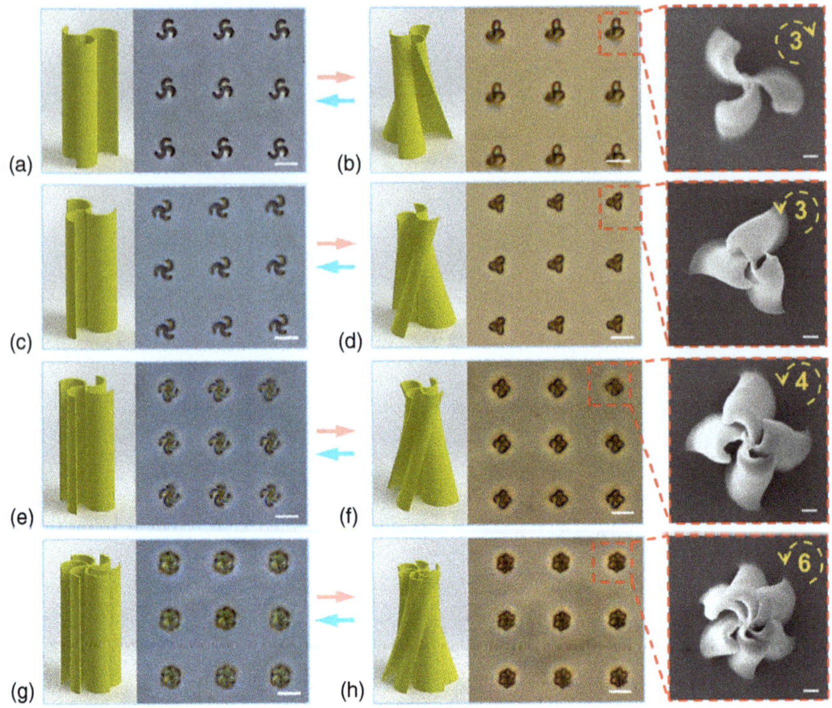


**Figure 4.12** Chiral structures with torsional deformation capability fabricated by splicing processing. (a and b) Schematic illustration of the left-handed structure, fabrication results, deformation results, and SEM results under supercritical drying conditions. (c and d) Schematic illustration of the right-handed structure, fabrication results, deformation results, and SEM results under supercritical drying conditions. (e and f) Schematic illustration of the right-handed structure composed of four microtubules, fabrication results, deformation results, and SEM results under supercritical drying conditions. (g and h) Schematic illustration of the right-handed structure composed of six microtubules, fabrication results, deformation results, and SEM results under supercritical drying conditions. Source: Li et al. [33]/Reproduced from American Chemical Society.

complex deformation patterns can be formed by placing more intermediate segments in the structure, as shown in Figure 4.11e–h. Microtubules containing two intermediate segments can form four deformation modes, indicating that the dynamic holographic strategy can precisely control the exposure dose distribution of microtubules along the height direction.

In addition, torsional deformation of the responsive structure can be achieved by splicing method, as shown in Figure 4.12. Microtubules processed by a semi-circular Bessel beam deform toward the gap in response to solvent stimulation. Figure 4.12a shows a left-handed structure consisting of three semicircular microtubules with different gaps. It undergoes chiral torsion when changing the pH of the solution (less than 9). Further, the twisting direction can also be adjusted by changing the gap direction. More complex chiral structures (containing four and six semicircular structures, respectively) were also fabricated to demonstrate the actuation modes.

### 4.2.4 Reconfigurable Soft Microdevices Fabricated by Complex Laser Scanning Strategy

The processing parameters of femtosecond laser direct writing, including the number of scanning repetitions, laser exposure time and laser power, can adjust the deformability of hydrogels [6]. With the increase of the exposure time, or the increase of the laser power and the number of scanning repetitions, the swelling ratio of the hydrogel gradually decreased. As shown in Figure 4.13, the stimuli-responsive polymer is processed by a reciprocating scanning method of the DLW system to construct a structure that mimics the shape of a plant leaf. Attributed to the outstanding processing capability of the DLW system, the size and number of blades can be easily adjusted. The printed leaf-like structure is attached to a glass substrate via a slender cylinder. The scanning electron microscope (SEM) image can clearly reflect the printed blade structure, and traces of scanning lines in the blade structure can be observed (the thickness of the blade: 2.9 μm). By adjusting the pH value of the liquid, the leaf structure can undergo reversible shape transformations, such as isotropic expansion in an alkaline environment.

The distortion deformation of blade structure can also be realized by adjusting the scanning mode, as shown in Figure 4.14. For the blade structure obtained by

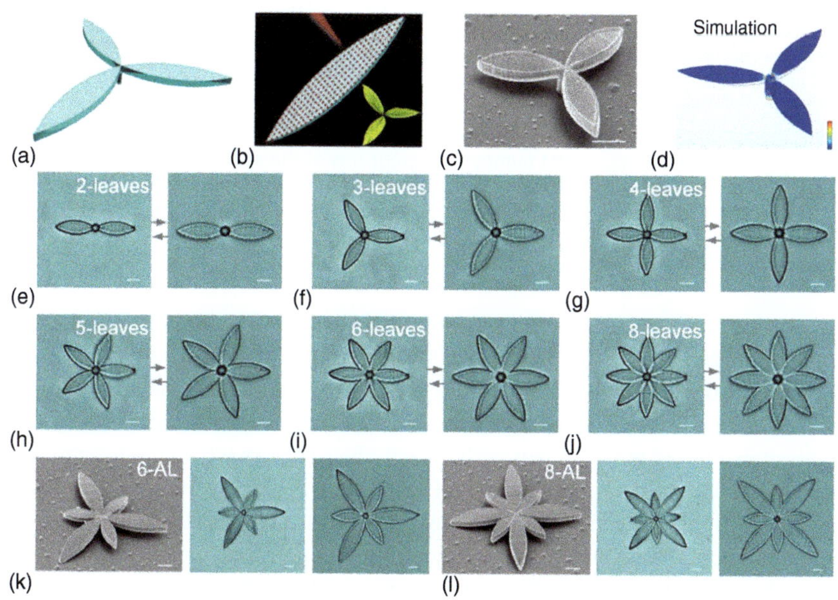


**Figure 4.13** pH-responsive leaf structures prepared by reciprocating scanning. (a) Schematic of the leaf structure, which consists of three leaves and an elongated cylinder as the base. (b) The reciprocating scanning strategy of the laser focus. (c) SEM image of the DLW-printed leaf structure. (d) Expansion and deformation of the printed leaf structure in an alkaline environment. (e–j) Expansion and contraction behavior of structures with different number of leaves. (k and l) Structures of lobes with inconsistent size. AL indicates asymmetrical leaf. Source: Hu et al. [6]/Reproduced from John Wiley & Sons, Inc.

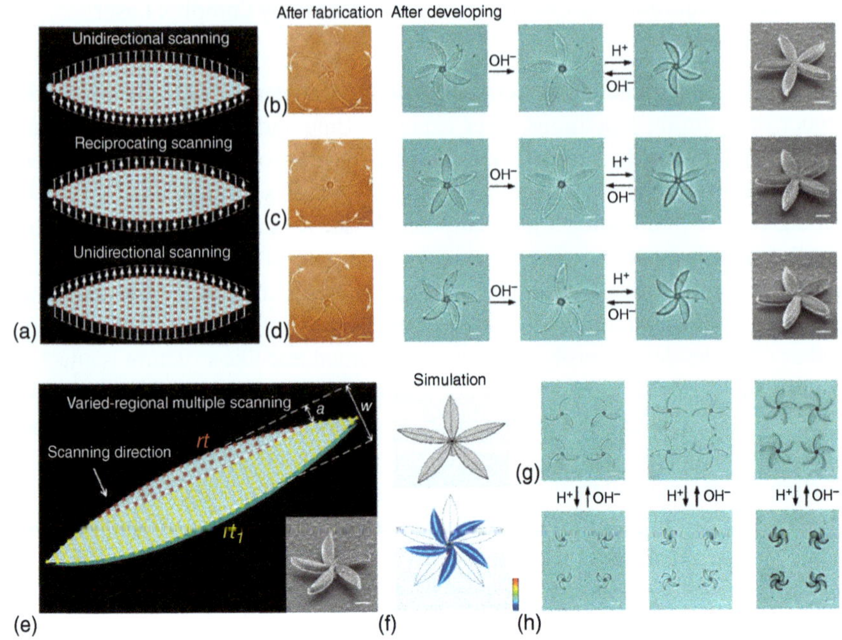


**Figure 4.14** Chiral distortion of leaf structures prepared by unidirectional scanning and multi-area multi-scanning strategies. (a) Schematic illustration of reciprocal scanning and two unidirectional scanning strategies. (b, d) Expansion, contraction, and twist deformation of the leaf structure fabricated by the unidirectional scanning strategy. (c) Expansion and contraction deformation of the leaf structure fabricated by the reciprocating scanning strategy. (e) Schematic illustration of different scanning strategies for different regions. The parameters *rt* and $rt_1$ represent the number of repetitions of the scan process. (f) Deformation simulation of a leaf structure fabricated with the complex scanning strategy. (g and h) Experimental results of the leaf structures constructed with the complex scanning strategy. Source: Hu et al. [6]/Reproduced from John Wiley & Sons, Inc.

reciprocating scanning, it will show uniform expansion or contraction behavior in the liquid environment. If the direction of the scan is unidirectional, as shown in the top schematic diagram in Figure 4.14a, the scan always starts from the upper side of the blade and ends at the lower side of the blade. When the printed leaf structure is immersed in a liquid environment with $pH < 9$, the structure will exhibit shrinkage behavior and chiral distortion. For the leaf structure prepared by the opposite unidirectional scanning method, the twist direction of the structure will also be reversed when it shrinks. The distortion caused by unidirectional scanning is related to the motion characteristics of the DLW system. When the laser spot is scanned from one side of the blade structure to the other, due to the buffer stage at the end of the movement, structures near the end of the scan usually experience longer exposure times, resulting in increased crosslink density and decreased expansion rates. When the leaf shrinks in a $pH < 9$ environment, the structure is deflected toward the side where the scan was initiated. In order to control the distortion of the leaf structure more finely, we can adopt different scanning strategies and scanning parameters in different regions, such as different repetition times (Figure 4.14e–h).

## 4.3 Modular Design of Reconfigurable Soft Robots Based on 4D Microscale Building Blocks

### 4.3.1 Designing Microscale Building Blocks

The 4D printing technique needs to solve complex forward and inverse problems, which are important for the precise match between the design and the realistic shape transformations. In this respect, simulation studies and theoretical analysis provide powerful strategies to deal with these problems. However, the simulation study of sophisticated 3D models would be limited by the great computational burden and become time-consuming. In this section, a modular design method based on DH matrix was developed. Inspired by the construction of LEGO toys, multiple discretized and small building blocks were adopted to assemble into an arbitrary 3D structure. In this manner, a reconfigurable micromachine can be obtained by the assembly of deformable building blocks. As shown in Figure 4.15, the traditional 4D building block was a bilayer structure consisting of a passive material and an active material. With the shrinking of the active material, the bilayer structure would bend toward the active layer. Thanks to the ultrahigh precision of the DLW technique, an articulated mechanism was introduced to improve the programmability and deformability of the building block. Multiple thin bilayers, rigid supports, and compliant hinges were included in a single articulated microbuilding block. The usage of compliant hinge could extend the morphing freedom of the 4D building block.

The classical Timoshenko beam theory was used to derive the force equilibrium equation for the building block in the traditional 4D printing strategy, as shown in Eq. (4.6).

$$\frac{(P_1 + P_2)h}{2} = \frac{E_1 I_1 + E_2 I_2}{\rho} \tag{4.6}$$

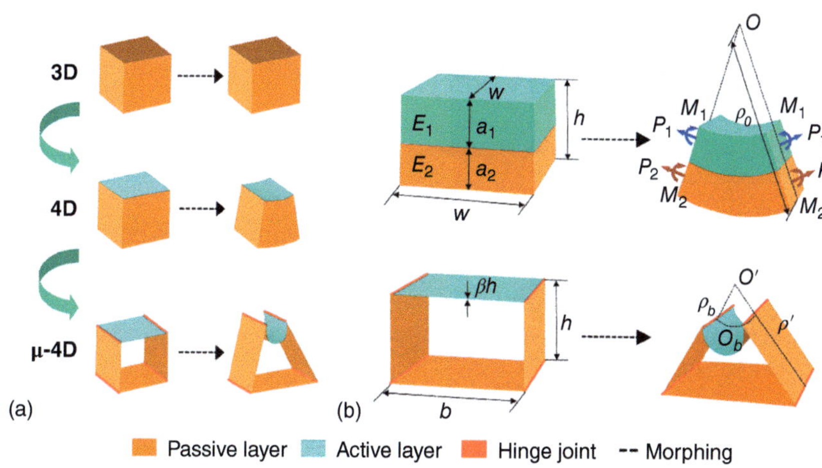


**Figure 4.15** Schematic illustration of the transformations of traditional and articulated building blocks. (a) The evolution of articulated building blocks from conventional building blocks. (b) Geometrical parameters and deformation forms of conventional building block and articulated building block. Source: Adapted from Huang et al. [81].

$P_i$, $E_i$, and $I_i$ ($i = 1, 2$) represented the axial forces, moduli, and moments of inertia of the bilayer structures, respectively. $h$ represented the total thickness. For simplicity, $h$ was assumed to be 1. $\rho$ represented the curvature radius of the middle layer. The geometric compatibility in this configuration required:

$$\lambda_1 - \lambda_{01} + \frac{P_1}{E_1 a_1} + \frac{a_1}{2\rho} = \lambda_2 - \lambda_{02} - \frac{P_2}{E_2 a_2} - \frac{a_2}{2\rho} \tag{4.7}$$

$a_1$ and $a_2$ represent the thicknesses of the active layer and the passive layer, respectively. $\lambda_i$ and $\lambda_{0i}$ were the final stretch and the initial stretch, respectively. If the width was assumed to be 1 for simplicity, $I_i$ and $E_i$ can be calculated as: $I_i = \frac{a_i^3}{12}$, $E_i = 3N_i kT$. $T$ and $k$ represented the absolute temperature and the Boltzmann constant, respectively. $N_i$ was the number of effective polymeric chains. With the following definitions: $\frac{a_1}{a_2} = m$, $\frac{E_1}{E_2} = \frac{N_1}{N_2} = n$, the curvature of the bilayer structures can be obtained as:

$$\kappa_0 = \frac{1}{\rho_0} = \frac{6[(\lambda_2 - \lambda_{02}) - (\lambda_1 - \lambda_{01})](1+m)^2}{\left[3(1+m)^2 + (1+mn)\left(m^2 + \frac{1}{mn}\right)\right]} \tag{4.8}$$

In Eq. (4.8), $m$ and $n$ are determined by the structural design and the exposure dosage, respectively. Figure 4.16 shows the impacts of exposure dosages and $m$ on the curvature of the bilayer structure, while for the building block designed based on articulated mechanism, the rotation of the rigid hinges around the joints would be activated by the bending of the bilayer unit. It is assumed that the two kinds of building blocks possess the same size, and the thickness of the bilayer unit was $\beta h$. The chord length $l$ and angle $\theta$ of the building block can be calculated as:

$$l = \sqrt{2{\rho_b}^2 - 2{\rho_b}^2 \cos\frac{b\lambda_1}{\rho_b \lambda_{01}}} \tag{4.9}$$

$$\theta = \frac{b\lambda_1}{\rho_b \lambda_{01}} \tag{4.10}$$

$\rho_b$ and $b$ represented the curvature radius of the bilayer unit and the length of the building block, respectively. The curvature of the building block can be calculated as:

$$\kappa' = \frac{1}{\rho'} = \frac{\frac{b\lambda_1}{\lambda_{01}} - l}{bh\frac{{\lambda_{01}}^2}{{\lambda_1}^2}} \tag{4.11}$$

$\rho'$ and $\kappa'$ represented the curvature radius and the curvature of hinges, respectively.

Based on the boundary condition that there is no intersection between the left and right hinges, the following requirement should be fulfilled: $\theta \leq 2\pi$. In addition, the maximum shape-morphing result of the building block should be a triangle shape. Thus, it can be concluded that:

$$\frac{b\lambda_1}{\lambda_{01}} < 2\frac{h\lambda_1}{\lambda_{01}} \tag{4.12}$$

Figure 4.16 illustrates the influence of exposure dosages, slenderness ratio ($b' = b/h$), on the curvature of the building block under different $\beta$.

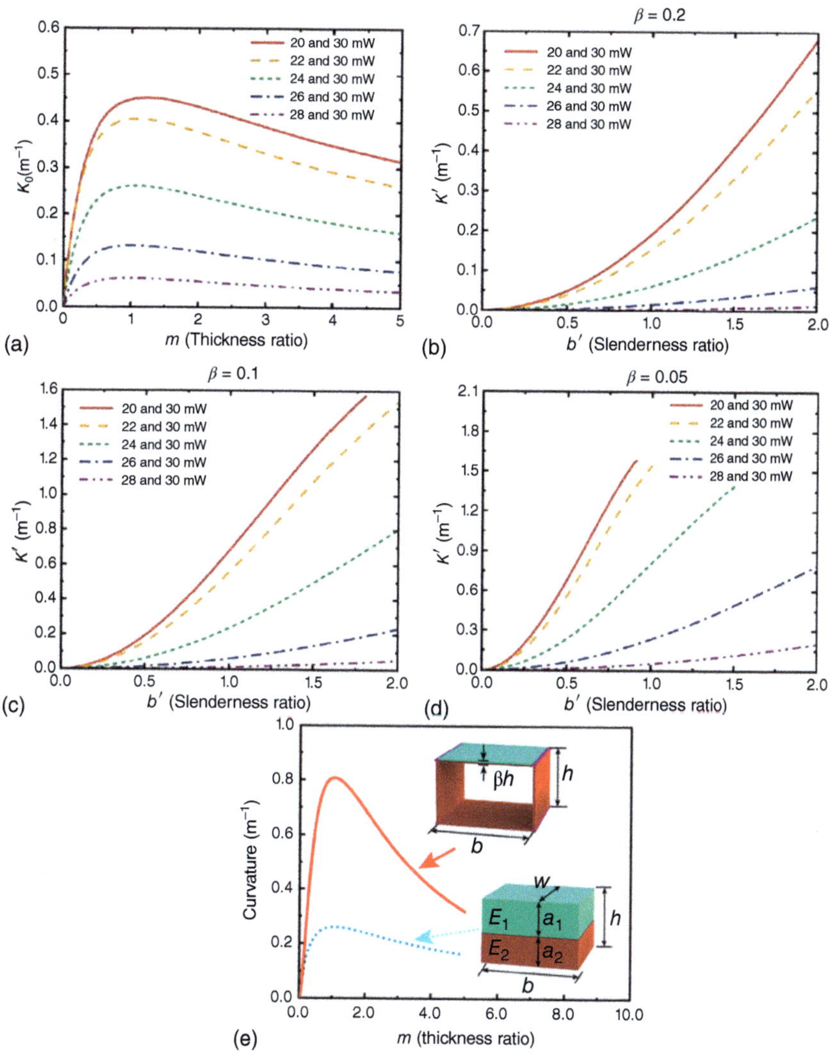


**Figure 4.16** Quantitative characterization of the curvatures of the deformed building blocks. (a) The relationship between the curvature of deformed bi-layered structures and the thickness ratio under different laser powers. (b–d) The impact of the slenderness ratio on the curvature of the articulated building block with different thickness coefficients and laser powers. (e) Comparison of the deformation (curvature) of conventional and articulated building blocks. Source: Adapted from Huang et al. [81]/American Association for the Advancement of Science.

As shown in Figure 4.16e, the deformation curvature of the articulated building block could attain at least two times larger than the curvature of the traditional building block, which confirmed the superior deformation ability derived from the articulated mechanism. In addition, since the crosslinking density in the fabricated structure was temporally and spatially well controlled, the morphing freedom of the building block can be extended by the arrangement of hinge joints and bilayer

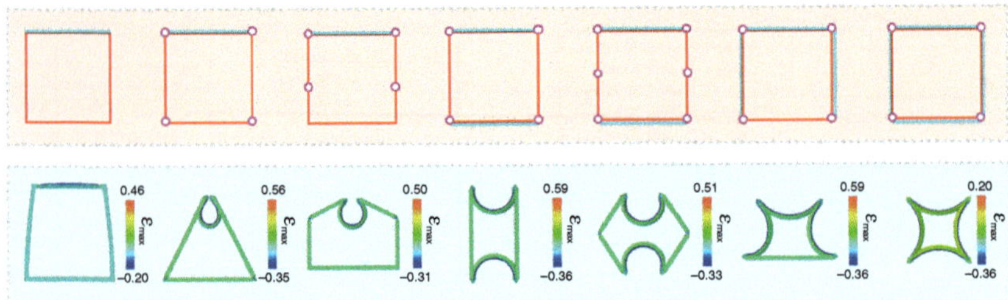

**Figure 4.17** Morphological transformations of articulated building blocks with varied bilayer structures and hinge joints. Source: Huang et al. [81]/American Association for the Advancement of Science.

units. Figure 4.17 exhibits several kinds of the articulated building blocks resulting in entirely different configurations with the varying distribution of hinge joints.

### 4.3.2 Modular Assembly of Building Blocks Based on DH Matrix

To fabricate a more sophisticated morphing modular system with the superposition principle, an octagonal prismatic microcylinder was introduced as the building block, as shown in Figure 4.18. From the side view, the new building block consisted of three pairs of passive layers, hinge joints, and a pair of active layers. The simulation study of the shape morphing of the building block was performed. A rotation of the assembled chain blocks was activated by the shrinkage of the prismatic building block, which can play a role of rotational joint. Desired 3D movements can be generated by the assembly of multiple building blocks like a robotic arm. The simulation study also works well to estimate the locomotion of assembled chains consisting of a finite number ($n < 60$) of building blocks, while due to the divergence of the numerical results, it cannot be applied to a large number of 4D building blocks ($n > 60$). To solve this problem, DH parameters, which fulfill the superposition principle, were introduced for the computation of the deformation of a robotic arm consisting of rigid bars and multiple joints. Any complex deformation can be described by four physical parameters with a closed analytical form. To obtain the kinematics equations for the modular system, a transformation $[Z]$ and a separate transformation $[X]$ were used. The relative movement allowed by each joint was characterized by the transformation $[Z]$, and the dimensions of each link were defined by the separate transformation $[X]$. Therefore, a sequence of transformations, which contain alternating link and joint transformations from the bottom to the end of the chain, was obtained as Eq. (4.13).

$$[T] = [Z_1][X_1][Z_2][X_2]\cdots[X_{n-1}][Z_n][X_n] \tag{4.13}$$

$$[Z_i] = \begin{bmatrix} \cos\theta_{z_i} & -\sin\theta_{z_i} & 0 & 0 \\ \sin\theta_{z_i} & \cos\theta_{z_i} & 0 & 0 \\ 0 & 0 & 1 & d_i \\ 0 & 0 & 0 & 1 \end{bmatrix} \tag{4.14}$$

$$[X_i] = \begin{bmatrix} 1 & 0 & 0 & R_{i,i+1} \\ 0 & \cos\alpha_{i,i+1} & -\sin\alpha_{i,i+1} & 0 \\ 0 & \sin\alpha_{i,i+1} & \cos\alpha_{i,i+1} & 0 \\ 0 & 0 & 0 & 1 \end{bmatrix} \tag{4.15}$$

$[X_i]$ and $[Z_i]$ $(i = 1, 2, \dots n)$ represented the dimensions and the relative movement of each link. $[T]$ represented the transformation for the link of the last building block. $d_i$, $\theta_{z_i}$, $R_{i,\,i+1}$, and $\alpha_{i,\,i+1}$ are the DH parameters, which define the motion planning and assembly rules of the building block (Figure 4.18). Five assembled building blocks encoded by different $\theta_z$ and $R$ and their simulated rotations are shown in Figure 4.18b,c. To avoid the deformation interaction of each building block, a rigid bar was used to connect these building blocks with each other. The parameter $\theta_{z_i}$

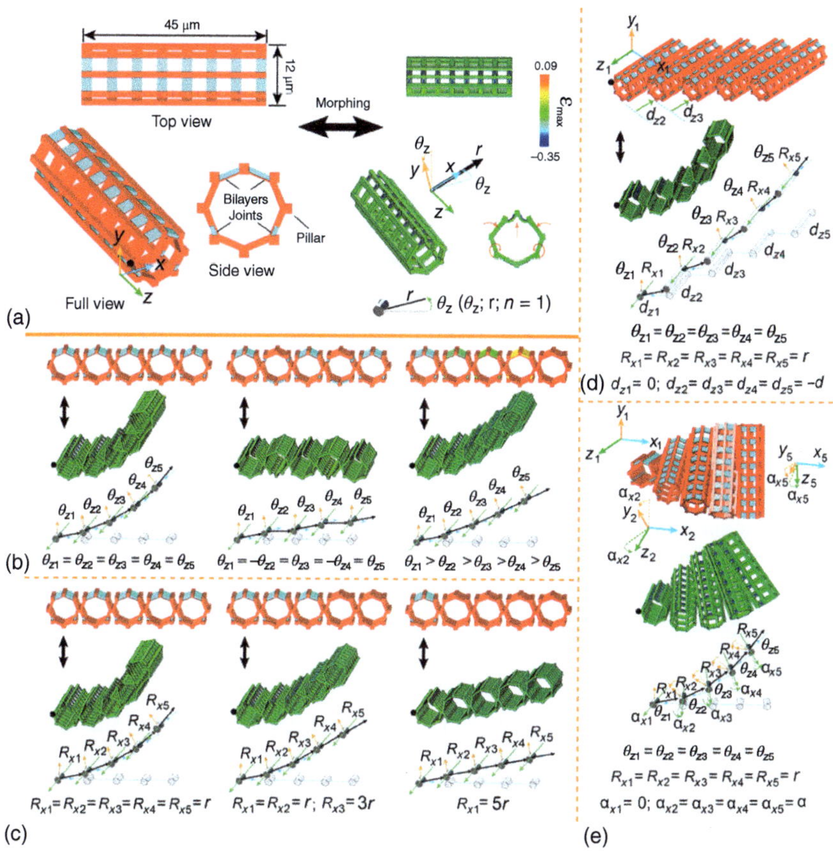


**Figure 4.18** Schematic illustration and simulation study of the modular system assembled with different DH parameters. (a) Schematic illustration of the geometry design and the morphological transformation of the articulated building block. (b–e) Rotational movements of modular systems with controlled orientation and amplitude activated by the shrinkage of active layer. Source: Huang et al. [81]/American Association for the Advancement of Science.

could determine the orientation and amplitude of the building block which rotates along the $Z$ axis, which can be tuned by the arrangement of the active layer and the laser power, respectively. As shown in Figure 4.18d, there is a morphable structure assembled by building blocks with offsets ($d_i$) along the opposite direction of the $Z$ axis. The screw displacement along the $Z$ axis is represented by Figure 4.18b,d. According to Figure 4.18c, the connection of the rotational joints by inactive building blocks could extend the rotational radius $R_{i,i+1}$ with respect to the $X$ axis. The morphable structure in Figure 4.18e was fabricated by the assembly of building blocks with angles $\alpha_{i,i+1}$ rotating with respect to the $X$ axis. Thus, the screw displacement along the $X$ axis was achieved by Figure 4.18c,e together.

### 4.3.3 Inverse and Forward Design of Modular Morphing System

We could convert arbitrary targeted 3D shapes to a discrete counterpart consisting of a limited number of joints. Then, inverse kinematics were performed to obtain the DH parameters and the morphological transformation between the initial and target modular system. For example, Figure 4.19 shows a modular system whose initial design was a roll shape and target design was a wave shape. The transformation between the initial and target shapes can be found by the DH parameters. In this respect, the only variable parameter was $\theta_z$. It should be noted that the building blocks in the modular system and the discretized joints of the target have the same number. As shown in Figure 4.19b, planar stripes composed of 60 building blocks with identical initial geometry but encoded by different parameters ($\theta_z$, $d$, and $R$) were fabricated. These printed structures could deform into a variety of targeted shapes including roll, wave, and honeycomb.

Typically, a reverse of the inverse design as a forward design could also provide effective solution, while for the "Googles" shape in Figure 4.19b, these two kinds of design strategies (the forward design and the reverse of the inverse design) are distinct and different. If a collision exists during the shape-morphing process, the forward kinematics cannot achieve its target shape. Figure 4.19b demonstrates that a planar stripe consisting of building blocks encoded by the same $\theta_z$ can deform into the goggle shape. Experimental results in Figure 4.19b (V and VI) illustrated that for building blocks with a constant Z axis offset ($d_i$), they could transform into a helix and spiral shape. Furthermore, more complex 3D-to-3D shape-morphing behaviors can be achieved by the fabrication of 3D structures encoded with four parameters ($\theta_z$, $d$, $R$, and $\alpha$). Following the forward design principle, a micrometer-scale transformer with 3D motion planning was developed. Figure 4.20 shows a modular system consisting of five functional sections and their connections, e.g. legs, neck, backbone, shoulder, and arms. A series of DH parameters can be used to separately capture the transformations of each section. The demonstration of the transformer could achieve a shape-morphing between a humanoid robot and a race car.

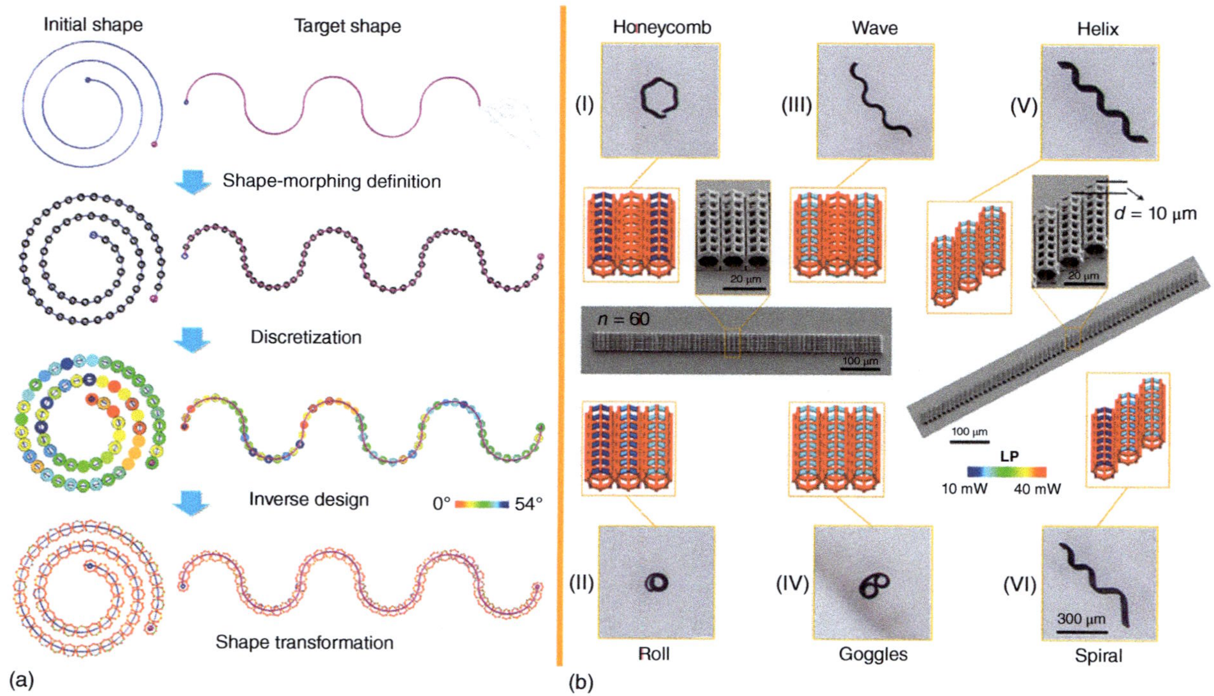


**Figure 4.19** Morphing modular systems developed using the inverse and forward design strategy. (a) A roll-shaped modular system deformed into a wave shape via the inverse design. (b) Experimental results of modular systems encoded by different DH parameters. Source: Huang et al. [81]/American Association for the Advancement of Science (AAAS).

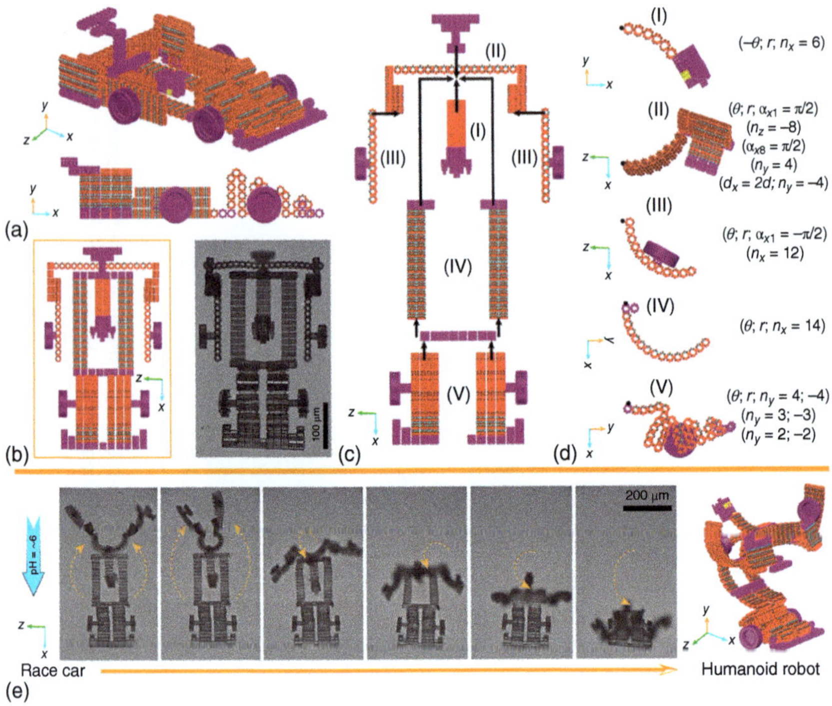


**Figure 4.20** Demonstration of a microtransformer that morphed from a race car to a humanoid robot. (a and b) Schematic of the shape of the race car and image of the fabricated race car. (c and d) Detailed connections and encoded DH parameters of the race car. (e) Experimental results of the transformation process of the microtransformer. Source: Huang et al. [81]/American Association for the Advancement of Science (AAAS).

## 4.4 Application of Reconfigurable Soft Robots

### 4.4.1 pH-Responsive Microparticle and Cell Gripper

Recently, stimuli-responsive microactuators have been widely used for the manipulation of microparticles and cells. Firstly, a pH stimuli-responsive device consisting of several bendable hydrogel microtubes was fabricated to perform *in situ* trapping. The maximum size of the hydrogel actuator was 200 μm × 200 μm × 200 μm and the resolution could reach 300 nm.

As shown in Figure 4.21, the microgripper device consists of four microtubes as fingers. Under open state (swelling), these "fingers" are straight, while in closed state (shrinking), they would bend toward the same point. The diameter and height of each microtube were 10 and 60 μm, respectively. There is a 20 μm distance between adjacent microtubes, which could match the size of microparticle. Nonuniform Bessel beam with a local gray value of 150 and laser power of 60 mW was used to fabricate the microgripper device. A drop of polystyrene (PS) microspheres solution (10 vol % and 13 μm diameter) was added to the open microstructures to verify the particle trapping capability. The gripper in Figure 4.21 could capture a

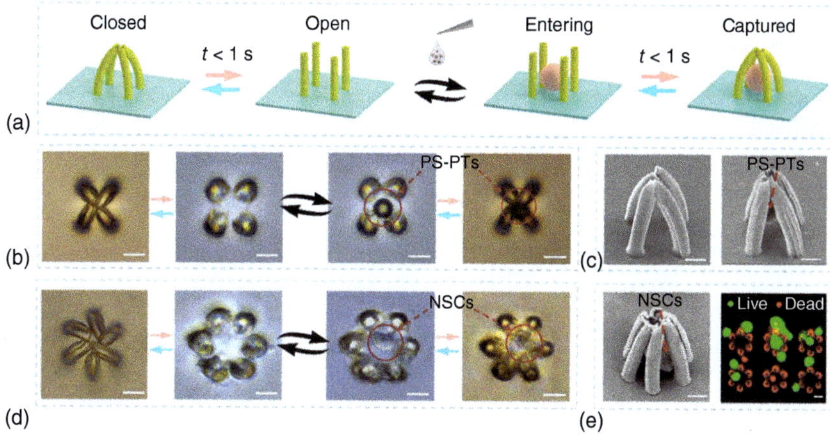


**Figure 4.21** pH-responsive microgripper for cell and cargo manipulation. (a) Schematic illustration of the design and actuation of the microgripper. (b) Experimental demonstration of the capture of a 13 μm microparticle. (c) Experimental demonstration of the trapping of 10–15 μm cells. Source: Li et al. [33]/Reproduced from American Chemical Society.

single PS microsphere. By adding a dilute acid solution, the bending of the gripper was realized to prevent the microparticle from being washed away. Subsequently, a dilute alkali solution was added to realize the unfolding of the gripper so that the microparticle can be easily removed. Therefore, with the pH-responsive gripper, microspheres can be captured and released within a second upon the varying environmental conditions. In addition, microspheres with different diameters can be captured by tuning the microtube height and the distance of adjacent microtubes.

Apart from microparticles, the microgripper can also trap a single cell. A six-finger microgripper is developed to enhance the success rate of capturing and lower the escape rate due to the existence of interspace between adjacent microtubes (Figure 4.21). The microgripper initially stayed in the open state to prevent dangerous mutations or cell death. A cell medium containing NE-4C neural stem cells was added to the microgripper. The distance between adjacent microtubes and the diameter of the middle space is 18 and 20 μm, respectively. The size of the cells is 10–15 μm. After the cell was fallen into the gripper, dilute hydrochloric acid was added to create a neutral environment (pH = 7), and the cell would be captured due to the closing of the gripper. The cell states were identified by live and dead cell assay on the following day. The red (hollow), red (solid), and green colors represent the hydrogel microgripper, dead cells, and alive cells, respectively. ~75% cell survival rate on the following day demonstrated that little harm was done to the cells by the capturing process. A SEM image that confirms the cell captured inside the microgripper is shown in Figure 4.21.

On the other hand, a 4D-printed microcage is demonstrated as shown in Figure 4.22. Due to the difference in the pore size under the contracted and expanded states, the microcage with the shape of a regular hexagonal could capture microparticles [6]. In the contracted state, the diameter of the inscribed circle of printed microcage was 8 μm. Thus, a $SiO_2$ particle with the diameter of 10 μm

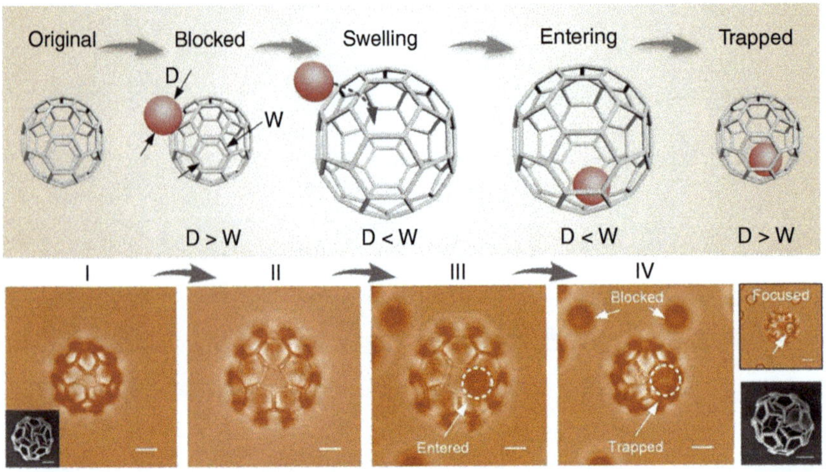


**Figure 4.22** Schematic illustration and experimental demonstration of pH-responsive microcage for selective particle trapping. Source: Hu et al. [6]/Reproduced from John Wiley & Sons, Inc.

could not enter the microcage, while under the swelling state, the pore size could reach 12 μm, allowing the entry of $SiO_2$ particles with size smaller than 12 μm. By introducing dilute hydrochloric acid, the capturing of the microparticles would be performed by the shrinking the microcage, while particles outside the microcage could move freely and be removed. Meanwhile, by changing pH value, the release of the microparticle can be realized by the expansion of the microcage. In addition, a microcage coated by a magnetic layer allows the remote transportation of the trapped microparticle. Due to the flexibility of the DLW technique, specific particles can be trapped by the microcage with well-designed pore size. The developed selective trapping method holds great promise for drug delivery and cell micromanipulation.

### 4.4.2 Localized Cancer Cell Treatment

Inspired by a natural fish with a deformable mouth (opening and closing states), a smart shape-morphing microfish (SMMF) was fabricated to controllably encapsulate and release doxorubicin (DOX) (Figure 4.23) [83]. The microrobots were printed by an Fs-DLW system with pH-responsive hydrogel. Under an alkaline environment (pH > 9), the hydrogel network expanded significantly due to electrostatic repulsion forces, and under a low pH environment, the shrinking of the hydrogel network would exhibit. Magnetic actuation exhibits advantages including remote control and fuel-free, which could facilitate the task execution of shape-morphing microrobots (SMMRs) in dynamic environments. After the 4D printing of the SMMRs, their magnetic driving capability was endowed by immersing them in a suspension of $Fe_3O_4$ NPs (Moment: 79.55 emu $g^{-1}$, diameter: [~200 nm]). To prevent a reaction with hydrochloric acid, silica shells were coated with $Fe_3O_4$ NPs. The developed SMMF loaded with $Fe_3O_4$ NPs presents good magnetic response

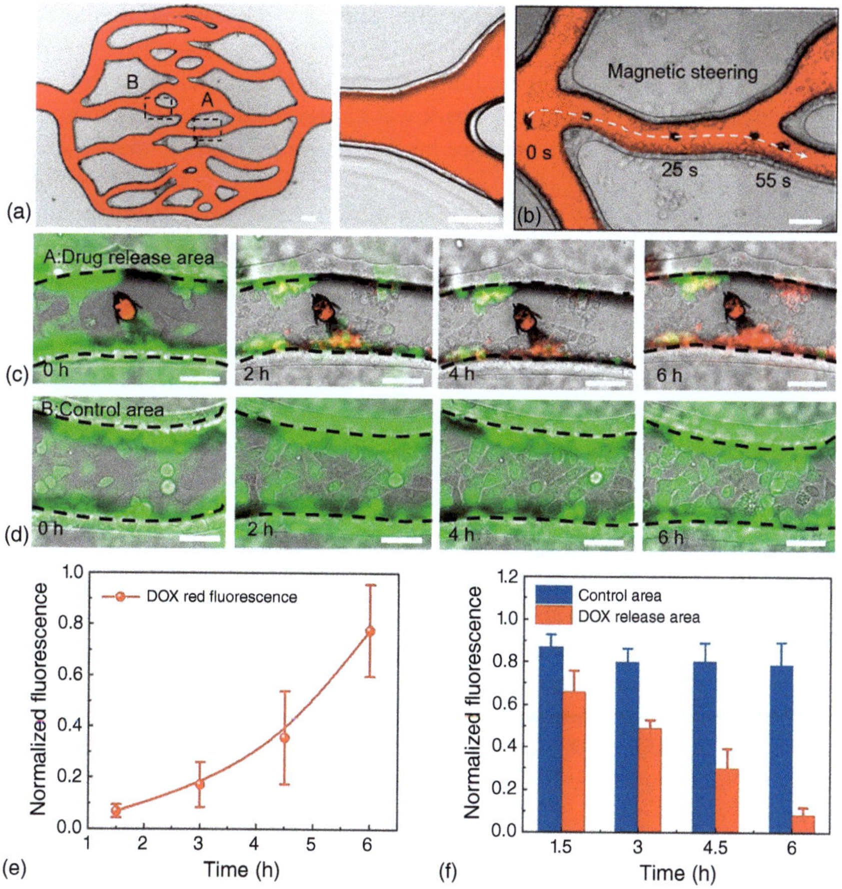


**Figure 4.23** Magnetic shape-morphing microrobots used for target drug delivery and localized cell delivery. (a) Image of an artificial vascular network. (b) The locomotion of magnetic microrobots in the artificial vascular network. (c and d) The cell viability in the drug-releasing region (c) and the control region (d). (e) The variation of fluorescence intensity of DOX with time. (f) The variation of fluorescence intensity of cells with time in control region and drug-releasing region. Source: Xin et al. [83]/Reproduced from American Chemical Society.

properties (Moment: 3.04 emu $g^{-1}$). Hydrogen bonds that build stable adhesion are formed due to the hydroxyl groups on the silica surface and the carboxyl groups in the microrobots. Furthermore, atomic force microscope (AFM) was used to measure the stable adhesion (~57 nN) between the $Fe_3O_4$ NPs and the surface of the SMMRs, which enables them to resist sonication vibration (100 W, five minutes) and tape peeling (3 M 610, five times).

To verify the capability of localized cancer cell treatment, magnetic SMMFs loaded with DOX were manipulated in a complex network for the localized treatment of HeLa cells. Mask UV lithography with polydimethylsiloxane (PDMS) replica molding was used to prepare the artificial vascular network. The width and depth of each microchannel were 70–150 and 80 μm, respectively. To enhance the bonding with the cover glass as well as the adhesion between HeLa cells and the channel, plasma treatment was performed on the network followed by the soaking in poly-L-lysine (PLL) solution for two hours. Nutrient medium with HeLa cells was added to the artificial vascular network. The entire network was almost filled with HeLa cells after two days of culture which would be used for drug delivery. Upon external magnetic stimulation, magnetic SMMFs were controlled inside the network until reaching a target location. After adding a mixture of hydrochloric acid (pH < 7), the mouth of the SMMF would be opened and DOX would be released. The area containing the SMMF was defined as the drug release area, and other regions were defined as the control area. DOX diffused rapidly from the body of the SMMF to the HeLa cells over six hours. The fluorescence signal (green) of the cells decreased during this period, while in the control channels, the growth and proliferation of HeLa cells were performed normally without no red DOX fluorescence signal exhibited. The variation of the signal intensity (red fluorescence) with the time was quantitatively characterized. The normalized intensity in HeLa cells rose from 7% to 78%. The signal intensity (green fluorescence) in the control area still remains ~80% after six hours. In comparison, the signal intensity in the drug release area drops to less than 10%. In this study, the feasibility of the developed microrobots was only verified in the artificial vascular network. The application in real blood vessels requires microrobots with a smaller size and feasible imaging method. Considering neutrobots (10–20 μm) that have been adopted for target delivery in vascular networks, the SMMRs with the optimized size less than 10 μm would meet the requirement of *in vivo* application in blood vessels. In addition, photoacoustic imaging and ultrasound imaging would be used for the targeting manipulation of the SMMRs.

## 4.5 Summary

In this chapter, reconfigurable soft micromachines based on stimuli-responsive hydrogels were discussed. The DLW technique could regulate the crosslinking density of hydrogel matrix in a point-by-point manner. The fabricated intelligent micromachines could perform 3D-to-3D morphological transformations with large deformations which may be amplified by various mechanical mechanisms. To improve the fabrication efficiency of the point-by-point DLW technique, the asymmetric Bessel beam with nonuniform energy distribution was adopted. Via the dynamic light modulation method, morphable microtubes that could respond to the variation of environment pH were fabricated. Inspired by the kinematic analysis problems of a robotic arm, an inverse and forward modular design method to achieve desired 3D-to-3D morphological transformation was proposed. Compared to traditional bilayer structures, the printed articulated building blocks exhibited better deformability. To investigate the shape-morphing process, the DH matrix and four DH parameters were used to guide the assembly of multiple articulated building blocks which also fulfills the superposition principle. The modular design method could lower the computational complexity for the construction of reconfigurable micromachines with complex morphological transformations. To verify the design strategy, a sophisticated transformer that could convert from a race car to a humanoid robot was demonstrated. The 4D printed responsive hydrogel machines hold great promises for a wide spectrum of biomedical applications, e.g. micromanipulation and target delivery. As demonstration, pH-responsive microgrippers and microcages were fabricated for *in situ* cell and particle manipulations. In addition, shape-morphing microrobots that could load and release drugs were fabricated and their functions were demonstrated in artificial vascular network. The 4D printing of stimuli-responsive hydrogel materials provides a programmable and facile strategy for the fabrication of reconfigurable intelligent devices.

## Abbreviations

| | |
|---|---|
| 4D printing | four-dimensional printing |
| Ace | acetone |
| AAc | acrylic acid |
| AFM | atomic force microscope |
| BLON | 1,4-butyrolactone |
| BG | blazed grating |
| CGH | computer-generated hologram |
| DI water | deionized water |
| DLW | direct laser writing |
| DMF | *N*,*N*-dimethylmethanamide |
| EL | ethyl lactate |
| EtOH | ethanol |
| IPA | isopropanol |
| LCST | lower critical solution temperature |
| LPM | local phase modulation |
| PDMS | polydimethylsiloxane |
| PNIPAAm | poly(*N*-isopropylacrylamide) |
| PLL | poly-L-lysine |
| SLM | spatial light modulator |
| SMMF | shape-morphing microfish |
| SMMRs | shape-morphing microrobots |
| SPP | spiral phase plate |

## References

1 Zhang, Y.S. and Khademhosseini, A. (2017). Advances in engineering hydrogels. *Science* 356: eaaf3627.

2 Liu, X., Liu, J., Lin, S., and Zhao, X. (2020). Hydrogel machines. *Materials Today* 36: 102–124.

3 Yuk, H., Wu, J., and Zhao, X. (2022). Hydrogel interfaces for merging humans and machines. *Nature Reviews Materials* 7: 935–952.

4 Ding, M., Jing, L., Yang, H. et al. (2020). Multifunctional soft machines based on stimuli-responsive hydrogels: from freestanding hydrogels to smart integrated systems. *Materials Today Advances* 8: 100088.

5 Nojoomi, A., Arslan, H., Lee, K., and Yum, K. (2018). Bioinspired 3D structures with programmable morphologies and motions. *Nature Communications* 9: 3705.

6 Hu, Y., Wang, Z., Jin, D. et al. (2020). Botanical-inspired 4D printing of hydrogel at the microscale. *Advanced Functional Materials* 30: 1907377.

7 Shin, Y., Choi, M.-Y., Choi, J. et al. (2021). Design of an electro-stimulated hydrogel actuator system with fast flexible folding deformation under a low electric field. *ACS Applied Materials & Interfaces* 13: 15633–15646.

8 Li, C., Iscen, A., Sai, H. et al. (2020). Supramolecular–covalent hybrid polymers for light-activated mechanical actuation. *Nature Materials* 19: 900–909.

9 Nava-Medina, I.B., Gold, K.A., Cooper, S.M. et al. (2021). Self-oscillating 3D printed hydrogel shapes. *Advanced Materials Technologies* 6: 2100418.

10 Lei, Z., Wang, Q., and Wu, P. (2017). A multifunctional skin-like sensor based on a 3D printed thermo-responsive hydrogel. *Materials Horizons* 4: 694–700.

11 Zhao, Y., Lo, C.Y., Ruan, L. et al. (2021). Somatosensory actuator based on stretchable conductive photothermally responsive hydrogel. *Science Robotics* 6: eabd5483.

12 Zhao, Y., Xuan, C., Qian, X. et al. (2019). Soft phototactic swimmer based on self-sustained hydrogel oscillator. *Science Robotics* 4: eaax7112.

13 Kim, Y.S., Liu, M., Ishida, Y. et al. (2015). Thermoresponsive actuation enabled by permittivity switching in an electrostatically anisotropic hydrogel. *Nature Materials* 14: 1002–1007.

14 Maiti, B., Abramov, A., Franco, L. et al. (2020). Thermoresponsive shape-memory hydrogel actuators made by phototriggered click chemistry. *Advanced Functional Materials* 30: 2001683.

15 Jeong, B., Kim, S.W., and Bae, Y.H. (2012). Thermosensitive sol–gel reversible hydrogels. *Advanced Drug Delivery Reviews* 64: 154–162.

16 Mahinroosta, M., Farsangi, Z.J., Allahverdi, A., and Shakoori, Z. (2018). Hydrogels as intelligent materials: a brief review of synthesis, properties and applications. *Materials Today Chemistry* 8: 42–55.

17 Liu, K., Zhang, Y., Cao, H. et al. (2020). Programmable reversible shape transformation of hydrogels based on transient structural anisotropy. *Advanced Materials* 32: 2001693.

18 Kim, J., Hanna, J.A., Byun, M. et al. (2012). Designing responsive buckled surfaces by halftone gel lithography. *Science* 335: 1201–1205.

19 Tang, L., Wang, L., Yang, X. et al. (2021). Poly (*N*-isopropylacrylamide)-based smart hydrogels: design, properties and applications. *Progress in Materials Science* 115: 100702.

20 Jochum, F.D. and Theato, P. (2013). Temperature-and light-responsive smart polymer materials. *Chemical Society Reviews* 42: 7468–7483.

21 Zhang, X., Chen, L., Lim, K.H. et al. (2019). The pathway to intelligence: using stimuli-responsive materials as building blocks for constructing smart and functional systems. *Advanced Materials* 31: 1804540.

22 Halperin, A., Kröger, M., and Winnik, F.M. (2015). Poly (*N*-isopropylacrylamide) phase diagrams: fifty years of research. *Angewandte Chemie International Edition* 54: 15342–15367.

23 Bokias, G., Staikos, G., and Iliopoulos, I. (2000). Solution properties and phase behaviour of copolymers of acrylic acid with *N*-isopropylacrylamide: the importance of the intrachain hydrogen bonding. *Polymer* 41: 7399–7405.

24 Ionov, L. (2014). Hydrogel-based actuators: possibilities and limitations. *Materials Today* 17: 494–503.

25 Hua, M., Kim, C., Du, Y. et al. (2021). Swaying gel: chemo-mechanical self-oscillation based on dynamic buckling. *Matter* 4: 1029–1041.

26 Shi, Q., Liu, H., Tang, D. et al. (2019). Bioactuators based on stimulus-responsive hydrogels and their emerging biomedical applications. *NPG Asia Materials* 11: 64.

27 Duan, J., Liang, X., Zhu, K. et al. (2017). Bilayer hydrogel actuators with tight interfacial adhesion fully constructed from natural polysaccharides. *Soft Matter* 13: 345–354.

28 Liu, X., Gao, M., Chen, J. et al. (2022). Recent advances in stimuli-responsive shape-morphing hydrogels. *Advanced Functional Materials* 32: 2203323.

29 Li, H., Go, G., Ko, S.Y. et al. (2016). Magnetic actuated pH-responsive hydrogel-based soft micro-robot for targeted drug delivery. *Smart Materials and Structures* 25: 027001.

30 Han, Z., Wang, P., Mao, G. et al. (2020). Dual pH-responsive hydrogel actuator for lipophilic drug delivery. *ACS Applied Materials & Interfaces* 12: 12010–12017.

31 Fan, X., Chung, J.Y., Lim, Y.X. et al. (2016). Review of adaptive programmable materials and their bioapplications. *ACS Applied Materials & Interfaces* 8: 33351–33370.

32 Shim, W.S., Kim, J.H., Park, H. et al. (2006). Biodegradability and biocompatibility of a pH-and thermo-sensitive hydrogel formed from a sulfonamide-modified poly ($\varepsilon$-caprolactone-*co*-lactide)–poly (ethylene glycol)–poly ($\varepsilon$-caprolactone-*co*-lactide) block copolymer. *Biomaterials* 27: 5178–5185.

33 Li, R., Jin, D., Pan, D. et al. (2020). Stimuli-responsive actuator fabricated by dynamic asymmetric femtosecond bessel beam for in situ particle and cell manipulation. *ACS Nano* 14: 5233–5242.

34 Zhang, J., Chu, L.-Y., Li, Y.-K., and Lee, Y.M. (2007). Dual thermo- and pH-sensitive poly(*N*-isopropylacrylamide-*co*-acrylic acid) hydrogels with rapid response behaviors. *Polymer* 48: 1718–1728.

35 Kamei, N., Morishita, M., Chiba, H. et al. (2009). Complexation hydrogels for intestinal delivery of interferon $\beta$ and calcitonin. *Journal of Controlled Release* 134: 98–102.

36 Erol, O., Pantula, A., Liu, W., and Gracias, D.H. (2019). Transformer hydrogels: a review. *Advanced Materials Technologies* 4: 1900043.

37 Park, N. and Kim, J. (2020). Hydrogel-based artificial muscles: overview and recent progress. *Advanced Intelligent Systems* 2: 1900135.

38 Li, C.Y., Zheng, S.Y., Hao, X.P. et al. (2022). Spontaneous and rapid electro-actuated snapping of constrained polyelectrolyte hydrogels. *Science Advances* 8: eabm9608.

**39** Lin, S.-B., Yuan, C.-H., Ke, A.-R., and Quan, Z.-L. (2008). Electrical response characterization of PVA–P (AA/AMPS) IPN hydrogels in aqueous $Na_2SO_4$ solution. *Sensors and Actuators B: Chemical* 134: 281–286.

**40** Morales, D., Palleau, E., Dickey, M.D., and Velev, O.D. (2014). Electro-actuated hydrogel walkers with dual responsive legs. *Soft Matter* 10: 1337–1348.

**41** Migliorini, L., Santaniello, T., Yan, Y. et al. (2016). Low-voltage electrically driven homeostatic hydrogel-based actuators for underwater soft robotics. *Sensors and Actuators B: Chemical* 228: 758–766.

**42** Jang, S. and Park, S. (2023). 4D printed untethered milli-gripper fabricated using a biodegradable and biocompatible electro-and magneto-active hydrogel. *Sensors and Actuators B: Chemical* 384: 133654.

**43** Kim, S.J., Kim, M.S., Kim, S.I. et al. (2006). Self-oscillatory actuation at constant DC voltage with pH-sensitive chitosan/polyaniline hydrogel blend. *Chemistry of Materials* 18: 5805–5809.

**44** Khan, Z.M., Wilts, E., Vlaisavljevich, E. et al. (2022). Electroresponsive hydrogels for therapeutic applications in the brain. *Macromolecular Bioscience* 22: 2100355.

**45** Ali, I., Xudong, L., Xiaoqing, C. et al. (2019). A review of electro-stimulated gels and their applications: present state and future perspectives. *Materials Science and Engineering: C* 103: 109852.

**46** Yang, C., Liu, Z., Chen, C. et al. (2017). Reduced graphene oxide-containing smart hydrogels with excellent electro-response and mechanical properties for soft actuators. *ACS Applied Materials & Interfaces* 9: 15758–15767.

**47** Li, T., Li, G., Liang, Y. et al. (2017). Fast-moving soft electronic fish. *Science Advances* 3: e1602045.

**48** Kwon, G.H., Park, J.Y., Kim, J.Y. et al. (2008). Biomimetic soft multifunctional miniature aquabots. *Small* 4: 2148–2153.

**49** Kwon, G.H., Choi, Y.Y., Park, J.Y. et al. (2010). Electrically-driven hydrogel actuators in microfluidic channels: fabrication, characterization, and biological application. *Lab on a Chip* 10: 1604–1610.

**50** Li, C., Xue, Y., Han, M. et al. (2021). Synergistic photoactuation of bilayered spiropyran hydrogels for predictable origami-like shape change. *Matter* 4: 1377–1390.

**51** Han, D., Li, Y., Liu, X. et al. (2020). Rapid bacteria trapping and killing of metal-organic frameworks strengthened photo-responsive hydrogel for rapid tissue repair of bacterial infected wounds. *Chemical Engineering Journal* 396: 125194.

**52** Wang, Z.J., Li, C.Y., Zhao, X.Y. et al. (2019). Thermo-and photo-responsive composite hydrogels with programmed deformations. *Journal of Materials Chemistry B* 7: 1674–1678.

**53** Wei, X., Chen, L., Wang, Y. et al. (2022). An electrospinning anisotropic hydrogel with remotely-controlled photo-responsive deformation and long-range navigation for synergist actuation. *Chemical Engineering Journal* 433: 134258.

**54** Lo, C.-Y., Zhao, Y., Kim, C. et al. (2021). Highly stretchable self-sensing actuator based on conductive photothermally-responsive hydrogel. *Materials Today* 50: 35–43.

**55** Li, L., Scheiger, J.M., and Levkin, P.A. (2019). Design and applications of photoresponsive hydrogels. *Advanced Materials* 31: 1807333.

**56** Wu, Y., Wu, S., Tian, X. et al. (2011). Photoinduced reversible gel–sol transitions of dicholesterol-linked azobenzene derivatives through breaking and reforming of van der Waals interactions. *Soft Matter* 7: 716–721.

**57** Kloxin, A.M., Kasko, A.M., Salinas, C.N., and Anseth, K.S. (2009). Photodegradable hydrogels for dynamic tuning of physical and chemical properties. *Science* 324: 59–63.

**58** Li, Q., Fuks, G., Moulin, E. et al. (2015). Macroscopic contraction of a gel induced by the integrated motion of light-driven molecular motors. *Nature Nanotechnology* 10: 161–165.

**59** Li, C., Lau, G.C., Yuan, H. et al. (2020). Fast and programmable locomotion of hydrogel-metal hybrids under light and magnetic fields. *Science robotics* 5: eabb9822.

**60** Griffin, D.R. and Kasko, A.M. (2012). Photodegradable macromers and hydrogels for live cell encapsulation and release. *Journal of the American Chemical Society* 134: 13103–13107.

**61** Klajn, R. (2014). Spiropyran-based dynamic materials. *Chemical Society Reviews* 43: 148–184.

**62** Gong, C., Wong, K.-L., and Lam, M.H. (2008). Photoresponsive molecularly imprinted hydrogels for the photoregulated release and uptake of pharmaceuticals in the aqueous media. *Chemistry of Materials* 20: 1353–1358.

**63** Park, C., Lee, K., and Kim, C. (2009). Photoresponsive cyclodextrin-covered nanocontainers and their sol-gel transition induced by molecular recognition. *Angewandte Chemie International Edition* 48: 1275–1278.

**64** Podual, K., Doyle Iii, F., and Peppas, N. (2000). Preparation and dynamic response of cationic copolymer hydrogels containing glucose oxidase. *Polymer* 41: 3975–3983.

**65** Kim, J.J. and Park, K. (2001). Modulated insulin delivery from glucose-sensitive hydrogel dosage forms. *Journal of Controlled Release* 77: 39–47.

**66** Lee, Y., Song, W., and Sun, J.-Y. (2020). Hydrogel soft robotics. *Materials Today Physics* 15: 100258.

**67** Chen, Q., Yan, X., Lu, H. et al. (2019). Programmable polymer actuators perform continuous helical motions driven by moisture. *ACS Applied Materials & Interfaces* 11: 20473–20481.

**68** Xue, B., Qin, M., Wang, T. et al. (2016). Electrically controllable actuators based on supramolecular peptide hydrogels. *Advanced Functional Materials* 26: 9053–9062.

**69** Qin, H., Zhang, T., Li, N. et al. (2019). Anisotropic and self-healing hydrogels with multi-responsive actuating capability. *Nature Communications* 10: 2202.

**70** Wei, S., Lu, W., Le, X. et al. (2019). Bioinspired synergistic fluorescence-color-switchable polymeric hydrogel actuators. *Angewandte Chemie* 131: 16389–16397.

**71** Song, Y., Wei, W., and Qu, X. (2011). Colorimetric biosensing using smart materials. *Advanced Materials* 23: 4215–4236.

**72** Ma, C., Li, T., Zhao, Q. et al. (2014). Supramolecular Lego assembly towards three-dimensional multi-responsive hydrogels. *Advanced Materials* 26: 5665–5669.

**73** Yao, C., Liu, Z., Yang, C. et al. (2015). Poly (*N*-isopropylacrylamide)-clay nanocomposite hydrogels with responsive bending property as temperature-controlled manipulators. *Advanced Functional Materials* 25: 2980–2991.

**74** Zhang, Z., Chen, Z., Wang, Y. et al. (2019). Bioinspired bilayer structural color hydrogel actuator with multienvironment responsiveness and survivability. *Small Methods* 3: 1900519.

**75** Lu, H., Wu, B., Yang, X. et al. (2020). Actuating supramolecular shape memorized hydrogel toward programmable shape deformation. *Small* 16: 2005461.

**76** Chen, Z., Zhao, D., Liu, B. et al. (2019). 3D printing of multifunctional hydrogels. *Advanced Functional Materials* 29: 1900971.

**77** Zhu, Q.L., Du, C., Dai, Y. et al. (2020). Light-steered locomotion of muscle-like hydrogel by self-coordinated shape change and friction modulation. *Nature Communications* 11: 5166.

**78** Zhu, Q.L., Dai, C.F., Wagner, D. et al. (2020). Distributed electric field induces orientations of nanosheets to prepare hydrogels with elaborate ordered structures and programmed deformations. *Advanced Materials* 32: 2005567.

**79** Gladman, A.S., Matsumoto, E.A., Nuzzo, R.G. et al. (2016). Biomimetic 4D printing. *Nature Materials* 15: 413–418.

**80** Huang, L., Jiang, R., Wu, J. et al. (2017). Ultrafast digital printing toward 4D shape changing materials. *Advanced Materials* 29: 1605390.

**81** Huang, T.-Y., Huang, H.W., Jin, D.D. et al. (2020). Four-dimensional micro-building blocks. *Science Advances* 6: eaav8219.

**82** Jin, D., Chen, Q., Huang, T.Y. et al. (2020). Four-dimensional direct laser writing of reconfigurable compound micromachines. *Materials Today* 32: 19–25.

**83** Xin, C., Jin, D., Hu, Y. et al. (2021). Environmentally adaptive shape-morphing microrobots for localized cancer cell treatment. *ACS Nano* 15: 18048–18059.

# 5

# Liquid Crystal Network and Elastomer-Based Miniature Soft Robots

## 5.1 Introduction

Liquid crystal (LC) is a form of existence with properties between those of liquids and those of solid crystals. Polymeric materials exhibiting liquid crystallinity are referred to as liquid crystal polymer (LCPs), liquid crystal networks (LCNs), and liquid crystal elastomers (LCEs), based on the polymer backbone and the crosslink density [1–4]. Thanks to their superior programmability and multi-stimuli responsiveness, LCNs and LCEs have attracted extensive research attention and demonstrated promising potentials in a wealth of different applications, especially soft robotics and novel actuating and sensing devices [5–13]. The morphological transformation of these materials is related to the mesogenic order–disorder transition induced by factors such as temperature [14–20], light [7, 21–29], humidity [30, 31], chemical solvent [32–35], or electric field [36–42]. The prominent feature of these materials, i.e. the large and reversible macroscopic deformation, stems from the phase transition of mesogens between ordered LC and disordered isotropic state [43–46].

Among the various stimuli, a changing temperature is a widely adopted triggering cue [47–51]. When temperature exceeds the phase transition temperature of the material, the mesogen alignment will be disrupted, inducing a shrinkage along the orientation direction and simultaneously an expansion in the transverse direction. In contrast, when temperature drops below the phase transition temperature, the mesogen alignment will restore itself to an ordered state and the material returns to its original shape. By coupling the programmed mesogen alignment with the material's geometric features, a variety of complex thermally induced reversible deformation could be achieved. As shown in Figure 5.1a, Li et al. controlled the orientation of LC mesogens through a magnetic field and achieved an unprecedented asymmetric deformation mode of the unit cell structure through the regulation of temperature [17]. The developed LCE structure with reconfigurable shape-morphing results exhibited switchable frictional properties.

In comparison to the aforementioned thermal triggering, the light-driven strategy shows an advantage of a high degree of controllability in time and space [52–54]. The photo-responsiveness of LCNs could be achieved through photochemical and photothermal mechanisms, see Figure 5.1b [7]. Using azobenzene derivatives as mesogens, LCN could undergo trans–cis photoisomerization under stimulation

*Untethered Miniature Soft Robots: Materials, Fabrications, and Applications*, First Edition.
Li Zhang, Jiachen Zhang, Neng Xia, and Yue Dong.

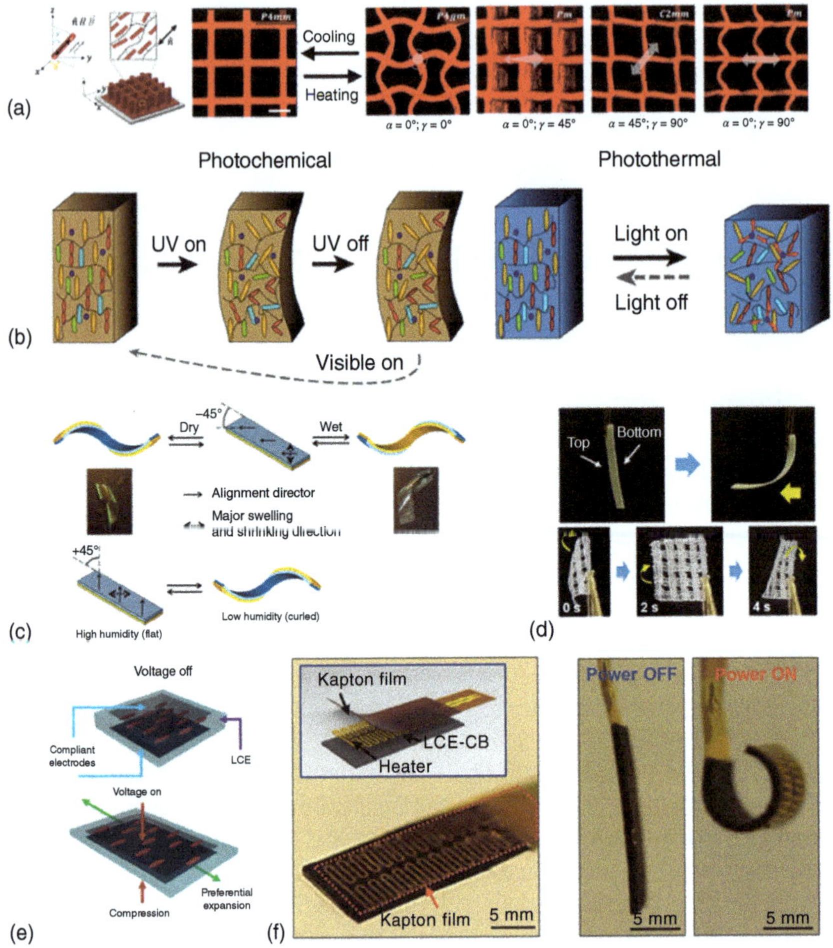


**Figure 5.1** Actuation stimuli types of LCNs and LCEs. (a) Temperature-responsive LCNs with programmable molecular alignments and symmetry breaking transformations of surface-attached LCE microlattice. Source: Li et al. [17]/Reproduced from John Wiley & Sons, Inc. (b) Schematic illustration of photo-responsive LCNs including photochemical and photothermal mechanisms. Source: Markus et al. [7]/Springer Nature/CCBY 4.0/Public Domain. (c and d) Transformations of humidity responsive LCNs with tunable molecular director. Source: (c) Haan et al. [30]/Reproduced from American Chemical Society and (d) Kim et al. [31]/Reproduced from John Wiley & Sons, Inc. (e and f) Schematic illustration and transformations of electro-responsive LCNs. Source: (e) Fowler et al. [40]/John Wiley & Sons and (f) Wang et al. [41]/Wiley-VCH Verlag GmbH & Co. KGaA, Weinheim.

(e.g. UV light), accompanied by a molecular configuration from a rod-like shape compatible with ordered LC, which leads to a change in the free volume. On the other hand, the photothermal effect was formed by doping photothermal conversion materials such as Au nanomaterials, graphene or organic dyes in the LCN to trigger the thermal responsive behavior [55].

Compared with other stimuli, humidity response is a relatively mild stimulation method. After the LCN was treated with an alkaline solution, the LCN film could absorb water and swell to form deformation behaviors such as folding or curling, as shown in Figure 5.1c,d [30, 31]. However, humidity stimulation is difficult to achieve precise control. In contrast, the electric field can provide a highly controllable driving mode, in which parameters such as amplitude, phase, and frequency can be programmed and adjusted, see Figure 5.1e,f [18, 40, 41]. Relying on the Joule heating effect, LCN can exhibit fast response and large deformation characteristics. Compared with the driving method based on the photothermal effect, the electric field can heat the LCN film more uniformly along the propagation direction. In addition, the orientation of liquid crystal molecules can also be regulated by an electric field.

## 5.2 Stimuli-Responsiveness Based on Programmed Director Field

The macroscopic shape-shifting of LCNs and LCEs are due to the order–disorder conversion of the material's microscopic mesogen alignment, which is denoted as the director field. Based on this working principle, the stimuli responsivenes of LCNs and LCEs can be designed via programming the director field together with the geometric features. Thus, the fabrication strategies to program the director field has been a paramount topic in this field, and a wealth of distinct approaches have been proposed. This section discusses some of the representative method to program a designated director field in LCNs and LCEs.

Conventionally, the director field patterns can be programmed using glass cells [56], optical patterning [57], and mechanical stretching [58–60]. These are widely adopted methods to fabricate LCNs and LCEs with programmed direction fields. They are relatively easy to perform and low-cost. As these methods have been extensively discussed in the literature and their capabilities have been fully investigated. Recent studies usually employ more advanced fabrication approaches to make complex director field patterns. Thus, these traditional method will not be further elaborated here.

### 5.2.1 Programmed Voxelated Director Fields via Two Surface-Patterned Confining Glasses

Powerful microfabrication techniques enable the control and encoding of director fields in 3D space inside LCN microstructures [61–64]. Two-photon polymerization (TPP) or direct laser writing (DLW) technique has significant advantages in preparing high-precision (nanoscale) 3D structures [65]. With this printing method, complex LCN structures with dimensions in microns or millimeters can be easily obtained, meanwhile making it possible to regulate the arrangement of molecular of LCs at high resolution [66].

Schematic illustrations of an exemplar fabrication process of 3D LCN microstructures with a programmed director field is shown in Figure 5.2 [66]. First, microchannels were fabricated on the top and bottom glass substrates of the chamber via TPP. The director field of LCNs could be regulated by these microchannels,

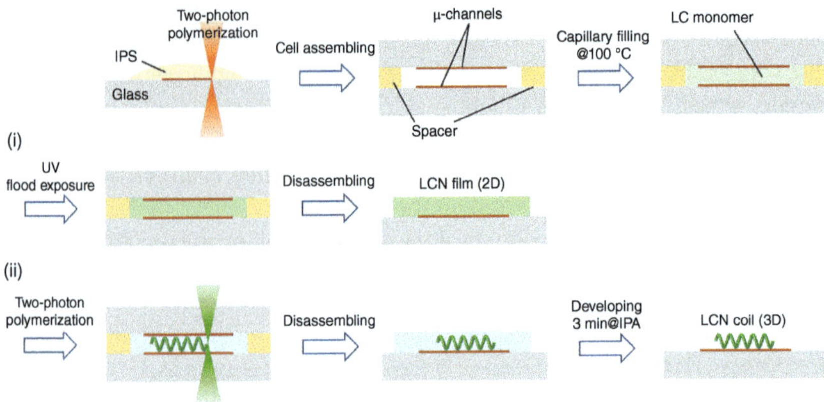


**Figure 5.2** Schematic illustration of the fabrication procedure for LCN microstructures with programmed director field. Source: Guo et al. [66]/John Wiley & Sons/CC BY 4.0.

as described in the following steps. A DLW system (Nanoscribe GmbH) using a 63× oil immersion objective (NA = 1.4) was adopted for the fabrication. After dispensing negative photoresist IP-S 780 (Nanoscribe GmbH) onto the cleaned glass, microchannels with the designed orientation were formed using a two-photon laser writing process. This approach is only effective if a programmed director field throughout the pre-polymerized voxels is successfully generated by the microchannels without any adverse effect from other counterparts. Then, the glass slides were immersed in propylene glycol monomethyl ether acetate for 30 minutes to dispose uncured photoresist.

For 2D alignment of LCNs, two glass substrates assembled to form a cell with mirror microchannel patterns on the two inner sides, followed by applying UV glue mixed with spacers of the desired size to the four corners of the assembled substrate. The two mirrored patterns on the two substrates were carefully positioned and aligned under a microscope. Then the glue was cured to ensure that the patterns are well aligned. Next, 3D LCN microstructures were fabricated in the cell. A 2 : 1 weight ratio mixture of RM006:RM257 was used as the printable material, which has been extensively adopted in TPP experiments. 1 wt% of photoinitiator Irgacure-369 (Ciba) was used as the photoinitiator of the mixture. The LCN precursor mixture was mixed in the molten state using a hot stage at 120 °C for one hour. The injection was later injected at 100 °C (above its clearing point) into the chamber with a thickness of approximately 40 μm to avoid flow alignment. Subsequently, the LC cells were cooled down to room temperature at a rate of 5 °C $h^{-1}$ for further operation. Finally, the photopolymerization of LCNs was performed by UV flood exposure (1 J $cm^{-2}$ 365 nm UV light) or by TPP.

### 5.2.2 Arbitrary Director Fields Formed via 3D Assembly

To relax the constrains posed on the complexity of director field's by conventional 2D fabrication methods, a LCE voxel-based bottom-up 3D assembly approach has been proposed to form arbitrary 3D programmed director fields, see

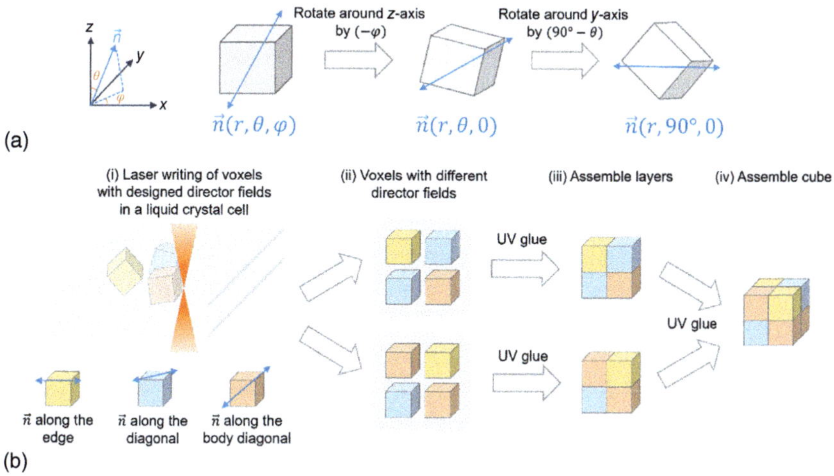


**Figure 5.3** Assembly of microscale LCE modules with heterogeneous director fields for the fabrication of 3D LCE structures. (a) Two-stage rotation enables the formation of LCE voxel with a director field along the *x*-axis. (b) Schematic illustration of 3D assembly of LCE voxels. Source: Guo et al. [67]/Springer Nature/CCBY 4.0/Public Domain.

Figure 5.3 [67]. The materials used in LC included disperse red 1 acrylate (dye), 1,4-Bis[4-(3-acryloyloxypropyloxy)benzoyloxy]-2-methylbenzene (ST3021) as the crosslinker, 4-methoxybenzoic acid 4-(6-acryloyloxyhexyloxy) phenyl ester (ST3866, LC monomer), and Irgacure 369 as the photoinitiator. A mixture of these materials was filled into an LC cell with a 150 μm cell gap, in which, a TPP process was applied to generate cubic voxels.

In Figure 5.3a, $\boldsymbol{n}(x, y, z)$ represents an arbitrarily uniform director field of a voxel, which can also be expressed as $\boldsymbol{n}(r, \theta, \varphi)$ in a spherical coordinate system. $r$ is the radial distance. $\theta$ and $\varphi$ represent the polar and azimuth angles, respectively. An LC cell was assembled in which TPP would be performed to fabricate the voxel. A director field was encoded in the LC cell along the $x$-axis, corresponding to $\boldsymbol{n}_p(r, 90°, 0)$ which was the predefined director field. Figure 5.3a shows the following two-step rotation to transform any 3D steering field $\boldsymbol{n}_a(r, \theta, \varphi)$ into $\boldsymbol{n}_p(r, 90°, 0)$. Firstly, $\boldsymbol{n}_a(r, \theta, \varphi)$ becomes $\boldsymbol{n}_t(r, \theta, 0)$ by rotating $-\varphi$ around the $z$ axis; Then, $\boldsymbol{n}_t(r, \theta, 0)$ becomes $\boldsymbol{n}_p(r, 90°, 0)$ by rotating $\theta - 90°$ around the $y$ axis. Therefore, Figure 5.3a illustrates that a new voxel can be formed by a two-stage rotation of a voxel together with the director field $\boldsymbol{n}_a(r, \theta, \varphi)$. Then, as shown in Figure 5.3b, with 3D TPP of the rotating voxels within the LC cell, a voxel with the desired director field $\boldsymbol{n}_a(r, \theta, \varphi)$ is obtained. Theoretically, the value ranges of $\varphi$ and $\theta$ are 0–360°, and 0–180°, respectively. For convenience, a Cartesian coordinate system is used in the following description, where $\boldsymbol{n}_a(r, \theta, \varphi)$ is equivalent to $\boldsymbol{n}_a(r\sin\theta\cos\varphi, r\sin\theta\sin\varphi, r\cos\theta)$. Finally, by 3D assembling the LCE voxels with arbitrary director fields, 3D structures with desired heterogeneous patterns of the director field $\boldsymbol{n}(x, y, z)$ could be formed. In comparison, the previous programmable alignment of LC was only applicable to in-plane 2D director field $\boldsymbol{n}(x, y, 0)$. The mixing of vertical alignment and in-plane surface alignment, or combining external electric fields and in-plane surface alignment can achieve director fields whose $z$-components are non-zero. While, only very limited variations to the $z$-component are allowed by these strategies.

### 5.2.3 3D Microstructures with Uniform and 2D Director Field Alignment

Guo et al. demonstrated the programmable director fields formed using fabricated microchannels [66]. As shown in Figure 5.4a, LC cells that have +1 defects on the bottom and top substrates were used to examine the alignment function of fabricated microchannels. According to the polarized optical microscope (POM) image in Figure 5.4a, there are four bright regions existing around the defect center. The successful alignment of LCN monomers based on the proposed design was confirmed by the signature of a defect with a charge of +1. Then, sinusoidal patterns on the top and bottom sides of the LC cell were applied to verify the alignment capability of pixelated arrays of microchannels. Along the vertical direction, there are alternating dark and bright patterns, confirming the capability of the developed method in the programming of LCNs with pixelated matching patterns on the top and bottom sides of the LC cell. To generate non-uniform director fields, substrates with +1 and +2 topological defects and uniform alignment, respectively, were used for the LC cell. Two and four bright regions with disclinations in the center exhibit in POM images of Figure 5.4c,d. The experimental results verify the programmability of the director fields via the fabricated microchannels.

After the TPP of microstructures, the induced molecular alignment can still be maintained. To confirm this capability, 3D microstructures (including rings, coils, and suspending membranes) with uniform molecular alignment are fabricated, as shown in Figure 5.5a–c. Compared with the design sketches, the SEM results in Figure 5.5 demonstrate the accuracy of the microfabrication. The POM results

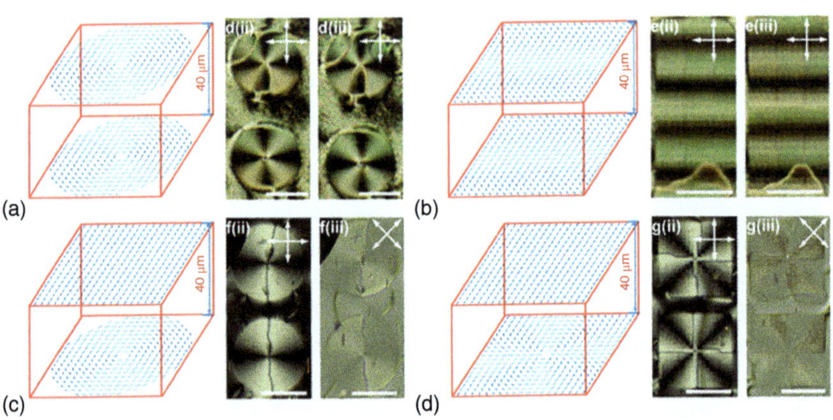


**Figure 5.4** Alignment design and polarized optical microscope images of liquid crystal (LC) monomers. (a) +1 defect pattern. (b) Sinusoidal pattern. (c) LC monomers with uniform pattern on the top substrate and +1 defect pattern on the bottom substrate. (d) LC monomers with uniform pattern on the top substrate and +2 defect pattern on the bottom substrate. Source: Guo et al. [66]/Reproduced from John Wiley & Sons, Inc./CC BY 4.0.

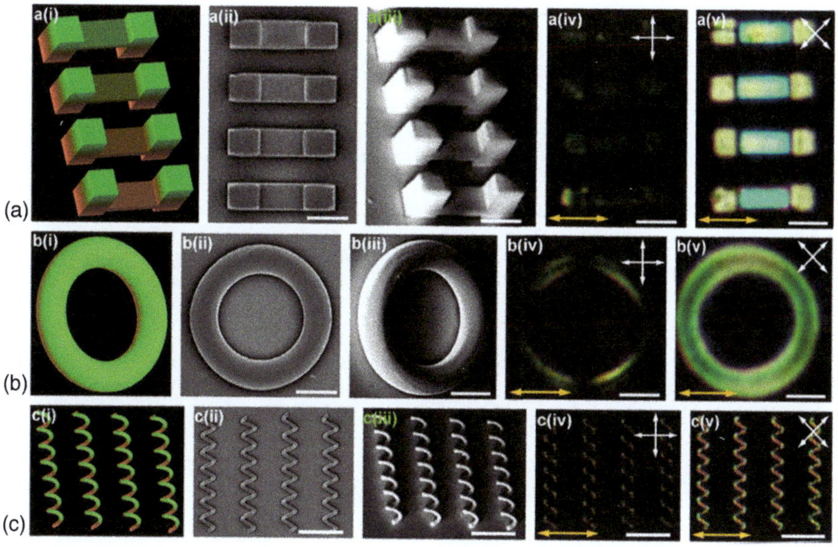


**Figure 5.5** Geometrical design, SEM and POM results of LCN microstructures with uniform director field. (a) Suspended films. (b) Rings. (c) Spiral coils. Source: Guo et al. [66]/Reproduced from John Wiley & Sons, Inc./CC BY 4.0.

of the director field along or at 45° with respect to the polarizer axis confirm the preservation of the designed uniform molecular alignment inside the fabricated LCN microstructures.

TPP and pixelated microchannel array were combined to facilitate the fabrication of 3D LCN structures with voxelated encoded director field. The in-plane 2D director fields were induced using microchannels with non-uniform patterns on both top and bottom substrates. The alternating dark and bright patterns in the POM result of Figure 5.6a illustrate that there are sinusoidal director fields formed in the DLW-printed film. The ring and coil microstructures with director fields in sinusoidal form were shown in Figure 5.6b,c. The POM images show that the dark and bright patterns (Figure 5.6b) are the same as the design of the director fields. In Figure 5.6c, the director field changes from ≈90° at the bottom to ≈45° in the middle and back to ≈90° at the top. Thus, there are two black and bright regions around the ring. Figure 5.6d,e shows the fabricated 2D structure and 3D ring geometry using LC cells with +1 topological defects on both sides. The signature director fields with designed defects were confirmed via four black and bright regions in the POM images of the LCN microstructures. Figure 5.6f presents a ring with the director field of +2 topological defect, which allows the generation of eight dark and bright regions under crossed polarized microscope.

### 5.2.4 3D LCN Microstructures with Encoded 3D Director Field and their 3D-to-3D Shape Transformation

Using cells with mismatched patterned substrates, 3D LCN devices that were encoded with non-uniform in-plane or through-thickness director fields were

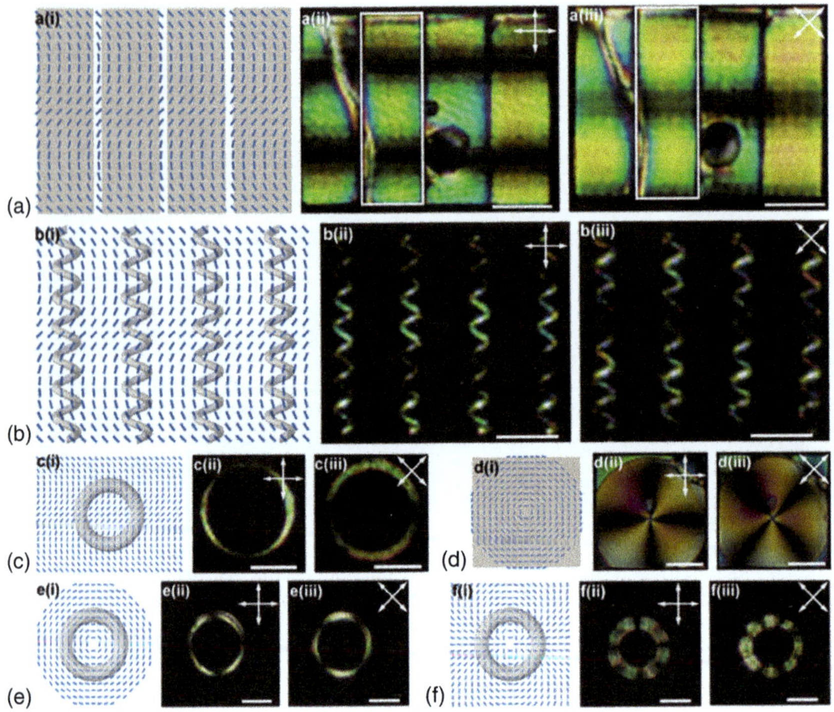

**Figure 5.6** LCN microstructures with 2D director field. (a–f) Director field patterns and POM images of films (a, d), coils (b), and rings (c, e, f). Source: Guo et al. [66]/Reproduced from John Wiley & Sons, Inc./CC BY 4.0.

fabricated. Firstly, substrates with uniform alignment on both sides but with ≈80° angle between them were used to form twist nematic distortion of the director field, as shown in Figure 5.7a. Four suspending LCN films at different heights from the bottom were fabricated to confirm the nematic order along the thickness, and they show different transmission intensities under the cross-polarized microscope. Similarly, in Figure 5.7b the top and bottom halves of a spiral coil present two groups of oblique lines with different angles under the cross-polarized microscope. Then, an in-plane non-uniform director field on one side was introduced, as shown in Figure 5.7c. The alternating dark and bright regions can be found on the POM image. Three rings with different 3D director fields using a sinusoidal alignment pattern, +1 topological defect, as well as +2 topological defect were fabricated (Figure 5.7d–f). For example, two dark and bright regions are shown in the POM image of Figure 5.7d and after the 45° rotation of the polarizer, the dark region disappears due to the complex 3D director field.

Various swelling experiments with organic solvent (*N,N*-dimethylformamide [DMF]) were performed to induce the anisotropic shape transition of the fabricated 3D LCN microstructures. As shown in Figure 5.8a, coils with similar initial geometric features (20 μm diameter, 150 μm length, and 20 μm pitch) but different director fields were fabricated. For the coil with uniform alignment, it experiences

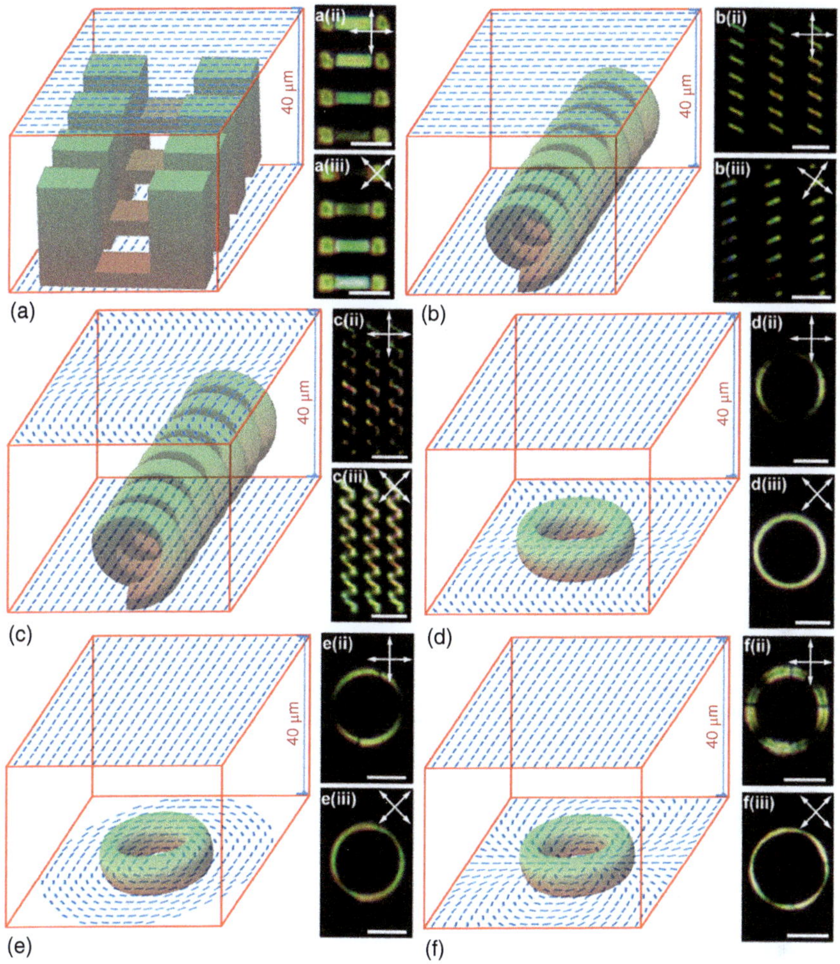


**Figure 5.7** LCN microstructures with 3D director field. (a–f) Director field patterns and POM images of suspended films (a), coils (b and c), and rings (d–f). Source: Guo et al. [66]/Reproduced from John Wiley & Sons, Inc./CC BY 4.0.

a significant swelling in the direction perpendicular to the director field. In comparison, coils with twist molecular alignment exhibit more complicated shape transformations such as bending along their helical axis. Similar experiments were performed for ring structures with different director fields (30 μm inner radius and 50 μm outer radius), as shown in Figure 5.8b. Swelling and out-of-plane bending would occur for both rings with global twist alignment.

### 5.2.5 Thermal Shape Transformation of LCN Structures Based on 1D and 2D Voxels Assembly

One of the characteristic advantages of LCEs is complex programmable reversible shape transformation based on the director fields and geometry of the LCE

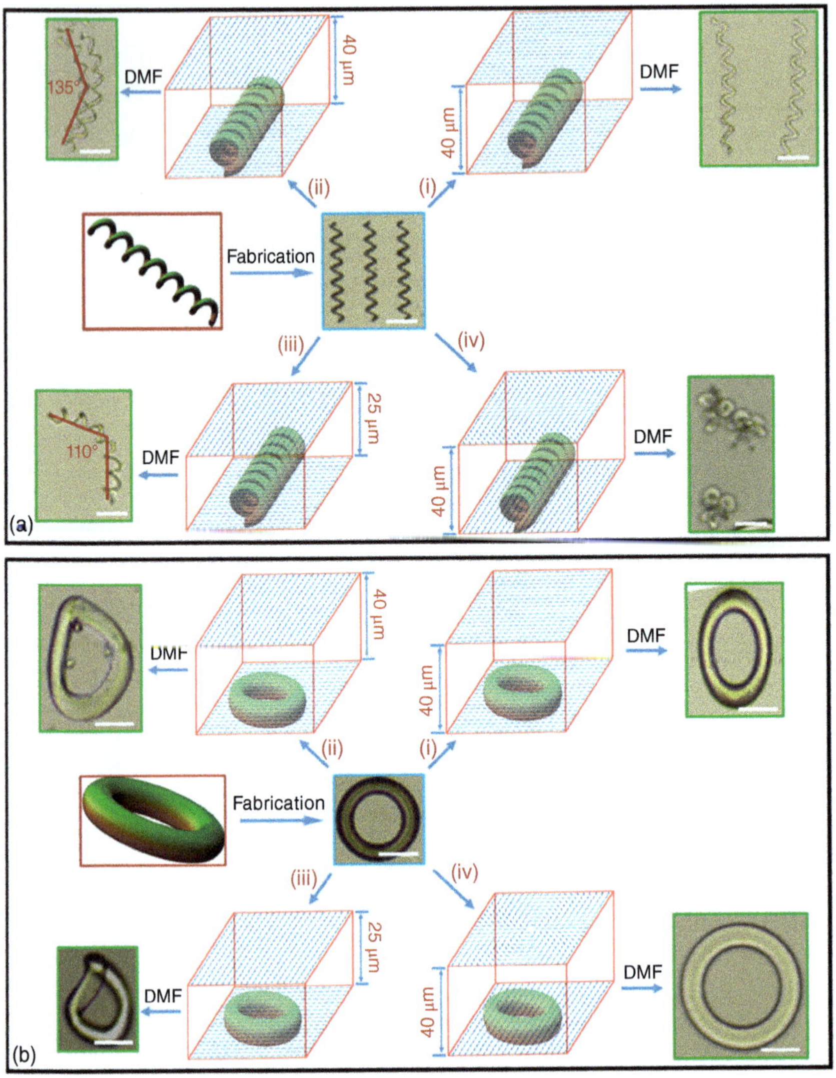


**Figure 5.8** 3D-to-3D shape transformation of LCN microstructures with encoded 3D director fields. (a) The shape-morphing result of coil geometry. (b) The shape-morphing result of ring geometry. Source: Guo et al. [66]/Reproduced from John Wiley & Sons, Inc./CC BY 4.0.

structure. The reported assembly strategy enables the fabrication of LCE structures with arbitrary director fields and shapes. Firstly, to demonstrate the assembly approach, LCE voxels assembled into 1D lines with three different encoded director fields were applied, as shown in Figure 5.9. After activating by heating source, these lines (width of 100 μm, height of 100 μm, and length of 2 mm) deformed into various geometries. A 1D LCE with four sections was shown in Figure 5.9a. When activated, sections 2 and 4 (from left) swell longitudinally while sections 1 and 3 swell transversely. For the 1D line in Figure 5.9b, sections 2 and 4 expand

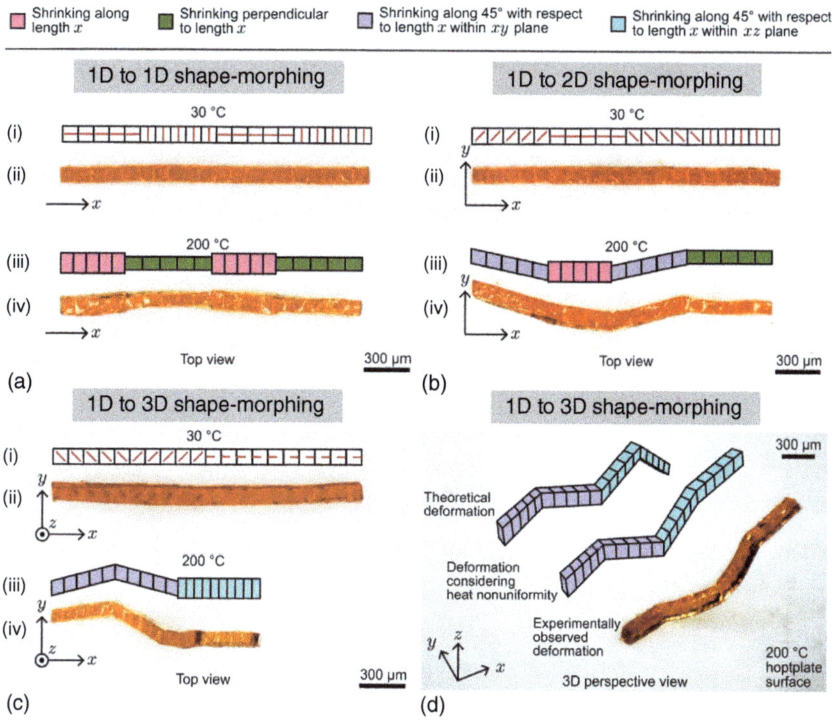


**Figure 5.9** LCE microstructures based on 1D cubic voxels assembly and their thermal responsive shape transformation. (a) 1D LCE microstructure transforming into 1D line with varying width. (b) 1D LCE microstructure transforming into 2D line shape. (c and d) 1D LCE microstructure transforming into 3D line shape. Source: Guo et al. [67]/Reproduced from Springer Nature/CC BY 4.0.

transversely and longitudinally, respectively. The deformation of section 1 and section 3 was oblique under the actuation. In Figure 5.9c, sections 3 and 4 of the LCE line deform in the $x$–$z$ plane, and sections 1 and 2 exhibit deformation in the $x$–$y$ plane. The results in Figure 5.9c were shown in Figure 5.9d from another perspective, taking into account the non-uniformity of heating.

Then, the assembly of 2D LCE structures was shown to be encoded with 3D-designed director fields. Previously, through strategies such as tips of atomic force microscopy, laser writing in pixel-by-pixel manner, plasmonic photopatterning, and magnetic regulations, various in-plane 2D director fields $\boldsymbol{n}(x, y, 0)$ can be formed into LCE films. In contrast, 2D structures with arbitrary 3D director field profiles such as nonzero $z$-component director field can be achieved by the developed assembly method. As shown in Figure 5.10, there are 2D LCE structures with four voxels and encoded director fields. Angles of voxels with $\boldsymbol{n}(1, 1, 0)$ director fields would change when driven by heating source: in the direction along and perpendicular to the director field, angles become larger and smaller, respectively. There are much smaller variations of angles for the other two voxels ($\boldsymbol{n}(1, 0, 0)$ and $\boldsymbol{n}(1, 1, 1)$). Therefore, the square 2D LCE structure in Figure 5.10a deformed into a kite shape after heating. Due to the expansion and contraction on both sides of

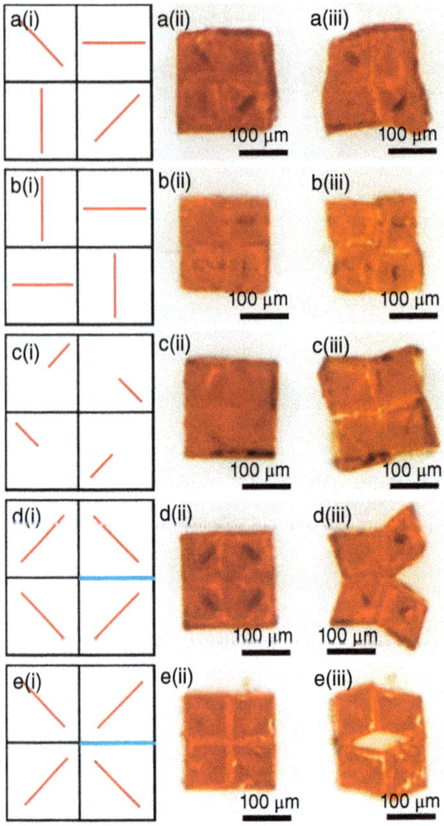


**Figure 5.10** LCE microstructures based on the 2D assembly of cubic voxels and their thermal responsive shape transformation. (a–c) 2D LCE microstructures with all surfaces bonded together. (d and e) 2D LCE microstructures with a non-glued surface. Source: Guo et al. [67]/Reproduced from Springer Nature/CC BY 4.0.

each line, spiral lines formed for structures in Figure 5.10b,c. For the structures in Figure 5.10d,e, an unglued edge was left in each case (marked with blue lines), resulting in the transformation into different 2D structures via thermal actuation. Through these examples, several advantages of the developed assembly method can be demonstrated: (i) the deformation mode could be enriched via the selection of un-bonded sections; (ii) nonzero $z$-component director fields can be programmed into 2D LCE microstructures; (iii) LCE 2D structures with large thickness, size ratios, and programmable director fields can be fabricated to explore complex morphing behaviors.

### 5.2.6 Reversible 3D-to-3D Shape Transformation of the Assembled 3D LCE Structures

3D LCE structures with encoded 3D director field profiles were fabricated. Because of the versatility of the assembly method, voxels that have arbitrary director fields can be assembled into 3D structures whose director field profiles are independent from their 3D shapes. As shown in Figure 5.11a, 80 voxels with a non-zero $z$-component director field are assembled into a cube frame with a dimension of $800 \times 800 \times 800\,\mu m^3$. All edges deformed toward the center of the cube body under

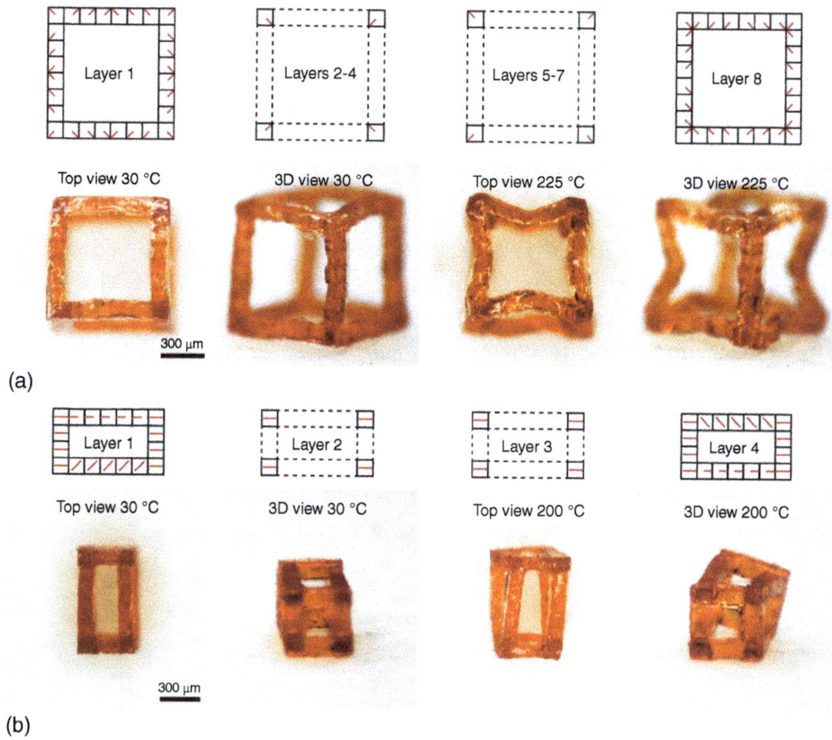


**Figure 5.11** LCE microstructures based on the 3D assembly of cubic voxels and their thermal responsive shape transformation. (a) Cuboid LCE structures. (b) Rectangular LCE structures. Source: Guo et al. [67]/Reproduced from Springer Nature/CC BY 4.0.

high temperature. The shape-morphing results can be potentially used for caging micro-objects. Figure 5.11b shows a rectangular frame with the dimension of $400 \times 700 \times 400\,\mu m^3$ assembled via 36 voxels. As shown in Figure 5.11, layers 1 and 4 in the rectangular frame are formed by voxels with non-zero $z$-components. Upon high temperature, the top layer of the assembled device would experience rotation compared to the bottom layer. In order to achieve such morphing mode, there are rotationally symmetric director field directions encoded in the two long edges in layers 1 and 4. The top and bottom surfaces of the LCE structure remain parallel after deformation, while angular displacements are generated between these two surfaces and four long sides of the geometry deform obliquely, which is consistent with the design.

## 5.3 Magnetic Liquid Crystal Elastomer Composites for Miniature Machines

Miniature machines cannot employ conventional actuation and control strategies due to their size limit. Instead, they utilize the intrinsic stimuli responsiveness of various smart materials to realize different functionalities. The integration of different advantages of multiple active ingredients into a robotic system is promising for the development of more versatile and intelligent miniature machines. LCEs could exhibit complex reversible shape deformation with high work density and large strain upon different environmental stimuli, e.g. light and temperature. Due to the fast and robust properties and deep penetration depth, magnetic actuation has been widely adopted for untethered miniature soft robots [68–72]. In addition, programmed magnetic inputs including magnetic field strength, direction, and spatiotemporal variation can be controlled by magnetic actuation systems including permanent magnet systems and electromagnetic coil systems [73]. The wide tunability of magnetic field inputs endows magnetic robots with unparallel versatility to exhibit more complex shape deformation behaviors. Integrating the respective stimulus responsiveness of LCE and magnetic agents can enhance the freedom of the design space and locomotion control, and is also expected to realize more complex behaviors such as local or sequential control.

### 5.3.1 Material Designs of Magnetic LCE Systems

Two approaches have been proposed to make magnetic LCE systems integrate the respective advantages of LCEs and magnetic soft materials. The first approach directly embeds magnetic micro- or nanoparticles (MMPs or MNPs) into the LCE matrix, while the second one stacks LCEs and magnetic soft materials together to form devices with distinct sections. Figure 5.12a schematically illustrates the concept of the first approach: a soft monolithic composite film consisting of LCEs as soft matrix and magnetic microparticles as remote-control agents [74]. The exemplar material samples shown in Figure 5.12b have a thickness of 40 μm and a circular director field. Because of the limited bonding strength of the rubbed and polyimide-coated surface, the largest thickness of the composite film was less than 100 μm. The volume ratio of MMP varied as 0%, 3.8%, 7.3%, 14%, and 24%, and the corresponding mass ratio of MMP to LCE is 0 : 1, 1 : 4, 1 : 2, 1 : 1, and 2 : 1.

Examining the composite films at different temperatures using optical microscopy and crossed polarizers, the observed images demonstrated that the director field of the LCE matrix was maintained until the light path was completely blocked by MMPs. When these samples were immersed in a high temperature (70 °C) water bath, out-of-plane morphological transformation would occur, even for the sample with the highest concentrations of MMPs. However, with an increasing MMP concentration, the deformation results of the films, i.e. mass ratios of 1 : 1 and 2 : 1, were distinctly different from the response of pure LCE (out-of-plane cone structure). Through the polarized optical microscopy images, it was found that the LCE director field of the LCE films with a moderate concentration of embedded MMPs (7.3% v/v) was maintained in regions without MMPs. In regions with MMPs, elastic repulsion between the surfaces and the MMPs was created by the programmed surface alignment, which would lead to the high concentration of particles near the middle of the film thickness. Even for the LCE composite with high concentration (24% v/v), the global director field remained intact to some extent, causing the out-of-plane

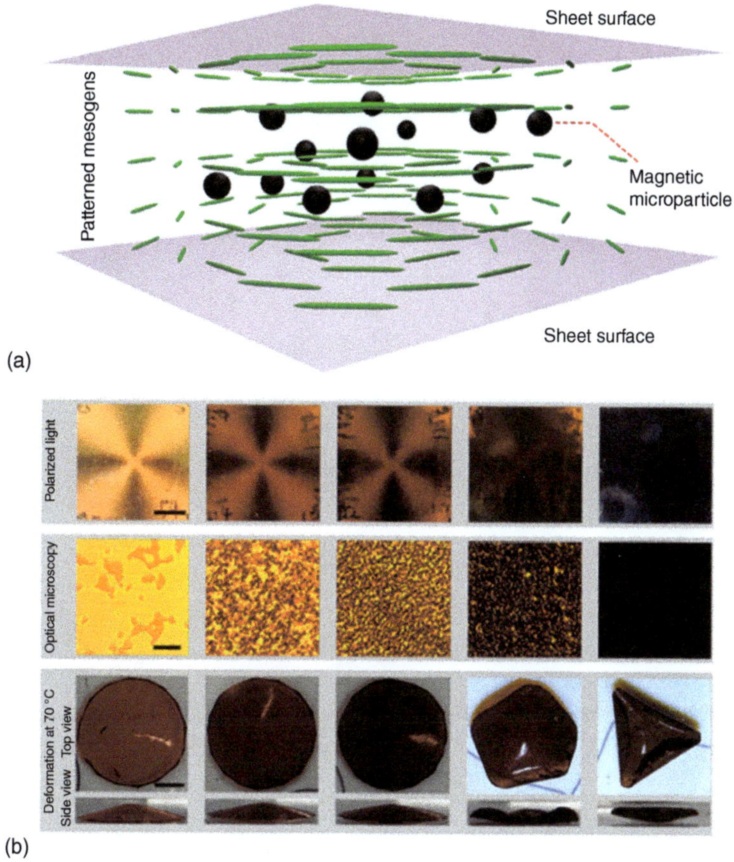


**Figure 5.12** Illustration and experimental observation of the composite film consisting of LCE base matrix and hard NdFeB magnetic microparticles. (a) Schematic illustration of the magnetic LCE film. (b) Experimental results of the composite film with a variety of magnetic microparticle concentrations (0 : 1, 1 : 4, 1 : 2, 1 : 1, and 2 : 1). Source: Zhang et al. [74]/Reproduced from John Wiley & Sons/CC BY-NC 4.0.

deformation activated by high temperature. In summary, the programmed alignment of LC and director field was not completely destroyed by the introduction of MMPs with moderate concentrations into the LCE matrix. Therefore, the magnetic LCE composites could exhibit comparable stimuli responsiveness to pure LCE.

Combining magnetically responsive and covalently adaptable networks in LCEs can endow actuators with multifunctional and reprogrammable properties [75]. As shown in Figure 5.13, magnetothermal responsive polyurethane (PU) LCE (M-PULCE) was developed with dynamic carbamate bonds. To endow the magnetothermal effect, $Fe_3O_4$ MNPs, that would present high magnetothermal conversion rates upon the stimuli of alternating magnetic field (AMF), were dispersed into PULCEs and covalent adaptable networks (CAN)-based LCEs with exchangeable dynamic carbamate bonds. LC oligomers with decreased crosslink density and prolonged chain length of the LC moiety were prepared to obtain M-PULCEs with

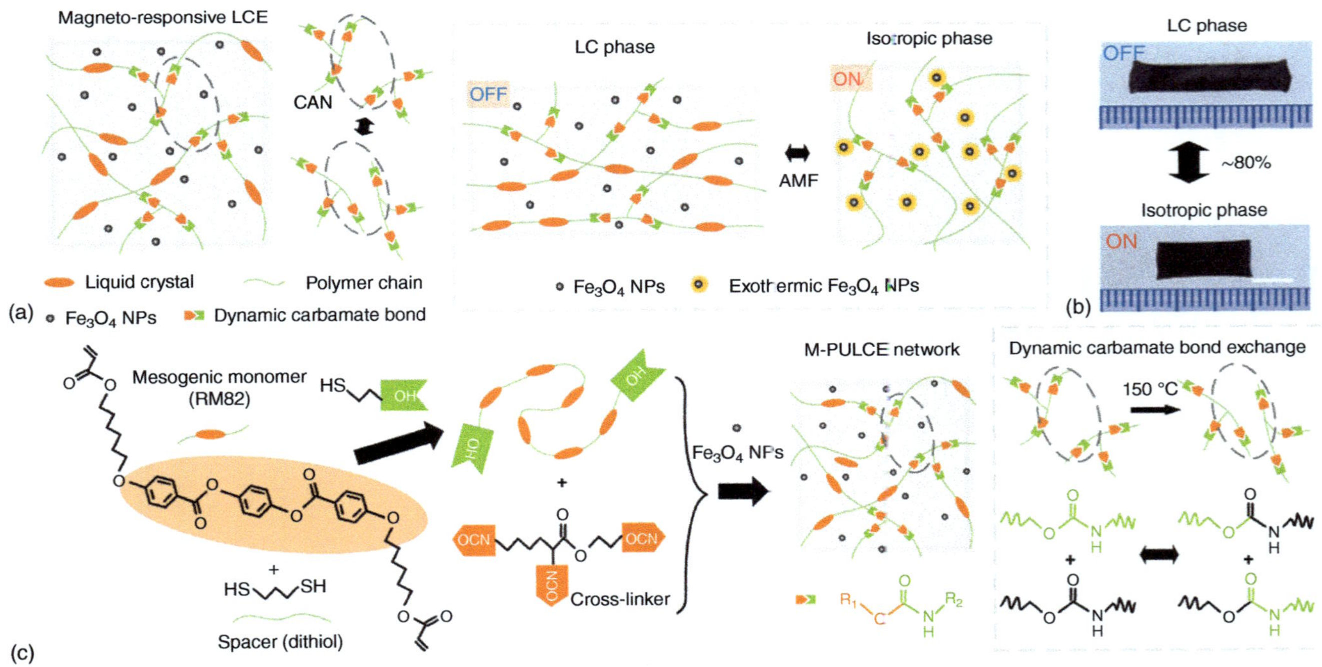


**Figure 5.13** Reprogrammable LCE systems doped with $Fe_3O_4$ nanoparticles. (a and b) Schematic illustration and experimental observation of the magnetic LCE system with magnetothermal responsiveness. (c) Synthesis of magnetothermal-responsive polyurethane LCE with dynamic carbamate bond exchange. Source: Wu et al. [75]/Reproduced from American Association for the Advancement of Science. (AAAS).

desired large morphological transformation. Commercially available LC monomer (RM82 (1,4-bis-[4-(6-acryloyloxyhexyloxy)benzoyloxy]-2-methylbenzene)), mercaptoethanol, and dithiol were used to prepare the hydroxyl-terminated LC oligomer whose average molecular weight was approximate 4700 g $mol^{-1}$ via thiol-Michael addition reaction. Dibutyltin dilaurate (DBTDL) and three-arm isocyanate were adopted as catalyst and crosslinker for polymerization and crosslinking, respectively. A classic reaction between isocyanates and alcohol groups was applied for the synthetization of PULCEs with exchangeable carbamate bonds. Superparamagnetic $Fe_3O_4$ nanoparticles with 20 nm diameter and the content of: 5, 10 and 15 weight percent (wt%) were added to PULCEs before the crosslinking process. The fabricated magnetic LCEs are abbreviated as M-PULCE-1 (5 wt% $Fe_3O_4$ NPs), M-PULCE-2 (10 wt% $Fe_3O_4$ NPs), and M-PULCE-3 (15 wt% $Fe_3O_4$ NPs), respectively. Due to the presence of CAN, the alignment and realignment of M-PULCE could be easily achieved via thermally induced network rearrangement. Transcarbamoylation would be activated under heating source and the catalysis of DBTDL. Considering the middling relaxation rate, the critical temperature for programming and reprogramming was chosen at 150 °C. The fabrication of blank PULCE and monodomain M-PULCE with reversible shrinkage was described as the following procedure. Uniaxial stretch to 180% of the original length was performed for the fabricated polydomain LCE sample. Network topological change at 150 °C enables the fix of the alignment, resulting in the formation of an actuator with reversible shrinking motion. Upon heating or cooling stimuli, the monodomain LCE sample experiences the LC-isotropic transformation or returns to the LC phase, respectively, resulting in reversible shrinkage and elongation processes. Due to the addition of $Fe_3O_4$ MNPs, the programmed deformation can be activated by AMF. Therefore, the combined effect of the LCE material as well as the magnetothermal capability of $Fe_3O_4$ particles leads to the magnetic responsive property of M-PULCEs.

Finally, multifunctional composites were also realized through a facile method of stacking LCEs and MREs, as shown in Figure 5.14 [76]. A splay director field was encoded in the LCEs, leading to the out-of-plane deformation upon environmental cues, e.g. the exposure of UV light or the change of temperature. A soft elastomer (SE) matrix (Ecoflex 0030) mixed with hard magnetic particles (NdFeB particle whose average diameter was 5 μm) with a given mass ratio (e.g. 2 : 1) was used for the MREs. The fabrication of bimorph LCE-MRE material was performed by arranging LCEs on top of the casted MREs or in the molds before the fabrication of MREs. The integration of MREs and LCEs was achieved via the polymerization of MREs. According to structure design in software, laser cutting (ProtoLaser, LPKF Laser & Electronics AG) was used to fabricate the bimorph devices. To program the magnetization profile of the bimorph devices, they would be deformed and magnetized under a strong magnetic field (1.8 T) generated by a vibrating sample magnetometer (VSM). The stimulus responsiveness of both MRE and LCE is integrated into the bimorph composites, and the activation of these responsive behaviors can be performed individually or jointly. The hard magnetic particles show high coercivity values and remanence. The strong magnetic responsiveness enables the bimorph device to respond to external weak magnetic fields, and the

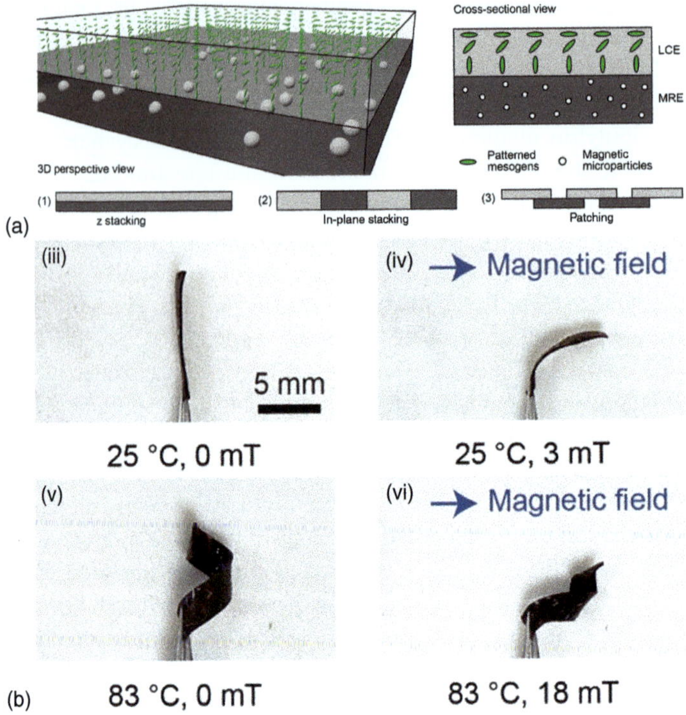


**Figure 5.14** Formation of bimorph material via the integration of LCEs and magnetic responsive elastomers (MREs). (a) Schematic illustration of the bimorph materials. (b) Experimental demonstration of the magnetic and thermal responsiveness of the bimorph material. Source: Zhang et al. [76]/Reproduced from John Wiley & Sons, Inc./CC BY 4.0.

magnitude and direction of permanent magnetization profiles can be programmed. As shown in Figure 5.14, the bimorph sample deformed toward the external magnetic field to ensure the alignment of the local magnetic moment with the external magnetic field direction. As the increase of the environment temperature (83 °C), a spiral structure was obtained due to the local shrinkage and swelling along and perpendicular to the director field, respectively. When both stimuli were applied simultaneously, the sample bent toward the magnetic field and also exhibit a helix shape. The experimental results functionalities including the *in situ* reconfigurable shape morphing and additional degrees of freedom in the activation of response behavior could be achieved via the integration of MRE and LCE, which could facilitate the task execution in realistic and complex environments.

### 5.3.2 Multiple-Stimuli Responsiveness of Magnetic LCE Systems

A variety of reconfigurable shape-morphing behaviors could be achieved with the integration of magnetic soft materials and LCEs with the patching configurations, as shown in Figure 5.15 [76]. The first column in Figure 5.15 illustrates the schematic of devices with designed parameters. To ensure that the temperature could attain the

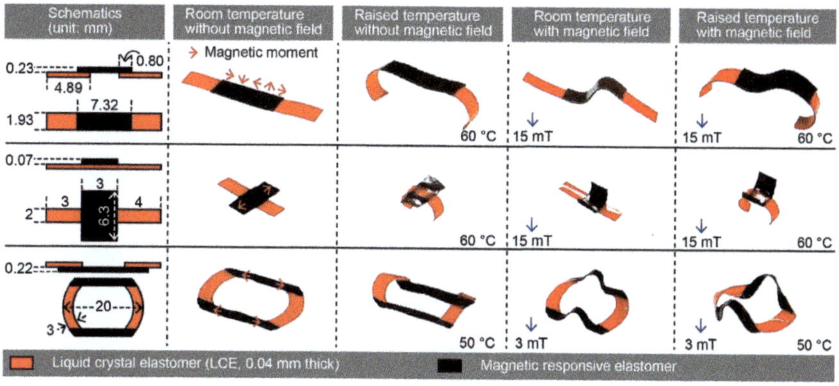


**Figure 5.15** Dual-responsive behaviors of the bimorph structures with three different configurations. Source: Zhang et al. [76]/John Wiley & Sons/CC BY 4.0.

desired value, the bimorph materials were placed on the surface of a hot plate, and to generate the desired actuation magnetic field, permanent magnets were adopted. These exemplary results demonstrated that the bimorph materials could be actuated by heating or external magnetic fields separately or be actuated conjunctly. Additional degree-of-freedom (DOF) in the design and actuation of samples toward in situ reconfigurable 3D shapes could be achieved by the multi-responsiveness of the bimorph materials. For instance, the control of the magnetic field (on/off) and temperature (high/low) enabled the realization of four distinct shapes of the strips. Figure 5.15 shows that any two of the four shapes could experience a reversible transition. Several possible shape reconfigurations were generated for the identical structure by the direct regulation of the magnetic field and temperature. The capabilities of bimorph structures could be greatly enhanced through the rich in situ reconfigurability.

Then, the integration of magnetothermal effect and LCEs for local and sequential actuation is demonstrated. Light-actuation strategies are widely adopted for reported sequential or localized deformations [77]. Nevertheless, the light stimulus is limited by its penetration depth for robotic actuation. For example, the penetration depths of ultraviolet light and near-IR (NIR) light are at the micrometer scale and millimeter scale, respectively. Therefore, light is hardly applied in the enclosed space, while magnetic field allows the penetration in any environment deeply, but is difficult to achieve the local or sequential actuation. A simple strategy using magnetic LCE system shown in Figure 5.13 was developed to realize the magnetically sequential and local control [75]. Under the stimulus of AMF that occupies the same $H$ value, the eventual stable temperatures of M-PULCEs vary with the content of doped $Fe_3O_4$ NPs. Meanwhile, the critical temperature of LC-isotropic phase transition (Ti) is independent with the concentration of $Fe_3O_4$ particles. Therefore, different magnetic field intensities are needed for the activation of different M-PULCEs. Interference welding was performed for the arrangement of different M-PULCEs in the desired regions. The movements of local positions in the assembled actuator can be controlled by tuning the intensity of AMF.

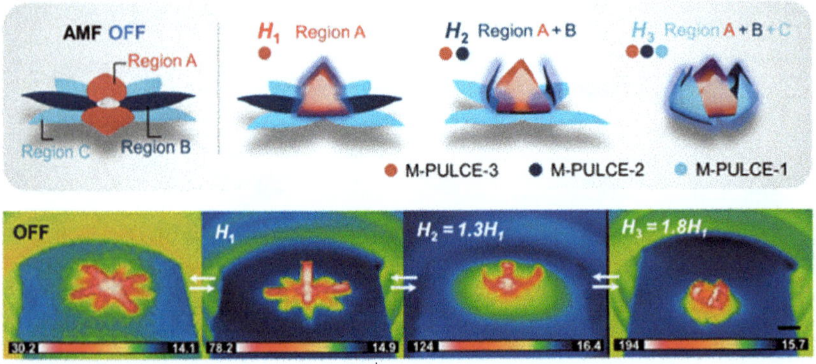


**Figure 5.16** Shape-morphing behaviors of eight-petal flower-shape M-PULCE actuator using AMF. Source: Wu et al. [75]/Reproduced from American Association for the Advancement of Science. (AAAS).

For example, to induce the deformation of the aligned M-PULCE-1, M-PULCE-2, and M-PULCE-3, the strength of AMF ($f = 320$ kHz) should attain 283.15 Gs ($H_3$), 219.27 Gs ($H_2$), and 163.08 Gs ($H_1$), respectively. With the $H_1$ intensity, region A in Figure 5.16 (made of M-PULCE-3) was independently controlled since there are two closed petals. With the increase of intensity to $H_2$ and $H_3$, regions B (made of M-PULCE-2) and C (made of M-PULCE-1) responded sequentially, resulting in an effect similar to a series of logic switches.

A hand-like actuator with locally controllable capability was fabricated, as shown in Figure 5.17. "Fingers" shape structures would respond and bend upon direct thermal actuation, while a series of localized locomotion can be generated by setting AMF with different intensities. Suitable design of the concentration of magnetic particles holds the promise to enhance the hierarchy of actuation. In this study, LCE matrix with a fixed Ti was adopted. Ti of the LCEs can be tuned by controlling the cross-linking density and the molecular weight of the LC oligomer. It can be envisioned that LCEs with diverse Ti would facilitate the realization of more complex sequences of motion and an enriched degree of freedom.

Because of the integration of the supercritical property and the magnetothermal effect inside the M-PULCEs, another unique performance that can be demonstrated was the stepwise magnetic control. The supercritical property endows the nematic LCE with a series of intermediate deformation states and continuous strain variation, which in turn enables stepwise control under given temperatures. The step-by-step controllability of M-PULCEs was confirmed by cooling and heating in the temperature range from 30 to 130 °C at 10 °C intervals. Based on the properties of M-PULCEs, a magnetically actuated robotic arm was demonstrated. When an alternating magnetic field is turned on, the temperature of the material rises rapidly and reaches the critical temperature, and the robotic arm will swing upwards. When the alternating magnetic field is turned off, the arm moves downward as the temperature decreases. Furthermore, we could gradually control the lifting of the robotic arm by modulating the alternating magnetic field, as shown in Figure 5.18a. When the intensity is 132.59 Gs, the material temperature reaches 80 °C, and the

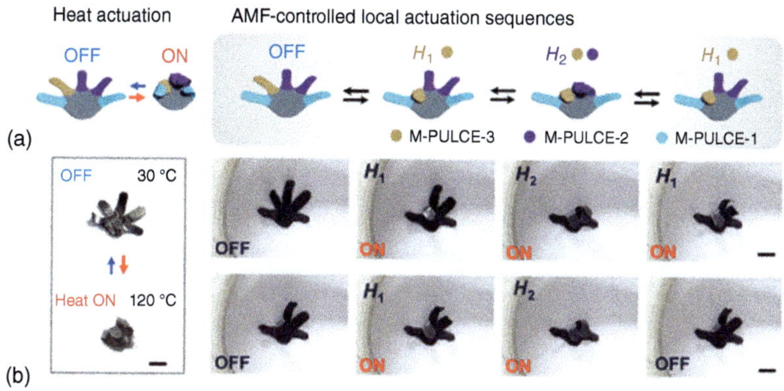


**Figure 5.17** Local and sequential actuation of hand-shape M-PULCE actuator. Schematic illustration (a) and experimental observation (b) of the actuation with the control of thermal source or AMF. Source: Wu et al. [75]/Reproduced from American Association for the Advancement of Science. (AAAS).

inclination angle of the robotic arm is 14°. At an intensity of 141.80 Gs, the temperature and tilt angle reach 100 and 31 °C, respectively. At an intensity of 158.27 Gs, the temperature and tilt angle reach 120 and 60 °C, respectively. By precisely adjusting the intensity of the alternating magnetic field, the interval of temperature change can be further refined such as 5 or 2 °C. This control method can be used to simulate the multi-level control strategy of traditional mechanical equipment, and meanwhile does not rely on complex circuit design or sensing devices, and can realize wireless actuation. Due to the existence of $Fe_3O_4$ nanoparticles, M-PULCEs can be actuated not only by the magnetothermal effect, but also by the photothermal effect such as NIR light, thereby enabling local actuation and reprogramming function. In addition, M-PULCEs have self-healing ability. Under the stimulation of alternating magnetic field or NIR light, material healing can be achieved through the exchange of dynamic carbamate bonds, and the healing efficiencies based on alternating magnetic field and NIR are 91.1% and 35.7%, respectively. As shown in Figure 5.18b, the healing and reconfigurable abilities of M-PULCEs under the action of magnetic field are demonstrated. First the material was cut into two separate parts, which were then restored under the external magnetic field. The healed samples could exhibit both elongation and bending deformations, and they can also be fabricated into various new thin film structures.

### 5.3.3 Adaptive Locomotion of Magnetic LCE-Based Robots

The locomotion of miniature robots in the low Reynolds number regime is particularly relevant for biomedical applications [78, 79]. In this regime, helical swimming behavior presents an efficient and effective strategy, and has been adopted by many reported microrobots [80], while due to the requirements of adjusting different locomotion methods based on working environments, few of these helical robots exhibit multi-modal locomotion schemes on multiple terrains, e.g. air surfaces and viscous

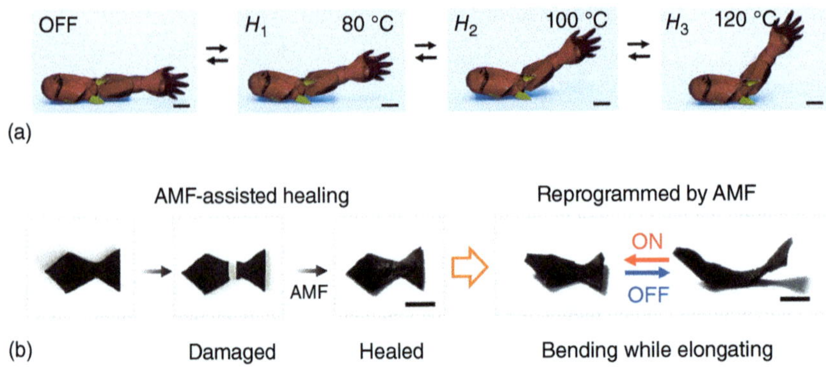


**Figure 5.18** Magnetic field-assisted multi-level control and self-healing of M-PULCE. (a) The step-by-step lifting process of the robotic arm under different magnetic field strengths (132.59, 141.80, 158.27 Gs). (b) Demonstration of self-healing and reconfigurable functions of M-PULCE under magnetic field assistance. The healed structure could exhibit both elongation and bending deformations. Source: Wu et al. [75]/Reproduced from American Association for the Advancement of Science. (AAAS).

liquids. A recent study reported the *in situ* shape reconfigurability of magnetic LCE and the formation of self-adaptive machines that are suitable for locomotion in different environments [74].

As shown in Figure 5.19, a spiral structure and a semicircle shape were adopted for swimming locomotion mode in viscous environment and walking locomotion mode in air, respectively. The magnetic LCE shown in Figure 5.12 was used for the fabrication of the film machine (with the size of 25 μm [thickness] × 1 mm × 9 mm) with a splay director field and 1 : 2 mass ratio. A sinusoidal magnetization profile was programmed into the soft machines by deforming it over the surface of a glass rod. It remained flat or deformed into the desired semi-circle curve by the removal or application of an external magnetic field. Once the semi-circle shape was formed, an oscillating magnetic field varying horizontally was applied, which would actuate the machine to move toward the edge of the substrate. When the machine fell into heated liquid (glycerol at 70 °C), the composite film deformed into helical shape. Once in this helical shape, a rotating magnetic field in the horizontal plane was applied, leading to the rotation of the machine with respect to its long axis and the generation of translational displacement. Once the external magnetic field was removed, the LCE machine sank to the container bottom.

A magnetic LCE-based structure with the shape of slender filament was adopted to demonstrate the capability of environmental adaption (Figure 5.20a). The mass ratio of LCE matrix and magnetic particles is 1 : 2. The size of the LCE filament was 17 mm (length), 130 μm (width), and 25 μm (thickness). A 1.7 T magnetic field was applied to magnetize the structure, resulting in a uniform magnetic domain along the length direction. A splay field was encoded in the LCE filament, where the aligning direction and the planar surface are parallel. One end of the magnetic LCE filament was attached to a glass substrate and the other end can be manipulated freely. A Halbach magnet array connected to a step motor was used to generate a dynamic magnetic

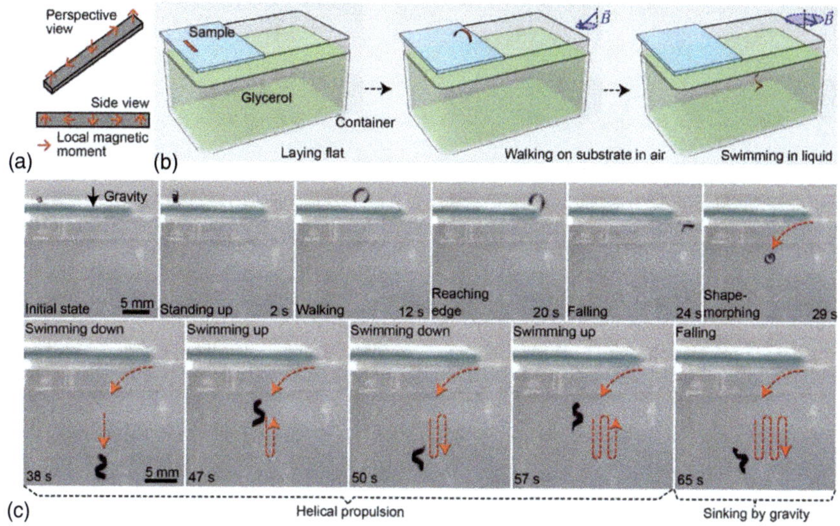


**Figure 5.19** Adaptive locomotion of a magnetic LCE-based soft robots in air and viscous fluids. (a) Magnetization profile of the soft machine. (b and c) Schematic illustration and experimental observation of multimode locomotion in different environments. Source: Zhang et al. [74]/Reproduced from John Wiley & Sons/CC BY-NC 4.0.

field (40 mT). Under the activation of the rotating field, the LCE filament would move back and forth, while when a hot needle (surface temperature: ~80 °C) was brought close to the LCE filament, it experienced a large shape transformation and the curled structure would capture the hot needle, which could mimic the twining behavior of vines.

The developed magnetic LCE material was also applied to deformable miniature springs which could exhibit tunable mechanical behaviors (displacement–force curve) based on magnetic interactions and resultant magnetic force. Moreover, via environmental stimulation, the developed LCE spring could also present *in situ* reconfigurability. A rectangular magnetic LCE film with the size of 16 mm (length), 8 mm (width), 40 μm (thickness) was fabricated. The mass ratio of LCE matrix and MMP is 1 : 2. A 1.8 T magnetic field was applied to magnetize the structure, resulting in a uniform magnetic domain perpendicular to the surface of the LCE film. A splay field was encoded in the LCE film across the thickness direction, leading to an out-of-plane bending upon heating. The developed LCE film was adhered to a glass substrate (at the center position) and subsequently fixed to a 25 g load cell. A ~40 mT magnetic field was generated by a cylindrial permanent magnet (N42, diameter: 35 mm, thickness: 4 mm). The distance between the permanent magnet and the LCE film was 18 mm, and the direction of the magnetization profile of the sample was antiparallel to the external magnetic field. As the distance between them decreases, the repulsive force applied to the LCE film would increase. When applying UV exposure to the LCE film (one minutes), it will deform due to the doping of the Disperse Red 1 inside the film and the photothermal effect. Due to the existence of magnetic interactions, this deformation can still be retained after

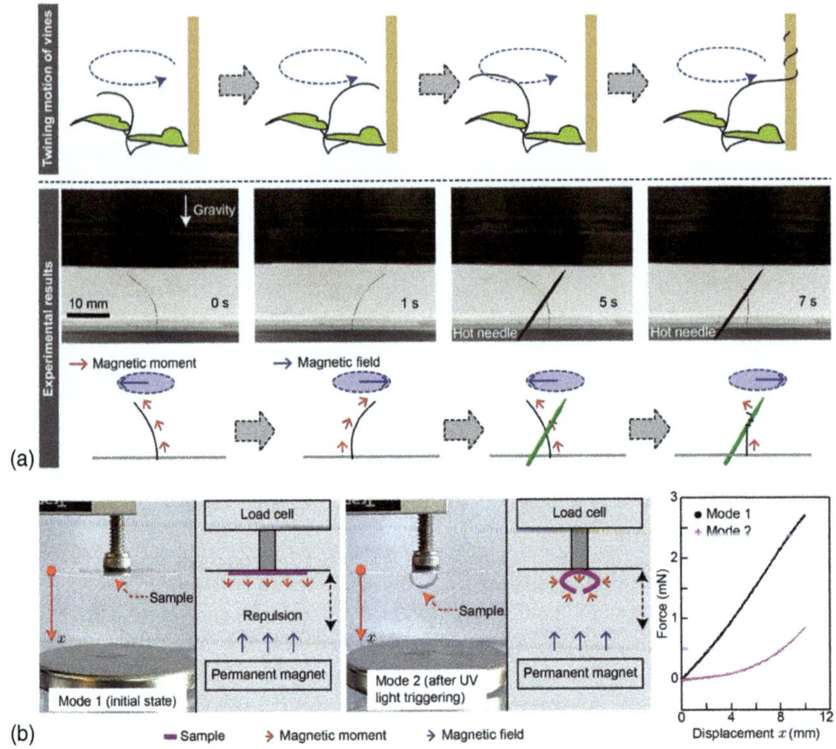


**Figure 5.20** Adaptability and reconfigurability of two kinds of magnetic LCE-based structures (magnetic filament and spring). (a) Bioinspired twining behavior of LCE filament. (b) Deformable miniature springs with in situ tunable mechanical properties. Source: Zhang et al. [74]/Reproduced from John Wiley & Sons/CC BY-NC 4.0.

the removal of UV exposure. Moreover, because of the reduction of net magnetization moment, the resultant magnetic force applied to the LCE film would also decrease. The measured displacement–force curves demonstrate that the developed magnetic LCE could exhibit *in situ* reconfigurable mechanical properties upon environmental cues (UV light).

An untethered miniature millipede-like robot fabricated by the bimorph materials shown in Figure 5.14 is presented in Figure 5.21 [76]. With the stimulus of a rotation magnetic field, the robot performed a millipede-inspired gait in water and in air. The central body of the robot was constructed by LCE and the 12 legs of the robot were constructed by MRE. Figure 5.21a shows the schematic illustration of the robotic structure. Local magnetic moments of the 12 legs of the robot were formed along alternating directions. Each leg was deformed to achieve the alignment between its free end and the direction of the external magnetic field. By applying a rotating magnetic field, the robot mimicked the locomotion of an oar with the synchronized rotation of the free end of each leg. Therefore, the leg bent downward to lift the robot body up, bent backward to push the robot forward, bent upward to disengage from the substrate, bent forward to move back to the starting pose, and then repeated the periodical locomotion. No less than six legs supported the robot at any time due to

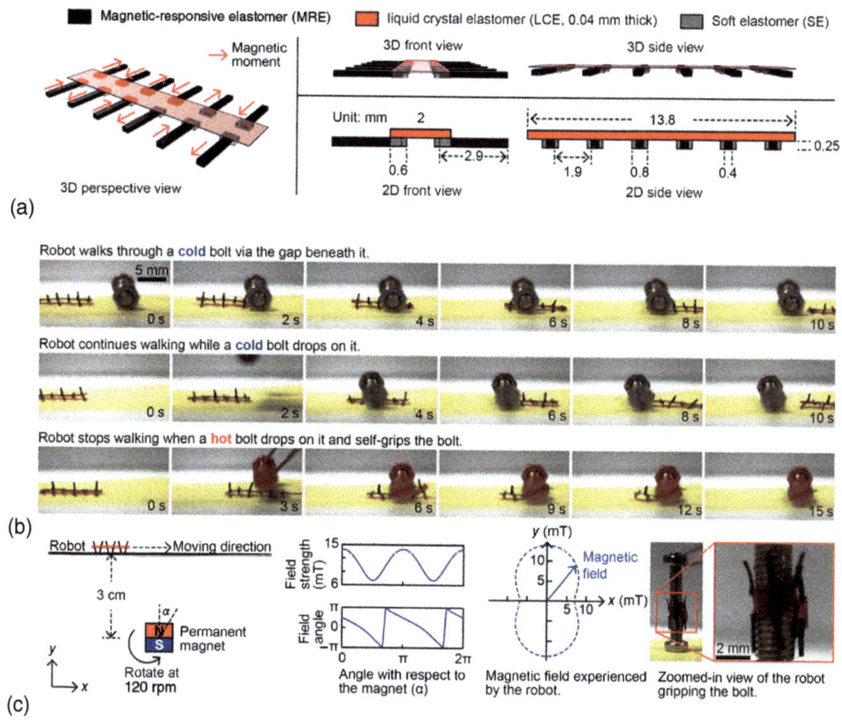

**Figure 5.21** Untethered miniature 12-legged robot made of bimorph material. (a) Schematic illustration of the structure of the multilegged robot. (b) Actuation of the multilegged robot. (c) Experimental setup and magnetic field design for the actuation. Source: Zhang et al. [76]/Reproduced from John Wiley & Sons, Inc./CC BY 4.0.

the alternating magnetic moments. Figure 5.21b shows snapshots of the robot during experimental demonstrations in different scenes. For the first case in Figure 5.21b, the robot passed through a preset cold bolt through the gap below. For the another case, during the walking process of the robot, a cold bolt fell on the robot, while the robot kept to walk without any disturbance. For the third case, during the walking of the robot, a hot bolt (approximate 100 °C surface temperature) fell on the robot. The walking process was terminated and self-gripping was performed by curling the robot body around the bolt.

## 5.4 Summary

This chapter discusses some newly proposed strategies to fabricate complex pre-programmed director fields of LCNs and LCEs, as well as the recently emerged magnetic LCE systems with multiple-stimuli responsiveness. Firstly, microchannels with pixelated arrays could induce the formation of director fields in pre-polymerized LCNs with designated 2D and 3D patterns. Combining it with the technique of TPP, 3D LCN devices with encoded director fields could be fabricated.

Next, the heterogeneous 3D assembly of LCE voxels with encoded individual director field could allow for the formation of LCE 3D structures with arbitrary 3D director field patterns and geometries, whose shape-morphing behaviors were demonstrated upon thermal and solvent stimulation. The extended freedom in the fabrication of 3D LCE with arbitrary director fields would pave the way toward broad application of LCEs in soft robotics.

Magnetic LCE systems emerged as a result of the effort to integrate the respective advantages of multiple-stimuli responsiveness of LCEs and the swift and agile shape-morphing of magnetic soft materials. Magnetic controllability could be endowed by embedding hard-magnetic microparticles such as NeFeB or soft-magnetic nanoparticles such as $Fe_3O_4$ into LCE matrices, or a facile stacking of LCE and magnetic soft materials. The resultant multiple-stimuli-responsive magnetic composite material could be utilized to make miniature robots with DOFs that could be independently, selectively, and sequentially actuated, enabling precise control and sophisticated functionalities in genuine real-world applications.

The enrichment of fabrication strategies for LCNs and LCEs to make more complex and more sophisticated director fields and structures could lead to the further enhancement of their material advantages and extension of the device capabilities based on these materials. The newly emerged magnetic LCE composite material indicates a new director of combining the advantages of LC materials with other smart materials for better versatilities and unprecedented applicability to real-world applications. With this recent development, it is envisioned that LCNs and LCEs will play an increasingly important role in smart materials and the large variety of corresponding applications in the near future.

## Abbreviations

| | |
|---|---|
| 1D | one-dimensional |
| 2D | two-dimensional |
| 3D | three-dimensional |
| AMF | alternating magnetic field |
| CAN | covalent adaptable networks |
| DBTDL | dibutyltin dilaurate |
| DOF | degree of freedom |
| LC | liquid crystal |
| LCN | liquid crystal network |
| LCE | liquid crystal elastomer |
| MMP | magnetic microparticles |
| MRE | magnetic-responsive elastomers |
| M-PULCE | magnetothermal-responsive polyurethane (PU) LCE |
| NIR | near-infrared |
| NPs | nanoparticles |
| POM | polarized optical microscope |
| PU | polyurethane |

PULCE polyurethane (PU) LCE
SEM scanning electron microscope
TPP two-photon polymerization

## References


1 White, T.J. and Broer, D.J. (2015). Programmable and adaptive mechanics with liquid crystal polymer networks and elastomers. *Nature Materials* 14: 1087–1098.

2 Ohm, C., Brehmer, M., and Zentel, R. (2010). Liquid crystalline elastomers as actuators and sensors. *Advanced Materials* 22: 3366–3387.

3 Nie, Z.-Z., Zuo, B., Wang, M. et al. (2021). Light-driven continuous rotating Möbius strip actuators. *Nature Communications* 12: 2334.

4 Wang, Q., Yu, L., Yu, M. et al. (2017). Liquid crystal elastomer actuators from anisotropic porous polymer template. *Macromolecular Rapid Communications* 38: 1600699.

5 Zhao, T., Wang, J., Fan, Y., and Dou, W. (2022). Helical liquid crystal elastomer miniature robot with photocontrolled locomotion. *Advanced Materials Technologies* 7: 2200222.

6 Wie, J.J., Shankar, M.R., and White, T.J. (2016). Photomotility of polymers. *Nature Communications* 7: 13260.

7 Lahikainen, M., Zeng, H., and Priimagi, A. (2018). Reconfigurable photoactuator through synergistic use of photochemical and photothermal effects. *Nature Communications* 9: 4148.

8 Xiao, Y.Y., Jiang, Z.C., Tong, X., and Zhao, Y. (2019). Biomimetic locomotion of electrically powered "Janus" soft robots using a liquid crystal polymer. *Advanced Materials* 31: 1903452.

9 Liu, Z., Bisoyi, H.K., Huang, Y. et al. (2022). Thermo- and mechanochromic camouflage and self-healing in biomimetic soft actuators based on liquid crystal elastomers. *Angewandte Chemie International Edition* 61: e202115755.

10 Da Cunha, M.P., Debije, M.G., and Schenning, A.P. (2020). Bioinspired light-driven soft robots based on liquid crystal polymers. *Chemical Society Reviews* 49: 6568–6578.

11 Barnes, M. and Verduzco, R. (2019). Direct shape programming of liquid crystal elastomers. *Soft Matter* 15: 870–879.

12 Liu, H., Tian, H., Li, X. et al. (2022). Shape-programmable, deformation-locking, and self-sensing artificial muscle based on liquid crystal elastomer and low-melting point alloy. *Science Advances* 8: eabn5722.

13 Wang, M., Hu, X.B., Zuo, B. et al. (2020). Liquid crystal elastomer actuator with serpentine locomotion. *Chemical Communications* 56: 7597–7600.

14 Li, S., Bai, H., Liu, Z. et al. (2021). Digital light processing of liquid crystal elastomers for self-sensing artificial muscles. *Science Advances* 7: eabg3677.

15 Li, Y., Yu, H., Yu, K. et al. (2021). Reconfigurable three-dimensional mesostructures of spatially programmed liquid crystal elastomers and their ferromagnetic composites. *Advanced Functional Materials* 31: 2100338.

**16** He, Q., Wang, Z., Wang, Y. et al. (2021). Electrospun liquid crystal elastomer microfiber actuator. *Science Robotics* 6: eabi9704.

**17** Li, S., Librandi, G., Yao, Y. et al. (2021). Controlling liquid crystal orientations for programmable anisotropic transformations in cellular microstructures. *Advanced Materials* 33: 2105024.

**18** Wang, C., Sim, K., Chen, J. et al. (2018). Soft ultrathin electronics innervated adaptive fully soft robots. *Advanced Materials* 30: 1706695.

**19** Zhang, M., Shahsavan, H., Guo, Y. et al. (2021). Liquid-crystal-elastomer-actuated reconfigurable microscale kirigami metastructures. *Advanced Materials* 33: 2008605.

**20** Wu, J., Yao, S., Zhang, H. et al. (2021). Liquid crystal elastomer metamaterials with giant biaxial thermal shrinkage for enhancing skin regeneration. *Advanced Materials* 33: 2106175.

**21** Shahsavan, H., Aghakhani, A., Zeng, H. et al. (2020). Bioinspired underwater locomotion of light-driven liquid crystal gels. *Proceedings of the National Academy of Sciences* 117: 5125–5133.

**22** Liu, J., Gao, Y., Wang, H. et al. (2020). Shaping and locomotion of soft robots using filament actuators made from liquid crystal elastomer–carbon nanotube composites. *Advanced Intelligent Systems* 2: 1900163.

**23** Hu, J., Nie, Z., Wang, M. et al. (2023). Springtail-inspired light-driven soft jumping robots based on liquid crystal elastomers with monolithic three-leaf panel fold structure. *Angewandte Chemie International Edition* 62: e202218227.

**24** Wang, Y., Yin, R., Jin, L. et al. (2023). 3D-printed photoresponsive liquid crystal elastomer composites for free-form actuation. *Advanced Functional Materials* 33: 2210614.

**25** Li, S., Lerch, M.M., Waters, J.T. et al. (2022). Self-regulated non-reciprocal motions in single-material microstructures. *Nature* 605: 76–83.

**26** Liang, R., Yu, H., Wang, L., and Shen, D. (2022). Light-guided dynamic liquid crystalline elastomer actuators enabled by mussel adhesive protein chemistry. *Advanced Functional Materials* 33: 2211914.

**27** Yu, H. and Ikeda, T. (2011). Photocontrollable liquid-crystalline actuators. *Advanced Materials* 23: 2149–2180.

**28** Liu, W., Guo, L.X., Lin, B.P. et al. (2016). Near-infrared responsive liquid crystalline elastomers containing photothermal conjugated polymers. *Macromolecules* 49: 4023–4030.

**29** Wang, Z., Li, K., He, Q., and Cai, S. (2019). A light-powered ultralight tensegrity robot with high deformability and load capacity. *Advanced Materials* 31: 1806849.

**30** de Haan, L.T., Verjans, J.M., Broer, D.J. et al. (2014). Humidity-responsive liquid crystalline polymer actuators with an asymmetry in the molecular trigger that bend, fold, and curl. *Journal of the American Chemical Society* 136: 10585–10588.

**31** Kim, K., Guo, Y., Bae, J. et al. (2021). 4D printing of hygroscopic liquid crystal elastomer actuators. *Small* 17: 2100910.

**32** Lee, J., Guo, Y., Choi, Y.J. et al. (2020). Mechanically programmed 2D and 3D liquid crystal elastomers at macro-and microscale via two-step photocrosslinking. *Soft Matter* 16: 2695–2705.

**33** Li, S., Deng, B., Grinthal, A. et al. (2021). Liquid-induced topological transformations of cellular microstructures. *Nature* 592: 386–391.

**34** Boothby, J.M., Kim, H., and Ware, T.H. (2017). Shape changes in chemoresponsive liquid crystal elastomers. *Sensors and Actuators B: Chemical* 240: 511–518.

**35** Stumpel, J.E., Wouters, C., Herzer, N. et al. (2014). An optical sensor for volatile amines based on an inkjet-printed, hydrogen-bonded, cholesteric liquid crystalline film. *Advanced Optical Materials* 2: 459–464.

**36** Boothby, J.M., Gagnon, J.C., McDowell, E. et al. (2022). An untethered soft robot based on liquid crystal elastomers. *Soft Robotics* 9: 154–162.

**37** He, Q., Wang, Z., Wang, Y. et al. (2019). Electrically controlled liquid crystal elastomer-based soft tubular actuator with multimodal actuation. *Science Advances* 5: eaax5746.

**38** Zadan, M., Patel, D.K., Sabelhaus, A.P. et al. (2022). Liquid crystal elastomer with integrated soft thermoelectrics for shape memory actuation and energy harvesting. *Advanced Materials* 34: 2200857.

**39** Wang, Y., He, Q., Wang, Z. et al. (2023). Liquid crystal elastomer based dexterous artificial motor unit. *Advanced Materials* 35: 2211283.

**40** Fowler, H.E., Rothemund, P., Keplinger, C., and White, T.J. (2021). Liquid crystal elastomers with enhanced directional actuation to electric fields. *Advanced Materials* 33: 2103806.

**41** Davidson, Z.S., Shahsavan, H., Aghakhani, A. et al. (2019). Monolithic shape-programmable dielectric liquid crystal elastomer actuators. *Science Advances* 5: eaay0855.

**42** Wang, Y., Liu, J., and Yang, S. (2022). Multi-functional liquid crystal elastomer composites. *Applied Physics Reviews* 9: 011301.

**43** Jiang, Z.C., Xiao, Y.Y., and Zhao, Y. (2019). Shining light on liquid crystal polymer networks: preparing, reconfiguring, and driving soft actuators. *Advanced Optical Materials* 7: 1900262.

**44** Ma, J., Yang, Y., Valenzuela, C. et al. (2022). Mechanochromic, shape-programmable and self-healable cholesteric liquid crystal elastomers enabled by dynamic covalent boronic ester bonds. *Angewandte Chemie International Edition* 61: e202116219.

**45** de Haan, L.T., Sánchez-Somolinos, C., Bastiaansen, C.M. et al. (2012). Engineering of complex order and the macroscopic deformation of liquid crystal polymer networks. *Angewandte Chemie International Edition* 51: 12469–12472.

**46** Krause, S., Zander, F., Bergmann, G. et al. (2009). Nematic main-chain elastomers: coupling and orientational behavior. *Comptes Rendus Chimie* 12: 85–104.

**47** Mol, G.N., Harris, K.D., Bastiaansen, C.W., and Broer, D.J. (2005). Thermo-mechanical responses of liquid-crystal networks with a splayed molecular organization. *Advanced Functional Materials* 15: 1155–1159.

**48** Kim, H., Lee, J.A., Ambulo, C.P. et al. (2019). Intelligently actuating liquid crystal elastomer-carbon nanotube composites. *Advanced Functional Materials* 29: 1905063.

**49** Lv, P., Yang, X., Bisoyi, H.K. et al. (2021). Stimulus-driven liquid metal and liquid crystal network actuators for programmable soft robotics. *Materials Horizons* 8: 2475–2484.

**50** Yang, X., Chen, Y., Zhang, X. et al. (2022). Bioinspired light-fueled water-walking soft robots based on liquid crystal network actuators with polymerizable miniaturized gold nanorods. *Nano Today* 43: 101419.

**51** Yang, Y., Zhan, W., Peng, R. et al. (2015). Graphene-enabled superior and tunable photomechanical actuation in liquid crystalline elastomer nanocomposites. *Advanced Materials* 27: 6376–6381.

**52** Zeng, H., Wasylczyk, P., Wiersma, D.S., and Priimagi, A. (2018). Light robots: bridging the gap between microrobotics and photomechanics in soft materials. *Advanced Materials* 30: 1703554.

**53** Sitti, M. and Wiersma, D.S. (2020). Pros and cons: magnetic versus optical microrobots. *Advanced Materials* 32: 1906766.

**54** Camacho-Lopez, M., Finkelmann, H., Palffy-Muhoray, P., and Shelley, M. (2004). Fast liquid-crystal elastomer swims into the dark. *Nature Materials* 3: 307–310.

**55** Zuo, B., Wang, M., Lin, B.-P., and Yang, H. (2019). Visible and infrared three-wavelength modulated multi-directional actuators. *Nature Communications* 10: 4539.

**56** Gelebart, A.H., Jan Mulder, D., Varga, M. et al. (2017). Making waves in a photoactive polymer film. *Nature* 546: 632–636.

**57** Ware, T.H., McConney, M.E., Wie, J.J. et al. (2015). Voxelated liquid crystal elastomers. *Science* 347: 982–984.

**58** Ahn, C., Liang, X., and Cai, S. (2015). Inhomogeneous stretch induced patterning of molecular orientation in liquid crystal elastomers. *Extreme Mechanics Letters* 5: 30–36.

**59** Xia, Y., Zhang, X., and Yang, S. (2018). Instant locking of molecular ordering in liquid crystal elastomers by oxygen-mediated thiol–acrylate click reactions. *Angewandte Chemie* 130: 5767–5770.

**60** Finkelmann, H., Kim, S.T., Munoz, A. et al. (2001). Tunable mirrorless lasing in cholesteric liquid crystalline elastomers. *Advanced Materials* 13: 1069–1072.

**61** Shahsavan, H., Salili, S.M., Jákli, A., and Zhao, B. (2015). Smart muscle-driven self-cleaning of biomimetic microstructures from liquid crystal elastomers. *Advanced Materials* 27: 6828–6833.

**62** Zeng, H., Martella, D., Wasylczyk, P. et al. (2014). High-resolution 3D direct laser writing for liquid-crystalline elastomer microstructures. *Advanced Materials* 26: 2319–2322.

**63** Martella, D., Nocentini, S., Nuzhdin, D. et al. (2017). Photonic microhand with autonomous action. *Advanced Materials* 29: 1704047.

**64** Kotikian, A., Truby, R.L., Boley, J.W. et al. (2018). 3D printing of liquid crystal elastomeric actuators with spatially programed nematic order. *Advanced Materials* 30: 1706164.

**65** Xing, J.-F., Zheng, M.-L., and Duan, X.-M. (2015). Two-photon polymerization microfabrication of hydrogels: an advanced 3D printing technology for tissue engineering and drug delivery. *Chemical Society Reviews* 44: 5031–5039.

**66** Guo, Y., Shahsavan, H., and Sitti, M. (2020). 3D microstructures of liquid crystal networks with programmed voxelated director fields. *Advanced Materials* 32: 2002753.

**67** Guo, Y., Zhang, J., Hu, W. et al. (2021). Shape-programmable liquid crystal elastomer structures with arbitrary three-dimensional director fields and geometries. *Nature Communications* 12: 5936.

**68** Xia, N., Zhu, G., Wang, X. et al. (2022). Multicomponent and multifunctional integrated miniature soft robots. *Soft Matter* 18: 7464–7485.

**69** Xia, N., Jin, D., Iacovacci, V., and Zhang, L. (2022). 3D printing of functional polymers for miniature machines. *Multifunctional Materials* 5: 012001.

**70** Jin, D. and Zhang, L. (2021). Collective behaviors of magnetic active matter: recent progress toward reconfigurable, adaptive, and multifunctional swarming micro/nanorobots. *Accounts of Chemical Research* 55: 98–109.

**71** Zhou, H., Mayorga-Martinez, C.C., Pané, S. et al. (2021). Magnetically driven micro and nanorobots. *Chemical Reviews* 121: 4999–5041.

**72** Kim, Y. and Zhao, X. (2022). Magnetic soft materials and robots. *Chemical Reviews* 122: 5317–5364.

**73** Yang, Z. and Zhang, L. (2020). Magnetic actuation systems for miniature robots: a review. *Advanced Intelligent Systems* 2: 2000082.

**74** Zhang, J., Guo, Y., Hu, W. et al. (2021). Liquid crystal elastomer-based magnetic composite films for reconfigurable shape-morphing soft miniature machines. *Advanced Materials* 33: 2006191.

**75** Wu, Y., Zhang, S., Yang, Y. et al. (2022). Locally controllable magnetic soft actuators with reprogrammable contraction-derived motions. *Science Advances* 8: eabo6021.

**76** Zhang, J., Guo, Y., Hu, W., and Sitti, M. (2021). Wirelessly actuated thermo-and magneto-responsive soft bimorph materials with programmable shape-morphing. *Advanced Materials* 33: 2100336.

**77** Liu, Y., Shaw, B., Dickey, M.D., and Genzer, J. (2017). Sequential self-folding of polymer sheets. *Science Advances* 3: e1602417.

**78** Ng, C.S.X., Tan, M.W.M., Xu, C. et al. (2021). Locomotion of miniature soft robots. *Advanced Materials* 33: 2003558.

**79** Palagi, S. and Fischer, P. (2018). Bioinspired microrobots. *Nature Reviews Materials* 3: 113–124.

**80** Dong, Y., Wang, L., Iacovacci, V. et al. (2022). Magnetic helical micro-/nanomachines: recent progress and perspective. *Matter* 5: 77–109.

# 6

# Flexible Ferrofluid as Soft Robotic Agents

## 6.1 Introduction

Untethered miniature soft robots that can respond to external stimuli possess programmable structural changes, enabling their promising applications in intelligent electronics [1–4], biomimetics [5–9], smart grippers [5, 6, 10–13], and precision medicine [14–23]. Particularly, their deformation ability allows them to perform dexterous operations and navigate in some confined spaces of the human body, holding great potential in biomedical applications. However, these silicon-, hydrogel-, carbon-, and liquid crystal polymer-based soft robots developed by our group and other groups are composed of rigid materials, which endow soft robots with limited and predesigned deformation, making them only capable of entering the spaces comparable to the size of soft robots and specific functions. To achieve biomedical applications in very narrow, clustered, and constrained spaces and changeable functions by varying structural changes, developing liquid materials-based soft robots is necessary [24–26]. Ferrofluids with liquid properties and magnetic response have emerged and been recognized as a good choice to develop soft robots by employing themselves (ferrofluids droplet robots (FDRs)) or coating them on inanimate targets as magnetic clothes [27–36]. FDRs have been widely and systematically researched to investigate their behaviors (fission, fusion, jumping, deformation magnitude and direction, etc.) and applications (cargo delivery, object sorting, liquid cilia, cooperative fluid mixing, etc.). The behaviors of FDRs are mainly determined by the external magnetic field, including its type (rotating field, oscillating field, conical field, permanent magnet-produced field, etc.), magnitude, direction, and frequency. Additionally, the type, size, and wetting behavior with contacting substrate of FDR also influence its behavior and application field [37]. When the ferrofluids are used as magnetic clothes, the high wetting behavior of ferrofluids on inanimate targets enables the successful construction of soft robots, which shows potential in variable structural changes when releasing and further coating ferrofluids. Therefore, ferrofluids exhibit excellent capabilities in constructing soft robots. To let readers fully understand this field, we review all points concerning the behaviors and application of ferrofluids-based soft robots in this chapter.

*Untethered Miniature Soft Robots: Materials, Fabrications, and Applications*, First Edition.
Li Zhang, Jiachen Zhang, Neng Xia, and Yue Dong.

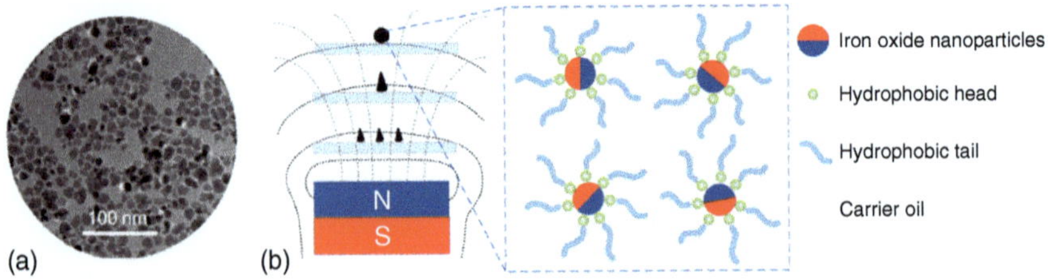


**Figure 6.1** a Transmission electron microscopy (TEM) image of the nanoparticles in the ferrofluid. b The composition of the oil-based ferrofluid. Source: Fan et al. [31]/Reproduced from American Association for the Advancement of Science. (AAAS).

## 6.2 Description of Ferrofluids

Ferrofluid, as a magnetic liquid material, is a stable colloidal suspension made of ferromagnetic or sub-ferromagnetic particles ($Fe_3O_4$ nanoparticles) with a diameter of ~10 nm, carrier fluids (usually oil-based organic solvent and water), and surfactants (Figure 6.1) [29, 31, 38]. These magnetic nanoparticles endow the ferrofluids with magnetic properties that are strongly affected by the content of nanoparticles [39]. The higher the concentration of magnetic nanoparticles, the stronger the magnetic properties of ferrofluids. The carrier fluids permit ferrofluids to keep fluid properties. The oil-based (or water-based) carrier fluids allow ferrofluids to form droplets in water-based fluids (or oil-based) due to the immiscibility and surface tension within different fluids [40]. As mentioned above, FDRs tend to be applied in water-based fluids that widely exists in microfluidic channels and biological environment for biomedical applications, so oil-based biocompatible ferrofluids are mainly used [41–46]. Each magnetic nanoparticle is thoroughly coated by a surfactant to inhibit aggregation of nanoparticles due to the formation of the van der Waals force, which makes ferrofluids stable for a relatively long time even though being magnetized by external magnetic fields [47, 48].

## 6.3 Deformation Behaviors of FDRs

FDRs have both active deformation and passive deformation [27–29]. The active deformation enables FDRs to freely adapt to the structures of the applied environment due to their liquid properties just like water. The passive deformations with unlimited deformation capabilities controlled by a variable magnetic field are widely researched in recent years. The related parameters of magnetic fields, FDRs, and contacting substrate have an influence on the deformation of FDRs. The deformations of FDRs are also linked with the motion of FDRs, so the motion styles are also included in the deformation behaviors. Current deformation behaviors include stretching, fission, fusion, as well as jumping, tumbling, rotating, transporting, kayaking, and wobbling motions.

### 6.3.1 The Influence of Magnetic Fields on the Deformation Behaviors of FDRs

There are two ways to produce magnetic fields using permanent magnets and electromagnetic coils [29]. The magnetic gradient fields are generated by permanent magnets and electromagnetic coils, while the magnetic uniform fields are only generated by electromagnetic coils. Magnetic gradient fields and uniform fields have differences in the control of the deformation behaviors of FDRs.

#### 6.3.1.1 Magnetic Fields Induced by Permanent Magnets

Permanent magnets are the typical representative of producing magnetic gradient fields (Figure 6.2a) [29]. Various complex deformation shapes (disk-, cube-, and ring-shapes) of FDRs have been achieved through programming magnetic field spatiotemporally by approaching permanent magnets with special shapes to FDRs from the bottom (Figure 6.2b,c). This is because magnetic nanoparticles within the FDRs will move to the location with the lowest magnetic potential energy. According to the simulation results about the spatial gradient of the induced magnetic potential energy, the morphology of the potential energy well is similar to the shape of permanent magnets, resulting in the similar deformation shapes of FDRs. By changing the deflection angles of permanent magnets, the deformation shapes of FDRs will be apparently changed. For example, the ring-shaped FDR produced by a ring-shaped magnet appears as a C-shape when the magnet is inclined. The FDRs move along the motion direction and speed of magnets. Combined the motion with the deformation, FDRs have a high possibility of realizing diverse manipulation tasks. The fission, fusion, and transportation of FDRs are also achieved by permanent magnets. The fission of FDRs happens when a stack of two cylindrical magnets with N, S-pole in the vertical direction is close enough to the FDRs underneath (Figure 6.2d,e) [49]. With the distance between magnets and FDR decreasing, the FDR is gradually elongated due to the increasing magnetic gradient field, leading to the reduction in the diameter of contacting area between FDR and substrate (Figure 6.2f). When the diameter is smaller than the calculated critical wavelength ($\lambda_c$), the FDRs are split by overcoming its surface tension. The critical wavelength is calculated as the flowing equation: $\lambda_c = 2\pi\sqrt{\frac{\sigma}{f_n}}$. $\sigma$ is the interfacial tension between ferrofluid and surrounding fluid, the normal force density ($f_n$) is the sum of magnetic force density and gravitational force density ($\Delta\rho$ g, $\Delta\rho$ is the density difference between the surrounding fluid and FDR, and g is the gravitational acceleration). With further decreasing distance, the FDR is split into smaller FDRs (Figure 6.2g).

Fusion of these smaller FDRs is achieved by rotating the cylindrical magnet by 90° that induces horizontal magnetization and mutual attraction between the droplets (Figure 6.3a,b) [49]. These smaller FDRs can be transported in a swarm with a vertical magnet. In this process, FDRs are still separated due to mutual repulsion. Additionally, immiscible fluid recognized as a lubricating layer prevents pinning and allows the moving flexibly of smaller FDRs. Under the cooperation of two magnets, individual smaller FDR is extracted from a swarm of FDRs (Figure 6.3c,d).

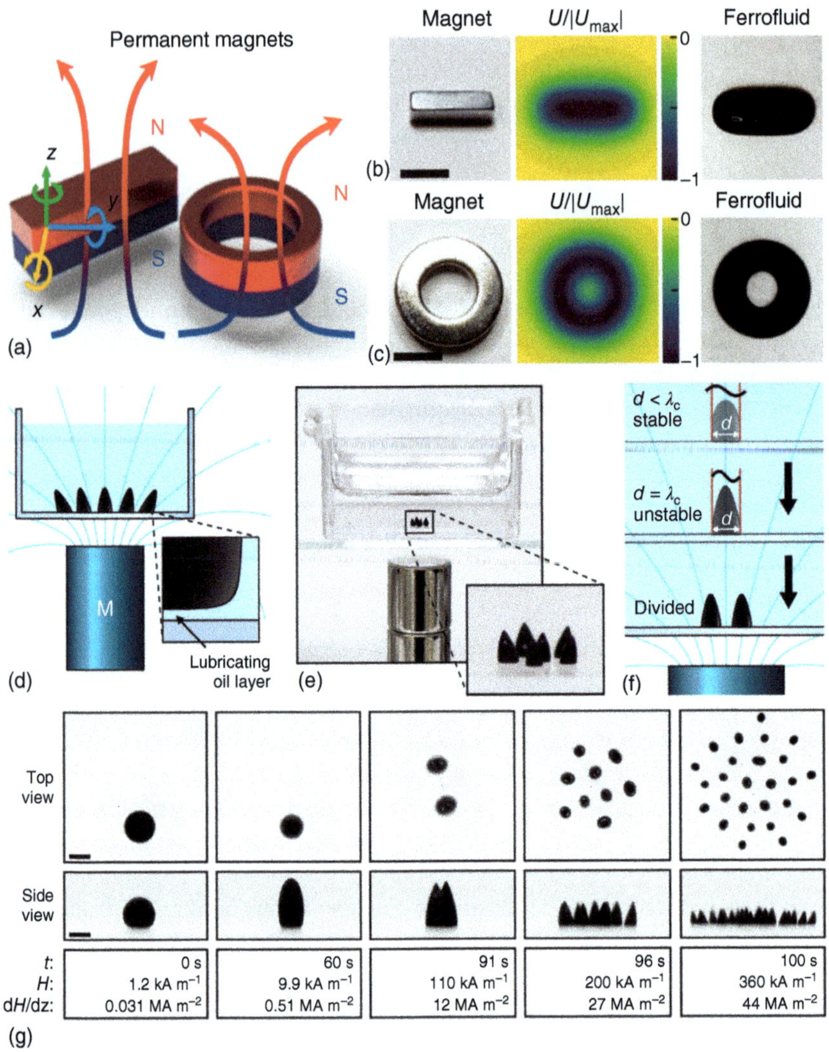


**Figure 6.2** (a)–(c) The schematic diagram, optical images, and magnetic field distribution of permanent magnets with cube- and ring-shapes, and the generated shape of FDRs. Source: Fan et al. [29]/Reproduced from National Academy of Science. (d)–(e) The schematic diagram and optical images of FDRs in a polystyrene container actuated by two stacked cylindrical magnets underneath. Source: Latikka et al. [49]/Reproduced from John Wiley & Sons, Inc./CC BY 4.0. (f) The schematic diagram of FDR splitting in an increasing magnetic field. Source: Latikka et al. [49]/John Wiley & Sons/CC BY 4.0. (g) Top and side views of FDRs splitting in silicone oil. Source: Latikka et al. [49]/Reproduced from John Wiley & Sons, Inc./CC BY 4.0.

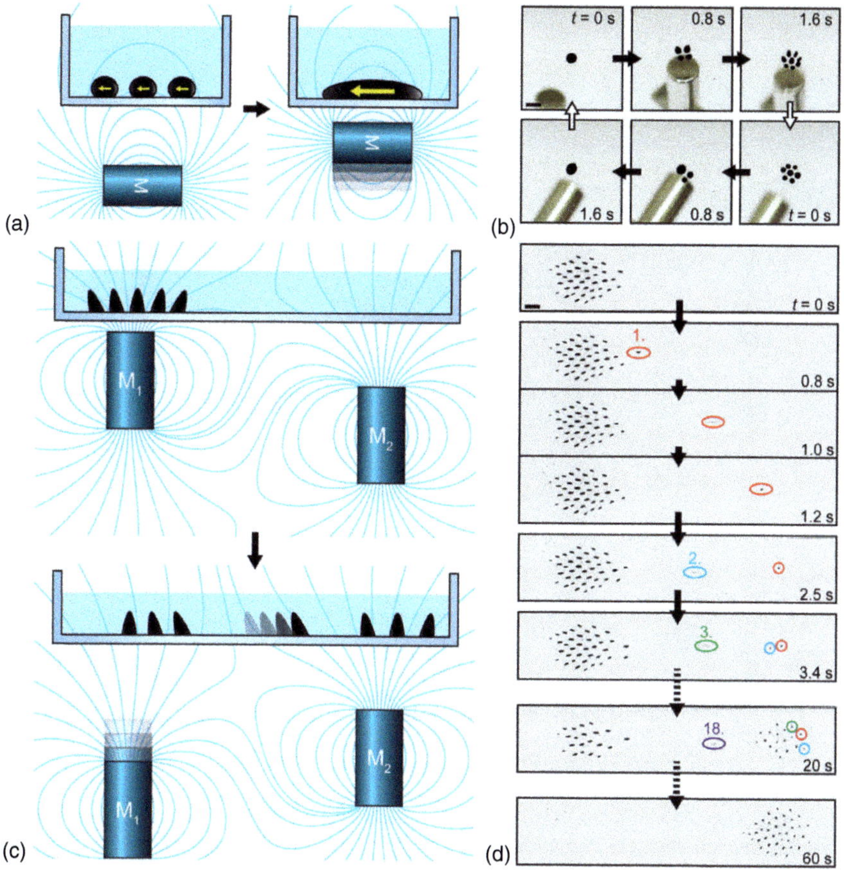


**Figure 6.3** (a) The schematic diagram of field-induced droplet combination. (b) The optical images of splitting (top row) and combining (bottom row) of FDR with a magnetic field. (c) The schematic diagram of sequential transport of FDRs. (d) Top-view images of sequential transport of FDRs. Source: Latikka et al. [49]/Reproduced from John Wiley & Sons, Inc./CC BY 4.0.

The controllable FDRs allow the development of more flexible manipulation in population-based digital microfluidics.

Considering largely changeable sizes of spaces ranging from centimeters to micrometers when FDRs are applied in the human body, Fan et al. designed a multiscale magnetic miniature robot actuation ($M^3RA$) system by combining electromagnetic coils and a spherical permanent magnet in one system, enabling to achieve trans-scale deformation, motion, and scale reconfiguration of FDRs [31]. In this $M^3RA$ system, electromagnetic coils are used to produce a weak background magnetic field ($\boldsymbol{B}_b$), which controls the position and/or orientation of a permanent magnet to generate a strong actuation magnetic field ($\boldsymbol{B}_a$) with a powerful gradient force and torque in a programmable way. Thus, the ultimate magnetic fields exerted on FDRs are generated by a permanent magnet. The largest $\boldsymbol{B}_a$ strength can reach 300 mT and the highest $\boldsymbol{B}_a$ gradient exceed $50\,T\,m^{-1}$ with this $M^3RA$ system. Even

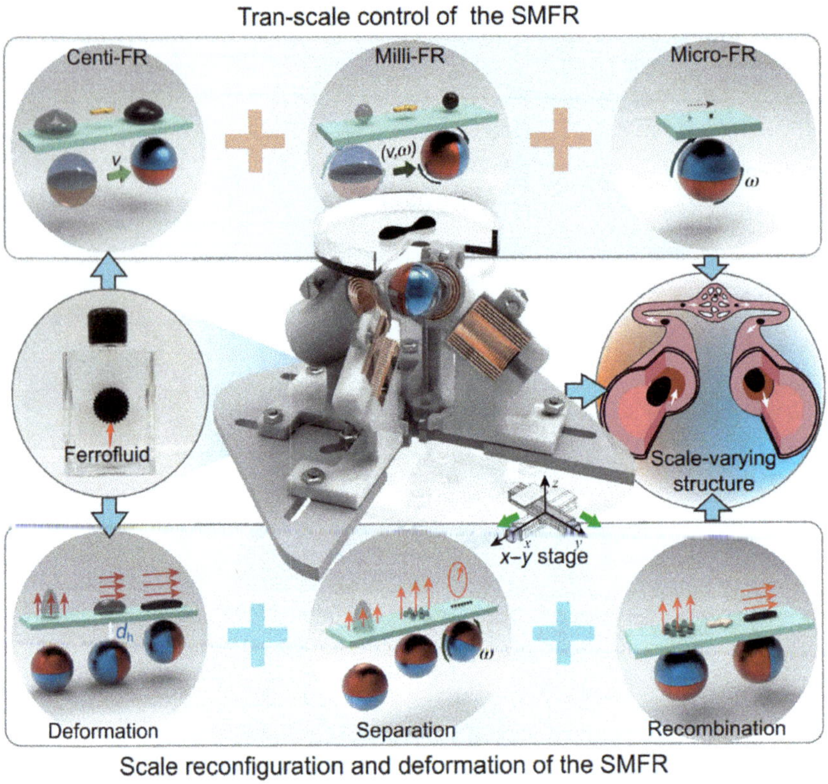


**Figure 6.4** Overview of trans-scale navigation and scale reconfiguration strategies of FDRs. Source: Fan et al. [31]/American Association for the Advancement of Science. (AAAS).

though the magnet is far away from FDRs (the distance between them [$d_h$] ranges from 10 to 20 cm), the $\boldsymbol{B}_a$ can still reach 30 mT or 2 T m$^{-1}$. The structural design of the M$^3$RA system is displayed in Figure 6.4. Four orthogonal solenoids, located at each bottom vertex of the cubic space, are arranged to point to the center of the spherical permanent magnet. The magnet is supported by a goblet-like cover to make its center always located at the focus of four coils. A 3D mobile platform increases the system's flexibility and allows programming magnetic fields in a large working space.

By controlling the magnet with its N, S-pole in the horizontal direction, the FDR is gradually stretched with $d_h$ decreasing (Figure 6.5a–c). The deformation degree is defined as the ratio (a/b) of the long axis (a) and short axis (b). Clearly, FDRs with different scales, including centimeter-scale (centi-FRs), millimeter-scale (milli-FRs), and microscale (micro-FRs), exhibit varying deformation degree as a function of $d_h$. The micro-FR with limited size displays slighter deformability than milli-FR which can stretch a quarter of its initial radius. The highest deformation degree of milli-FR is even larger than centi-FR. With the increase of $d_h$, the deformation of centi-FR first increases to a large value and then decreases. This is because degenerate deformation occurs when the magnet is close enough to the centi-FR. The stretching angle

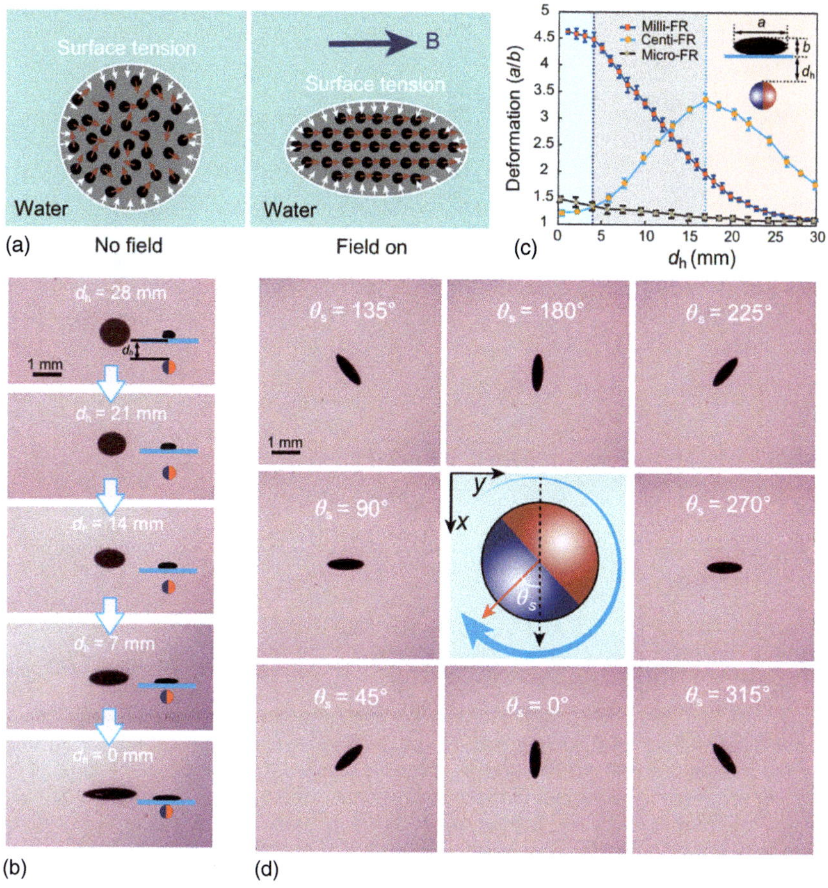


**Figure 6.5** (a) The schematic diagram of the stretch mechanism of the FDR under an external magnetic field. (b) Deformation magnitude control of a milli-FR. (c) The deformation of the FDRs with decreasing distance between the robot and the magnet. (d) Stretching angle control of a milli-FR. Source: Fan et al. [31]/American Association for the Advancement of Science. (AAAS).

of milli-FR is also indirectly regulated by adjusting the orientation of the permanent magnet (Figure 6.5d). The average and maximum error during this experiment is 1.32° and 5.7°, respectively, ensuring the stable control of FDRs.

The splitting and recombination of centi-FR and milli-FR are also achieved by this $M^3RA$ system. As mentioned above, the fission and fusion of centi-FR are similarly achieved by vertically applying the spherical magnet from the bottom of centi-FR and then rotating it. The high magnetic strength and vertical gradient induce repulsive forces within the inner parts of the centi-FR, causing them to separate from each other until the centi-FR is split into several milli-FRs. The resulting milli-FRs can be recombined together by rotating the magnet to make them align, collide, and merge into a centi-FR (Figure 6.6a,c). However, the above method cannot be used to continuously split milli-FR. The small volume of milli-FR only creates weak

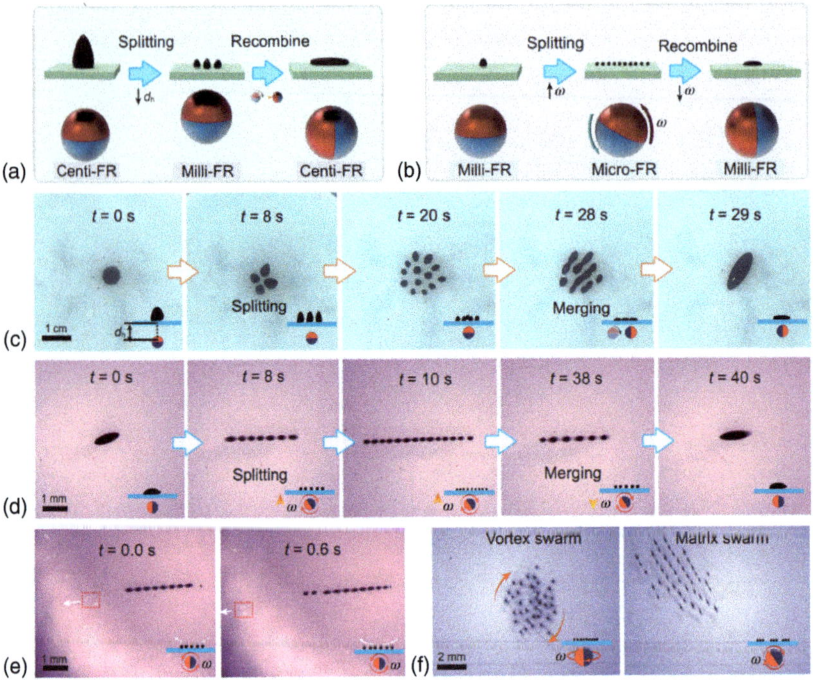


**Figure 6.6** (a) The schematic diagram for the splitting of centi-FR into milli-FRs. (b) The schematic diagram for the splitting of milli-FR into micro-FRs. (c) The optical images for the splitting process of centi-FR. (d) The optical images for the splitting and recombination of milli-FR. (e) Large-scope controlled fluid induced by chain-like micro-FR assemblies. (f) Emergency and control of the resulting vortex swarm and matrix swarm of micro-FRs. Source: Fan et al. [31]/American Association for the Advancement of Science. (AAAS).

repulsive forces, which is hard to overcome the surface tension and enrichment effect of magnetic gradient forces. An intelligent strategy is proposed to further split the milli-FR based on this $M^3RA$ system (Figure 6.6b). When the rotation speed of $\boldsymbol{B}_a$ is fast enough, the milli-FR is split into multiple micro-FRs. This is because of the anisotropic shear forces from the end to the top of milli-FR with an ellipsoid-like shape induced by $\boldsymbol{B}_a$. The shear forces at the end are larger than the shear forces at the top. Thus, the milli-FR breaks up from the end and is further split many times with an increasing rotation frequency of $\boldsymbol{B}_a$ from 1 to 120 Hz. Figure 6.6d shows that a milli-FR is gradually divided into 14 micro-FRs one by one to form a chain, which permits a large scope of controllable fluid motion along the chain direction, realizing efficient fluid stirring and noncontact cargo manipulation (Figure 6.6e). Apart from controlling micro-FRs in a chain-like state, a more practical and efficient driving mode of swarm control is proposed, achieving the matrix pattern and vortex pattern control by adjusting the rotating axis of $\boldsymbol{B}_a$ (Figure 6.6f). These micro-FRs are capable of being recombined into milli-FR by decreasing the rotation frequency from 120 to 1 Hz (Figure 6.6d). Therefore, the on-demand scale reconfiguration of FDRs has been achieved by simply turning the actuation strength and rotation frequency. The efficient motion control of centi-FR, milli-FR, and micro-FR is also achieved by this M$^3$RA system. The micro-FR is more sensitive to magnetic torque than magnetic force, so a rotating magnetic field is applied to actuate these robots in a rolling mode. Additionally, the micro-FR has been accurately controlled to track circular path and Chinese character-like path by controlling its rotating velocity and optimizing its position. The magnetic force and magnetic torque actuation of milli-FR are restricted by small actuation distance and limited reachable area, respectively. The most efficient method for actuating the milli- with an extensive working range is to apply magnetic force and magnetic torque simultaneously. In this way, the milli-FR can pass through a large and complex maze more flexibly. The centi-FR displays poor actuation efficiency under a rotating magnetic field, and its droplet-like shape is easily destroyed by a dynamic magnetic field. Therefore, magnetic torque is not suitable for actuating the centi-FR. However, the massive magnetic force of centi-FR brings by its large volume always enables more efficient motion, indicating that magnetic gradient force control of centi-FRs is preferable. The drag effect of the magnetic gradient force and the shape-shifting effect of the magnetic torque have been employed to enable a centi-FR squeezing through the narrow channel and crossing the barrier up.

#### 6.3.1.2 Magnetic Fields Induced by Electromagnetic Coils

Compared with permanent magnets that allow more complex shapes, the electromagnetic coils bring more flexibility in adjusting magnetic fields. In the previous studies, simple deformations and motion control of ferrofluids have been individually realized by experiments, simulations, and theoretical calculations, having potential in microfluids and assembly [50–56]. However, these researches mainly focus on the physical properties of ferrofluids and have no concern on their applications in robots. Considering these special properties of ferrofluids, Reza et al. combined both deformation and motion control in one control scheme with either four or eight electromagnetic coils system, developing miniature and deformable FDR [38]. The FDR enables higher deformation dexterity and is capable of splitting, recombination, squeezing through narrow gaps (Figure 6.7). The simultaneous control of

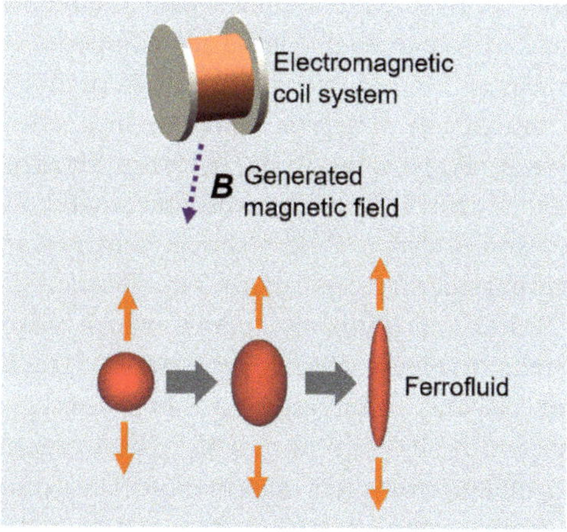


**Figure 6.7** Conceptualization of the electromagnetic coils for generating controlled magnetic fields to produce functional maneuvers.

deformation and motion is demonstrated. The FDR is elongated first along the horizontal direction and then along the vertical direction, while the FDR is controlled to move following a lemniscate path. The experiment is conducted twice to verify excellent controlling stability. Clearly, FDR moves the nearly same lemniscate path in these two experiments. Additionally, the experimental mean and $\pm 1$ standard error paths are also identical to specified path. When shape control scheme that produces a short pulse is applied at a specified angle, the FDR is first stretched to a large length and then split into two smaller FDRs. After applying shape control at the same angle but at a lower stretching length, these two smaller FDRs recombine via the mutual attraction and surface tension. The ability of FDR to squeeze through a narrow channel is accomplished by the closed-loop position control scheme combining position control and automated pulsing shape control. This scheme allows the movement of FDR to approach the channel entrance and squeeze itself through the channel. Certainly, the FDRs have the possibility to pass through the narrow channel by increasing its stretching length without pulsing shape control. However, the large stretching aspect ratio and resisting forces bring by the sharp edges of channel may cause FDR splitting.

Fan et al. develop a set of electromagnetic coils array with planar distribution to program the external magnetic fields on demand, which precisely realizes the splitting, merging, and controlled motion of FDRs (Figure 6.8a) [29]. The splitting behavior of an FDR is enabled by applying electric currents in the same direction in two coils located beneath the substrate. The splitting force on the FDR is determined by the spatial gradient of the induced magnetic potential energy. The magnetic force distribution in the deformation shape of FDR is illustrated using a simplified 2D model (Figure 6.8b,c). This simulation indicates that the magnetic forces exerted on an FDR point to the two regions with the minimum magnetic potential energy. Therefore, the elongation of FDR increases gradually due to the opposite magnetic forces and ends up when FDR is split into two smaller FDRs (Figure 6.8d). The merging behavior of two FDRs is realized by applying currents in opposite directions in two coils. Different from splitting, only one minimum magnetic potential energy exists in this case, so that the magnetic forces all point to this region, which drives two FDRs to approach each other, collide, and merge by the magnetic force and surface tension (Figure 6.8e). Superior to all the current methods, the volume ratios ($V_a/V_b$) of two smaller FDRs produced in the splitting process can be well controlled by tuning the actuation magnetic field ratios ($\boldsymbol{B}_a/\boldsymbol{B}_b$) in two coils. When the $\boldsymbol{B}_a/\boldsymbol{B}_b \geq 0.4$, the $V_a/V_b$ increases linearly as $\boldsymbol{B}_a/\boldsymbol{B}_b$ increases (Figure 6.8f). This allows an FDR to split into multiple smaller FDRs with required sizes, which benefits creating a team of FDRs for cooperative tasks or controlling microreactions in chemical engineering applications. The deformation of FDRs with different sizes (Volume: 50, 100, and 300 μl) is explored, which clearly increases in varying magnetic field strength (Figure 6.8g). For the largest FDR, its deformation is restricted by the size of coils and the largest deformation appears when the applied voltage in two coils exceeds 15 V. With the size of FDRs decreasing, the surface tension becomes more dominant over their deformation. The optimized sizes of FDRs range from 62 to 250 μl, which allows the maximum deformation with the minimum volume of FDRs.

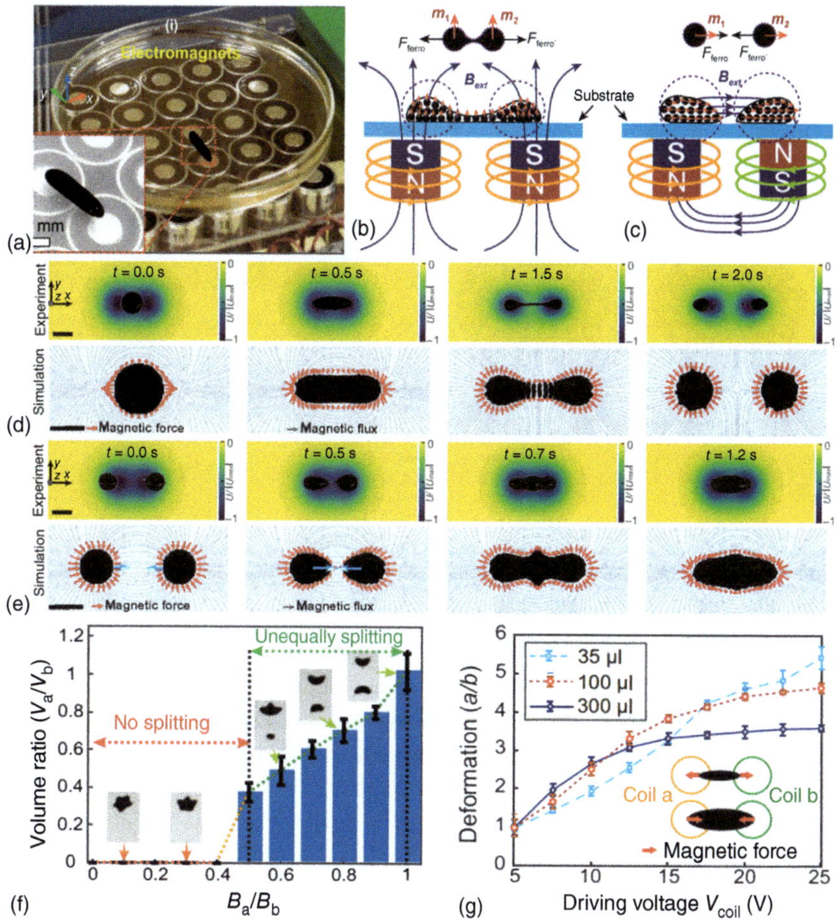

**Figure 6.8** (a) Experimental setup of a 2D electromagnet array for generating spatiotemporally distributed external magnetic fields. Illustration of the effect of interactive magnetic forces in the (b) splitting and (c) merging process of FDRs. Experimental and simulated time frames of the controllable (d) splitting and (e) merging behavior of an FDR. (f) The volume ratio ($V_a/V_b$) of two split droplets as a function of the relative magnetic field strength ($\boldsymbol{B}_a/\boldsymbol{B}_b$). (g) The deformation of an FDR as a function of different driving voltages. Source: Fan et al. [29]/Reproduced from National Academy of Science.

The motion control of an FDR is achieved by systematically regulating the current in neighbor coils, which shifts the minimum magnetic potential energy, enabling the controllable navigation of FDR (Figure 6.9a,b). By combining the splitting, merging, and motion of FDR, a programmable case is exhibited. The FDR first navigates along a spiral trajectory, and then splits into four smaller FDRs. Next, these smaller FDRs follow a programmable trajectory independently. Finally, smaller FDRs gather at the initial position and form a large FDR (Figure 6.9c).

The above-mentioned deformation behavior of FDRs is induced in the ununiform magnetic fields whatever produced by permanent magnets or electromagnetic

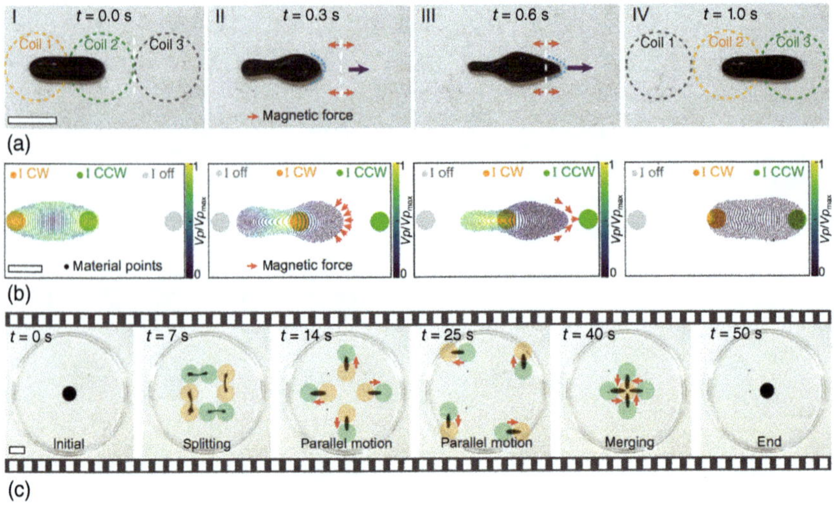


**Figure 6.9** (a) Snapshots of the dynamic motion of an FDR actuated in the double-coil mode. (b) Simulated velocity of an FDR under the spatiotemporal magnetic pulling forces. (c) Sequential snapshots of the splitting, coordinated navigation, and merging behaviors of FDRs. Source: Fan et al. [29]/Reproduced from National Academy of Science.

coils. Fan et al. investigate the deformation degree, motion control, alignment, and dispersion of FDRs in the uniform magnetic fields induced by a 3D Helmholtz coil system [30]. The deformation degree of FDR increases gradually with the magnetic field strength increasing when a uniform magnetic field is applied horizontally (Figure 6.10a). The rotating motion of FDR is achieved under the action of a rotating magnetic field circularly polarized in the (x, z) plane (Figure 6.10b). This is because of the cooperative effect of varying magnetic torques on the suspended magnetic nanoparticles generated by a rotating magnetic field and viscous coupling between particles and carriers of FDR. The frequency of the rotating magnetic field has an obvious influence on the motion velocity of FDR. The velocity of the FDR keeps an almost linear relationship with the rotating frequency if the frequency is smaller than the key frequency ($f_{\text{key}}^{\text{m}}$), while its speed reduces gradually if the frequency is larger than $f_{\text{key}}^{\text{m}}$ (Figure 6.10c). With the velocity of FDR increasing, the liquid-induced viscous torque is beyond the maximum torque produced by the magnetic field, which results in the asynchrony between the FDR and the magnetic field, further causing the reduction of velocity. The navigation velocity and direction of FDR is well controlled, and a complex taichi-shaped path has been achieved with an average tracking error of less than 7 μm, demonstrating its outstanding propulsion performance (Figure 6.10d).

Additionally, the surface tension of FDRs is adjusted to make them cannot merge. Thus, the alignment of several dispersed FDRs is automatically achieved by linking FDRs together to form a microstick when a uniform magnetic field or a rotating magnetic field parallel to the $(x, y)$ plane is applied (Figure 6.11a,b). Moreover, several gathered FDRs are decomposed into individual FDRs if the magnetic field perpendicular to the $(x, y)$ plane is applied. The deformation behaviors of FDRs are clearly

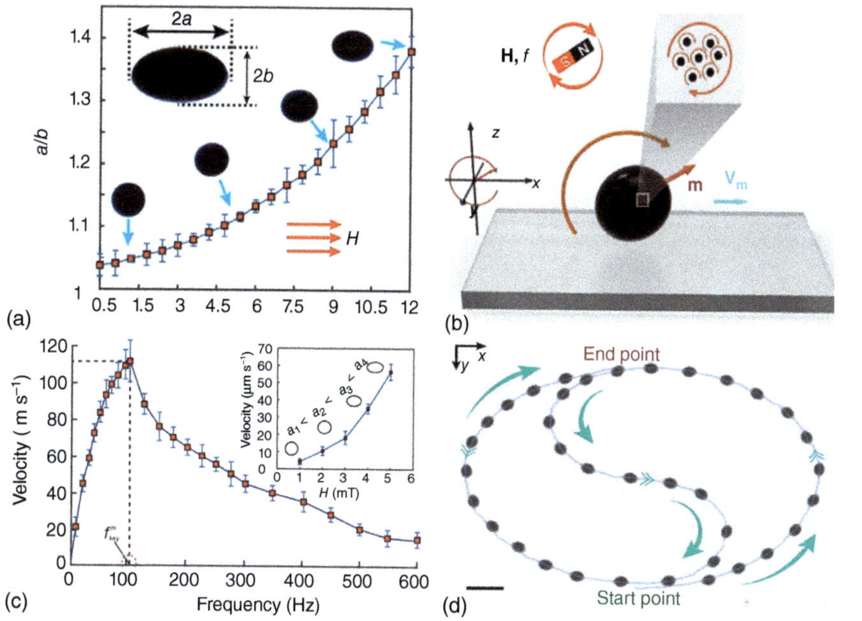


**Figure 6.10** (a) The deformability (a/b) of the FDR with increasing magnetic field strength. (b) The schematic diagram of the controllable motion of the FDR. (c) Average velocity of the FDR under oscillating magnetic field with different frequency. (d) The taichi-shaped path followed by the FDR. Source: Fan et al. [30]/Reproduced from John Wiley & Sons, Inc.

influenced by different magnetic fields. The deformation behavior of an as-formed microstick assembled by several FDRs is further investigated with the frequency of the rotating magnetic field increasing (Figure 6.11c,d). The microstick moves forward in a rolling state when a low-frequency magnetic field polarized in the $(x, z)$ plane is applied. The actuation force exerted on the microstick is produced by the alternating contact between its two heads and the surface of the substrate. The movement velocity of the microstick is far beyond a single FDR due to its larger step length in one turn at the same magnetic field. Additionally, its velocity increases linearly until the frequency is larger than $f_{\text{key}}^{\text{m}}$, where the microstick breaks into several parts. As the velocity of the microstick increases, the magnetic torque generated on the microstick cannot balance the increasing viscous resistance from the fluid. This results in the break of the stick for reducing the resistance. Subsequently, the divided parts of the microstick attract and reassemble again on the substrate to form a structure of micropie that is unfavorable to resistance. The velocity of micropie appears to a sudden change, which is explained by the reduction of displacement in one rotation because its length becomes smaller. The microstick can perform as a microtrain if a very large frequency of rotating magnetic field is directly applied instead of increasing it gradually. These FDRs included in the microtrain rotate asynchronously and individually between each other, resulting from the produced small torque, which is insufficient to overcome the viscous resistance to maintain the behavior of the microstick. In this case, every FDR only rotates around its

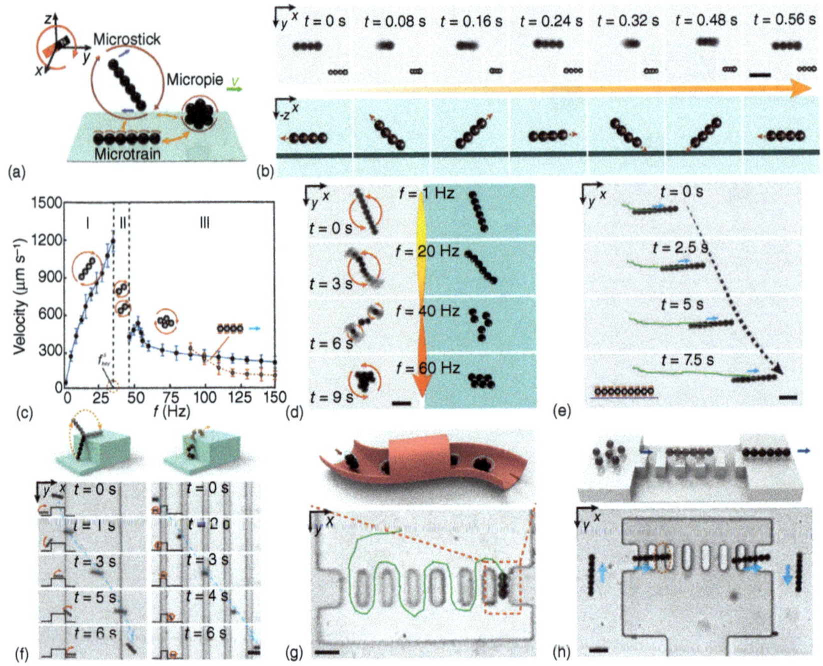

**Figure 6.11** (a) The schematic diagram of forming microstick, micropie, and microtrain in a rotating magnetic field. (b) The experimental (top view) and simulation (side view) results of a microstick in a motion cycle, respectively. (c) The average speed and formation of FDRs assemblies as a function of magnetic field frequency. (d) The experimental (left) and simulation (right) reshaping process of a microstick under increasing rotating frequency. (e) The motion of the microtrain. (f) Experimental processes of microsticks climbing vertical walls. (g) A micropie shuttles back and forth in a sealed microfluidic chip. (h) Snapshots of a microtrain to pass through gullies with deep grooves. Source: Fan et al. [30]/Reproduced from John Wiley & Sons, Inc.

geometric center on the substrate, follows its former FDR like a train with multiple cars, and generates attractions in its travel direction (Figure 6.11e). Just like the microstick and micropie, the velocity of microtrain reduces gradually when the frequency continues to increase. Based on the above motion behaviors of microstick, micropie, and microtrain, their capabilities of passing through different obstacles are further explored. The microstick can easily cross vertical obstacles on a 3D surface (Figure 6.11f). When the height of the obstacle is slightly higher than the length of the microstick, it will be jumped over with rotating less than a circle. However, when the height of the microstick is much lower than the height of the obstacle, the microstick passes through a higher obstacle with several circles of rotating. In this case, the elevated height of every circle of rotating accumulates continuously until the microstick reaches the top of the obstacle, and then it switches into a rolling mode to move forward. The interaction force between the two ends of the microstick and the wall of the obstacle provides climbing capability for the microstick. If the rotating speed of the microstick is fast enough, it cannot fall off from the wall by overcoming gravity. Obviously, the microstick has great potential to cross the vertical obstacle in the open environment. Owing to the small size of the micropie, it shows a special capability to pass through a narrow and sealed channel without collisions (Figure 6.11g). The stronger anti-flow ability is also explored, which is expected to resist blood flow. Additionally, the microtrain can cross surfaces with a series of adjacent deep grooves to achieve controllable locomotion. In the movement process, the microtrain cannot fall into these grooves due to its long and stable structure maintained by mutual magnetic attractions of FDRs (Figure 6.11h). Interestingly, the behaviors of these FDRs in microtrain behave like those of army ants that connect their bodies to form a bridge for passing through a gap, which indicates that the microtrain is strongly adaptable to complex environments.

Li Zhang's group further studies the deformation behavior of individual FDR under magnetic torque generated by a 3D Helmholtz coil system. By varying the types, magnitude, and direction of the external magnetic field, the stretching, jumping, splitting, rotating, tumbling, kayaking, and wobbling of FDR have been reported [27]. The FDR exhibits two modes of stretching and jumping when the one-dimensional (1D) oscillating magnetic field with different directions is applied. FDR (initial diameter: 2 mm) is stretched back and forth in situ when a 1D horizontal oscillating magnetic field is applied in the $x$–$y$ plane (upper part of Figure 6.12a). The stretching degree of FDR is dependent on the strength of the external magnetic field. The maximum length reaches to 7.5 mm at 9 mT with a corresponding aspect ratio of 9. Besides, if the pitch angle which is defined as the angle between the substrate and the FDR is increased, the FDR could generate translational motion in a reciprocal stretching pattern. The movement velocity of FDR first increases and then decreases as a function of the increasing frequency of the external magnetic field. Similarly, the FDR still is stretched at low frequency ($f < 50$ Hz) when a 1D vertically oscillating magnetic field is applied in the $x$–$z$ plane (lower part of Figure 6.12a). The maximum aspect ratio of FDR is about 5.5. Interestingly, the jumping behavior appears for FDR when the frequency of the applied magnetic field is larger than 50 Hz. The jumping height of FDR increases with the strength of the magnetic field, and the maximum jumping height can reach 4 mm at 9 mT ($f$ = 100 Hz). The jumping mechanism of FDR is explained that the high-frequency magnetic field induces the magnetization of FDR to form N–S poles, which cannot be changed quickly enough when the direction of the alternating magnetic field is reversed, so a repulsive force is exerted on FDR to make it jump up. The directional jumping is also achieved by changing the pitch angle of the magnetic field. The fission and fusion behaviors of FDR are achieved in the 2D oscillating and 3D wavy magnetic fields, respectively, which benefit from the magnetic responsiveness and fluidity of FDR. Actuated by the 2D oscillating magnetic field, the FDR (initial diameter: 2 mm) is split into eight smaller FDRs along a line (upper part of Figure 6.12b). The line orientation is controlled by the direction of the oscillating magnetic field. In this splitting process, the magnetic field first transforms FDR into a shuttle shape, which further starts to oscillate with the action of magnetic torque. As the oscillating frequency increases, the FDR cannot maintain its integrity and begins to split when the increasing viscous resistance from liquid is larger than the surface tension of the FDR. The fusion

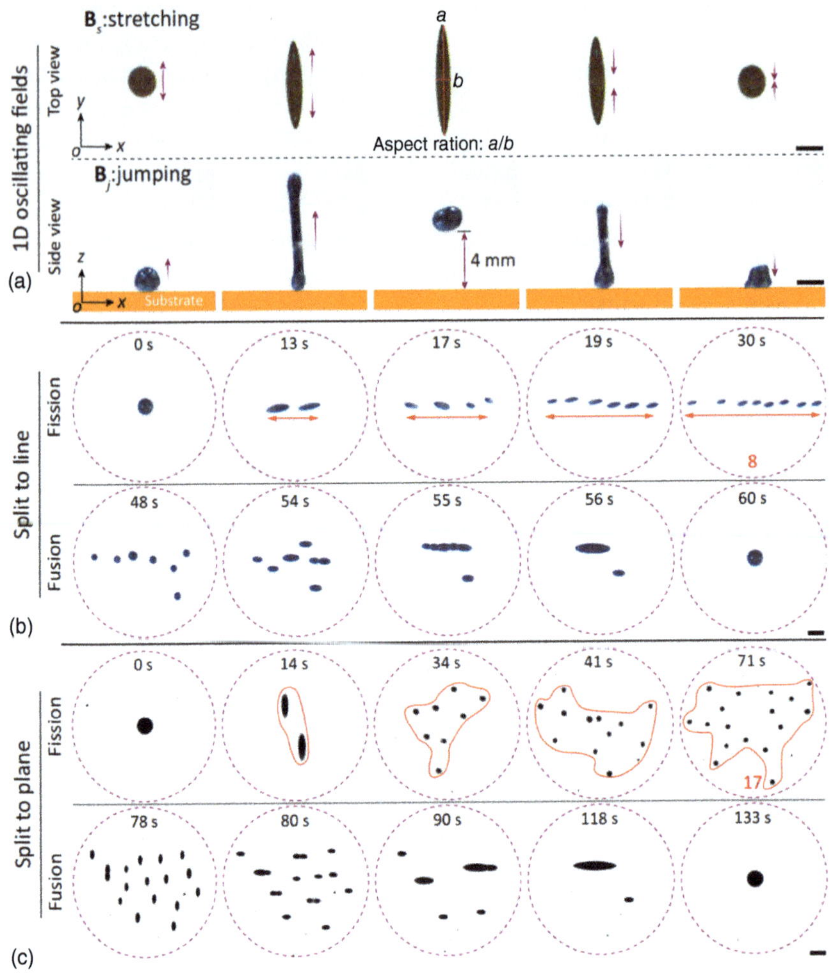


**Figure 6.12** (a) FDRs exhibiting stretching and jumping behaviors under 1D oscillating fields. The fission and fusion of FDR under (b) a 2D oscillating magnetic field and (c) a 3D wavy magnetic field. Source: Sun et al. [27]/Reproduced from Springer Nature/CC BY 4.0.

of smaller FDRs appears as the oscillating frequency decreases. In this case, the magnetic dipolar force on FDR occupies its dynamic behavior. The mutual magnetic attraction makes smaller FDRs approach each other and contact closely, as the interface flattens and the fluid film drains out (lower part of Figure 6.12b). With van der Waals intermolecular forces dominating, the fluid film between smaller FDRs is eventually ruptured, causing their coalescence. To achieve the planar splitting of FDR, a 3D wavy magnetic field composed of a 1D oscillating and a 2D rotating magnetic field is designed, which increases the splitting distance of smaller FDRs and prevents them from reassembling. In this magnetic field, the FDR with an initial diameter of 2 mm is split into 17 smaller FDRs on a plane (upper part of Figure 6.12c). The splitting number of smaller FDRs is controlled by modulating the distribution of the magnetic field. These smaller FDRs are then gradually fused

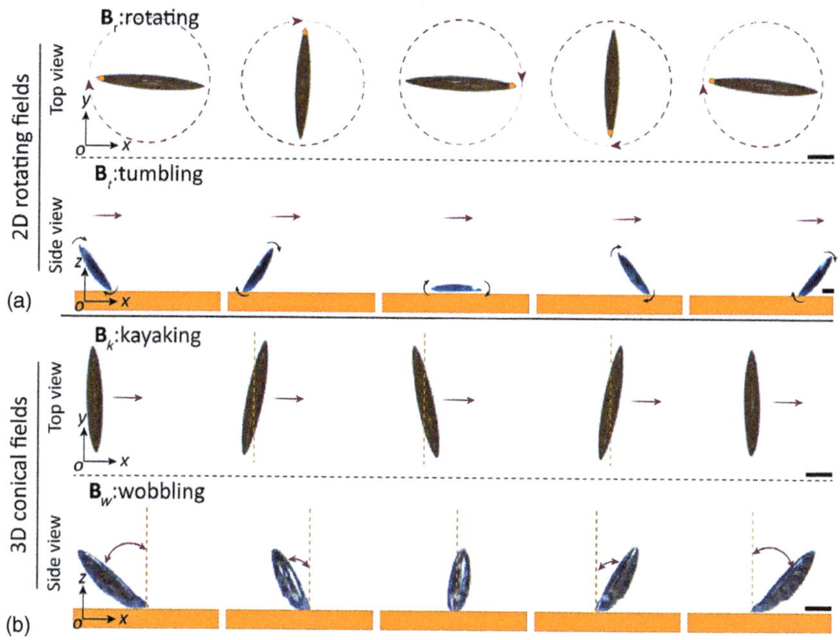


**Figure 6.13** (a) FDRs performing rotating and tumbling motion patterns under 2D rotating fields. (b) FDRs displaying kayaking and wobbling motion modes under 3D conical fields. Source: Sun et al. [27]/Reproduced from Springer Nature/CC BY 4.0.

to recover their initial state based on the same merging mechanism above when the input frequency reduces (lower part of Figure 6.12c). The fusion and fission of FDR enable its scale-reconfiguration capability to pass through narrow spaces with highly variable sizes.

The motion behaviors of rotating and tumbling are achieved when a 2D rotating magnetic field in the $x$–$y$ plane and $x$–$z$ plane is applied, respectively (Figure 6.13a). When a 3D conical magnetic field in the $x$–$y$ plane and $x$–$z$ plane is applied, the FDR exhibits two motion behaviors of kayaking and wobbling (Figure 6.13b). The FDR has high maneuverability using multimodal motion due to its higher degrees of freedom when compared to rigid robots and elastomer-based robots.

It's challenging for FDR to easily negotiate obstacles, textures, and variations in complex environments. Thus, a combination of multiple deformation and multimodal motion is employed to achieve controllable movement in complex artificial (continuous fences and wide gaps) and biological environments (uneven bladder, folded stomach, narrow vessel). The FDR can continuously override an array of three fences (peak-to-peak distance: 6 mm; the height of three peaks is 2.2, 3.6, and 4.5 mm, respectively.) when a directional 1D oscillating magnetic field (the angle of 80° to the $x$-axis) is applied. In the jumping process, there are four stages: elongation, contraction, descent, and landing (Figure 6.14a). The FDR is first stretched along the 80° direction, and then the FDR jumps from the substrate, contracts itself, and reaches the highest point. After that, the FDR starts to fall off under gravity. Lastly, the FDR successfully crosses a fence and lands on the gully. By repeating the

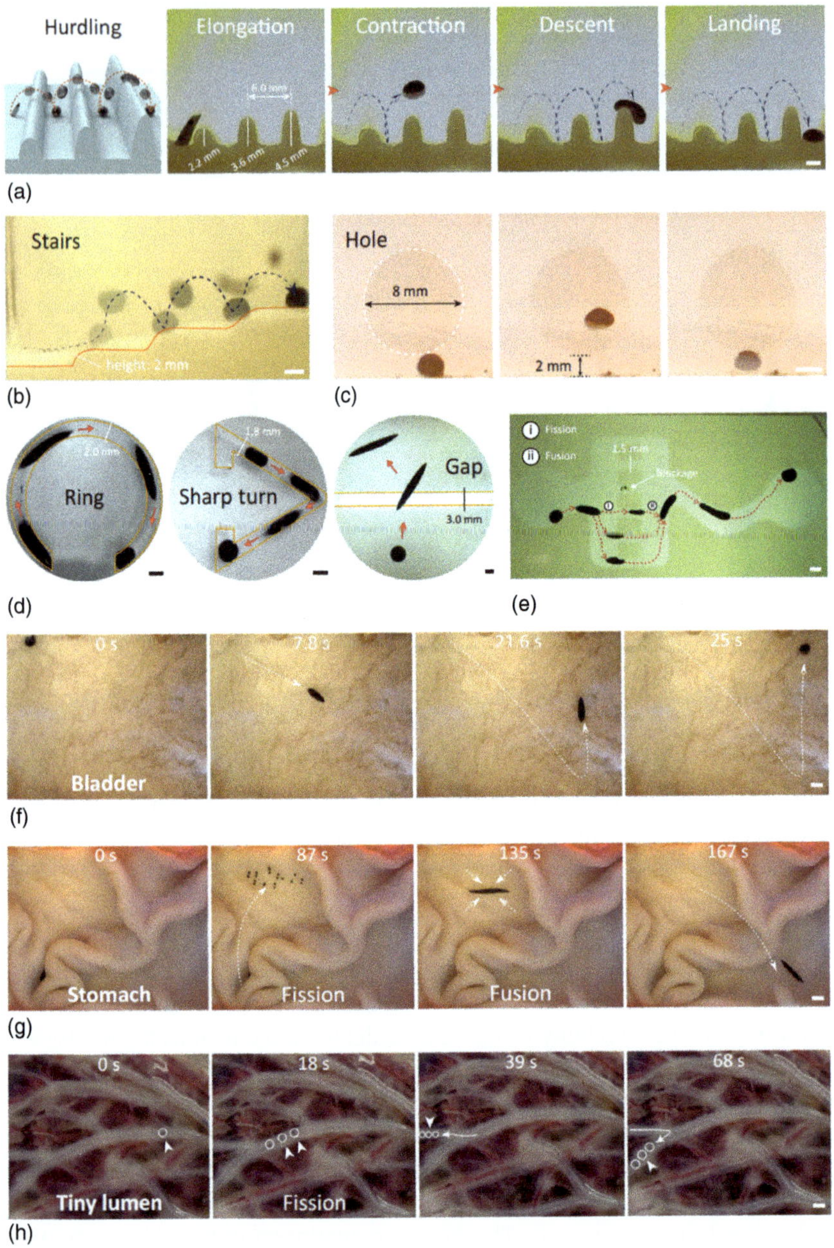


**Figure 6.14** (a) The directional hurdling of FDR over successive obstacles under a 1D oscillating magnetic field. (b) Jumping up the stairs. (c) Jumping over designated holes in the wall. (d) The FDR adaptively deform and actively pass through a circularly curved channel, a sharp turn and a gap. (e) The FDR splits and passes through the comb-constrained channel in a stretching motion mode. (f) The FDR tumbles over the uneven bladder lining. (g) The demonstration of an FDR maneuvering on the folded stomach lining. (h) FDR navigates within the unstructured narrow lumen. Source: Sun et al. [27]/Reproduced from Springer Nature/CC BY 4.0.

jumping process, the FDR reaches the final position. Additionally, the FDR has the ability to jump over three successive steps with each height of 2 mm (Figure 6.14b) and jump into a specified hole (inner diameter: 8 mm; height: 2 mm) above the floor (Figure 6.14c). The navigation capability in the confined space is critical for FDRs to realize their functions in lumens and cavities within the human body. The FDR can pass through the ring-shaped channel and sharp turns using its large deformation capability in a stretching mode (Figure 6.14d). The supporting force of the sidewall and passive adaptability to the shape plays a key role in this process. The FDR also can be actively elongated to span the gap with a width of 3 mm in the kayaking motion pattern. To pass through the comb-like narrow channels (inner diameter: 1.5 mm), the FDR is first split into three smaller FDRs with a 2D oscillating magnetic field, and then a 1D oscillating magnetic field is used to drive smaller FDRs to pass through different channels one by one. This is because the FDR (diameter: 3 mm) is hard to be driven to deform directly through the narrow channel. Eventually, these three smaller FDRs merge and then navigate to the target position (Figure 6.14e).

The navigation capability of FDR in the three real unstructured biological environments of the human body is further explored. The uneven bladder lining has continuous textured structures on its inner wall surface, which require the FDR to move in a controlled tumbling motion on it. Although the slipping behavior happens during navigation, the overall motion direction of FDR is controllable (Figure 6.14f). Unlike the bladder lining, the pig stomach has large folds (height: 10 mm) and many narrow traps on its inner wall. When the FDR is trapped in the recessed area, it is actuated to cross the trap by breaking into multiple smaller FDRs. After that, these smaller FDRs are merged at a 2D rotating magnetic field, and then the merged FDR is transformed into a shuttle shape, which will cross the large folds of the pig stomach by increasing the pitch angle of the rotating magnetic field (Figure 6.14g). The unstructured lumen of human placental vessels has variable inner diameters. The scale-reconfiguration capability of FDR has made it suitable to move inside the vessel. The FDR (diameter: 1.0 mm) first enters the main vessel with a diameter of 1.5 mm in the rotational translational motion, and then it is split into three smaller FDRs (diameter: 0.6 mm), allowing them to enter the branches of the vessel (inner diameter: 0.8 mm). After executing tasks, these three smaller FDRs are controlled to move back and return to the main vessel (Figure 6.14h). The above results demonstrate FDR can navigate freely in the biological tissues and lumens, holding great potential for biomedical applications.

After completing this work, the same group further proposes a coalescent collective system to achieve the splitting and coalescence properties of FDR in horizontal and vertical gravity-resistant collectives patterns [28]. The time-varying magnetic fields generated by the 3D Helmholtz coils system are used to induce the deformation behaviors of FDR. As shown in Figure 6.15a and the equation [1, 2], a dynamic magnetic field with continuously varying rotation planes (**B**h) is applied to split one FDR into multiple smaller FDRs and reconfigure them into horizontal swarms.

$$\boldsymbol{B}\text{h} = \begin{bmatrix} Bm\ \cos(\gamma)\cos(\alpha)\cos(2\pi f1t) - Bm\ \sin(\alpha)\sin(2\pi f1t) \\ Bm\ \cos(\gamma)\sin(\alpha)\cos(2\pi f1t) + Bm\ \cos(\alpha)\sin(2\pi f1t) \\ Bm\ \sin(\gamma)\cos(2\pi f1t) \end{bmatrix} \tag{6.1}$$

$$\alpha = \theta ceil(f2t) \tag{6.2}$$

In this equation, the rotating frequency ($f_1$) and switching frequency ($f_2$) determine the rotation speed and the switching speed of the magnetic field, respectively. The influence of the input frequency of the magnetic field on the patterns of the horizontal FDR swarm is obvious and displayed in the phase diagram (Figure 6.15b). By rationally adjusting the input frequency, irregular collectives (region 1), vortex-like patterns (region 2), chain-like patterns (region 3), and crystal-like patterns (region 4) have been achieved and well controlled (Figure 6.15c). The vortex-like pattern and crystal-like pattern can expand and contract periodically under the dynamic rotating magnetic field (Figure 6.15d). In the expansion or contraction process, these smaller FDRs inside and outside the vortex-like collectives migrate slowly and fast, respectively, which is verified by the distribution of the flow field (Figure 6.15e). The

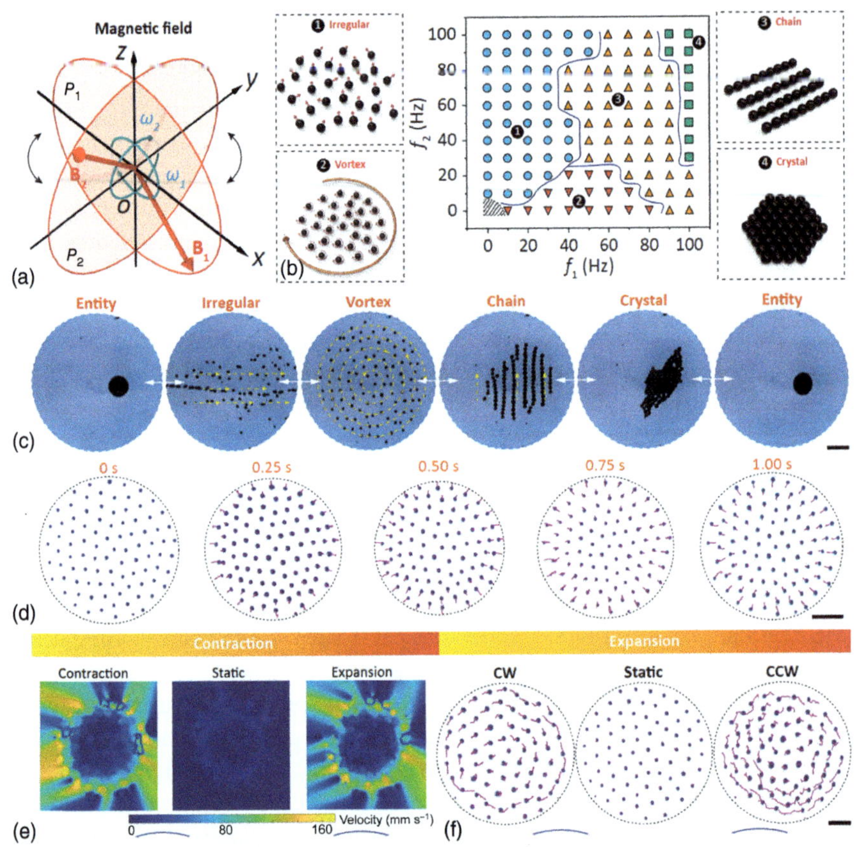


**Figure 6.15** (a) The schematic diagram of a dynamic magnetic field. (b) The relationship between collective patterns and the frequency. (c) The transition between different morphologies of horizontal droplet collectives. (d) Periodic expansion and contraction of vortex-like droplet collectives. (e) Fluid field distribution induced by droplet collectives. (f) The relationship between rotational direction of vortex-like droplet collectives and the input frequency value. Source: Sun et al. [28]/Reproduced from John Wiley & Sons, Inc.

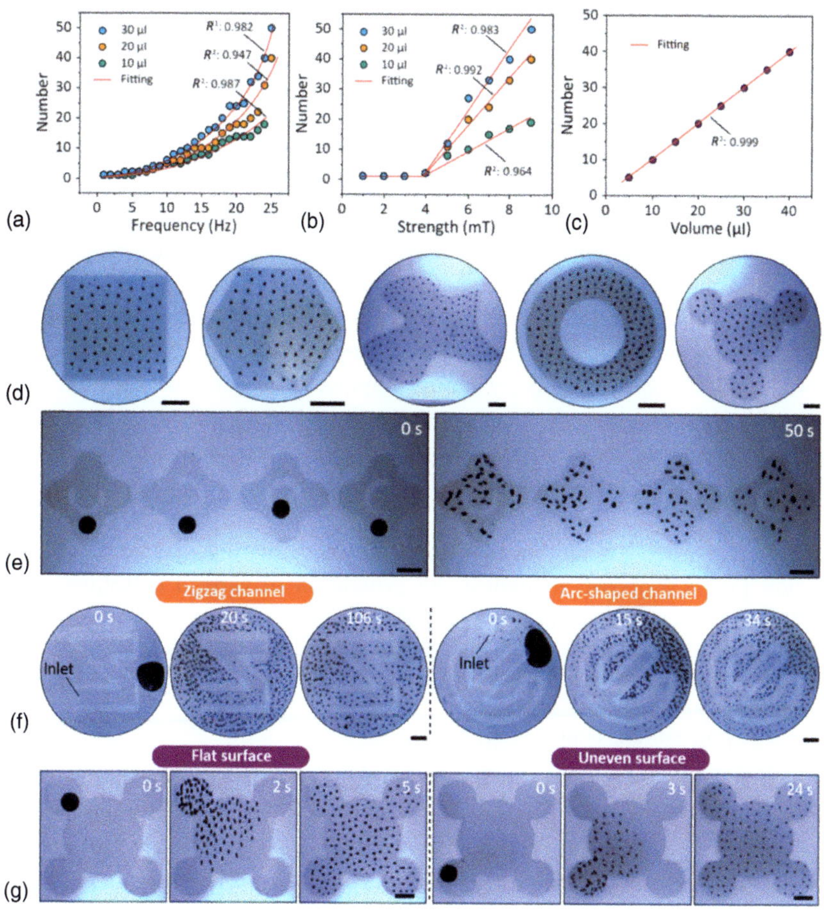


**Figure 6.16** The splitting numbers of FDR as a function of (a) the frequency f1, (b) the magnetic field strength, and (c) the volume of FDR. (d) The FDRs uniformly spread across various planar patterns. (e) The FDRs bypass obstacles and cover four different target areas simultaneously. (f) The smaller FDRs explore the zigzag channel and arc-shaped channel. (g) The FDRs self-assemble into arrays on flat and uneven surfaces. Source: Sun et al. [28]/Reproduced from John Wiley & Sons, Inc.

rotating direction of vortex-like collectives has a close relationship with frequency $f_1$ when the frequency $f_2$ is fixed to be 90 Hz. The vortex-like swarm rotates clockwise at a frequency $f_1$ of 20–30 Hz, while it rotates counterclockwise when the frequency $f_1$ is between 30 and 40 Hz (Figure 6.15f).

The splitting number of FDR is also well investigated as a function of the frequency $f_1$, the magnetic field strength, and the volume of FDR (Figure 6.16a–c). The splitting number increases exponentially with the frequency $f_1$ increasing. The splitting behavior of FDR cannot happen when the magnetic field strength is lower than 4 mT, while the splitting number is proportional to the magnetic field strength that is larger than 4 mT. In addition, the relationship between the splitting number and the volume of FDR displays a linear relationship with increasing volume. The splitting behaviors of FDR allow it to be distributed in largely confined regions with different shapes, including quadrilateral region, hexagonal region, irregularly shaped region, circular region, and a coupled region of four circles (Figure 6.16d). The distribution throughout the available area is quick and uniform, and these smaller FDRs spontaneously form periodic 2D arrays due to the magnetic repulsion. Upon forming the array patterns, they can be maintained under a vertical static magnetic field. The distance between smaller FDRs increases with the magnetic field strength. The mutual repulsion of smaller FDRs allows them to reach restricted regions by bypassing obstacles and covering all areas (Figure 6.16e). The zigzag and arc-shaped channels are also automatically explored, and then they assemble into periodic patterns (Figure 6.16f). The special property can be employed to reconstruct an unknown region with X-ray fluoroscopy imaging of distributed smaller FDRs. The surface roughness has an influence on the distribution time. It takes more time on uneven substrates when compared to flat terrain (Figure 6.16g).

The formation of FDR collectives along the vertical direction is a great challenge due to gravity. To address this problem, an oscillating magnetic field $\boldsymbol{B}\mathrm{v}$ (equation [3]) is proposed to induce the generation of vertical collectives and overcome the effects of gravity (Figure 6.17a).

$$\boldsymbol{B}\mathrm{v} = \begin{bmatrix} A_0 \sin(2\pi f\, t)\cos\theta + C_0 \sin\theta \\ 0 \\ A_0 \sin(2\pi f\, t)\sin\theta + C_0 \cos\theta \end{bmatrix} \tag{6.3}$$

In this equation, $A_0$ and $C_0$ represent the sinusoidal signal intensity and the constant component intensity, respectively. The amplitude ratio is proposed and defined as $\lambda = A_0/C_0$. The influence of $\lambda$ and the magnetic field frequency on the vertical patterns is displayed in Figure 6.17b. Clearly, different patterns form in these four specific regions. In region 1, the FDR cannot split and remains ellipsoidal. In region 2, the FDR is split into multiple smaller FDRs that aligns with the direction of the oscillating magnetic field. In region 3, these smaller FDRs stack together to form a vertical layer-upon-layer collective pattern. In region 4, the collective pattern starts to split as the increasing magnetic field strength in the vertical direction, forming several layer-upon-layer structures like vertical chains. As an example, the FDR (diameter: 5 mm) is split into ≈20 smaller FDRs by the oscillating magnetic field, which assembled into a vertical layer-upon-layer swarm with a height of ≈7 mm (Figure 6.17c). The vertical swarm is reconfigured by changing the input amplitude ratio $\lambda$ from 0.5 to 3.0, and the height of the swarm increases significantly from 7 to 14 mm (Figure 6.17d). If $\lambda$ is adjusted to the critical value of 3.0, the vertical swarm is split into multiple columnar collectives. The vertical swarms have been demonstrated to be stable when they are rotated or tilted, while maintaining their overall integrity in the total process (Figure 6.17e,f). No splitting and dropping has appeared, indicating the mutual magnetic attraction between smaller FDRs is sufficient. This special property of vertical swarms may be employed to mix fluids in the microfluidic chips.

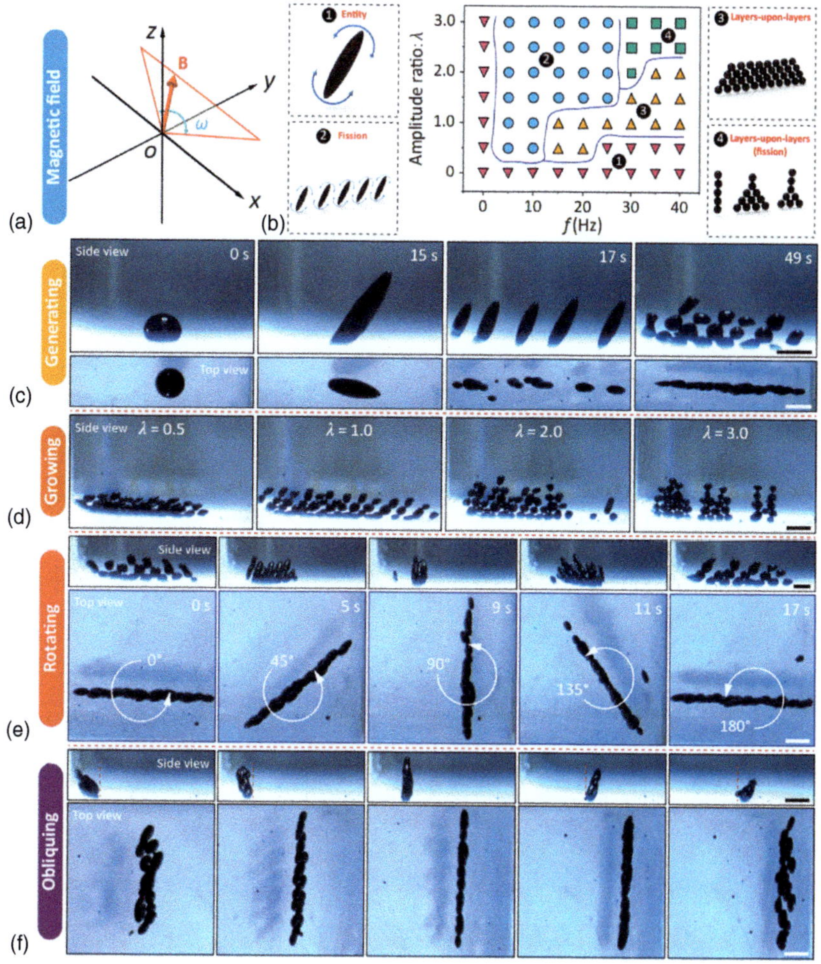


**Figure 6.17** (a) The schematic diagram of the external applied magnetic field. (b) The phase diagram showing the relationship between vertical collective patterns and the frequency. (c) Generation process of vertical FDRs collectives. (d) Reconfiguration of a vertical FDRs collective with growing height. (e) Rotation process and (f) inclination process of a vertical droplet collective. Source: Sun et al. [28]/Reproduced from John Wiley & Sons, Inc.

### 6.3.2 The Influence of FDRs' Parameters on their Deformation Behaviors

The parameters of FDRs consist of volume, viscosity, and the concentration of magnetic nanoparticles and surfactants. The volume of FDRs is closely linked with their deformation behaviors and the choice of different actuating magnetic fields, which has been discussed in detail in the above part. The effect of FDRs with different viscosities on their deformation is exhibited in Figure 6.18 [29]. The deformability degree of FDR is defined as the length ratio ($a/b$) of the long and short axis of the deformed FDR. Two types of high-viscosity and low-viscosity ferrofluids in

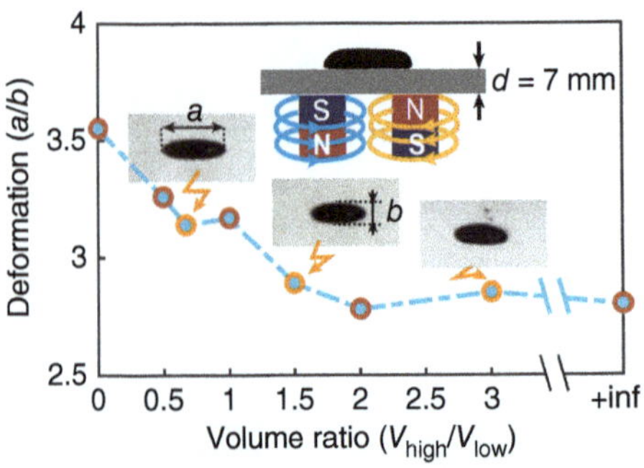


**Figure 6.18** The deformation of an FDR as a function of the mixed viscosity. Source: Fan et al. [29]/Reproduced from National Academy of Science.

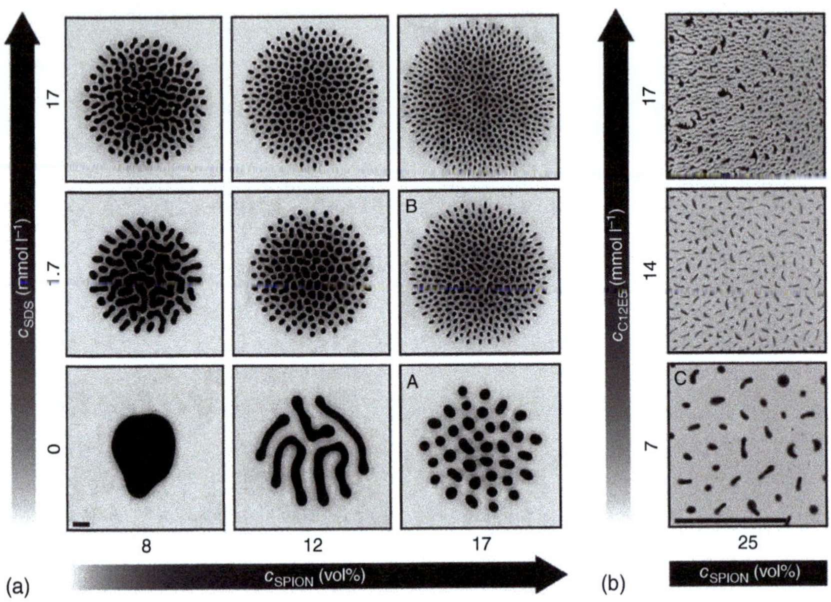


**Figure 6.19** (a) FDRs array created by field-induced splitting for different SDS and SPION concentrations. (b) The splitting of FDRs with different concentrations of $C_{12}E_5$. Source: Latikka et al. [49]/Reproduced from John Wiley & Sons, Inc./CC BY 4.0.

specific volume ratios are mixed to obtain different viscosities, while the total volume keeps identical for different mixing. Clearly, the lower the viscosity, the greater the deformation of FDR. However, the FDR with a higher viscosity displays better stability which is reflected by a smaller Bond number. The Bond number is defined as $\mathbf{B}_0 = (L/\lambda_c)^2$ ($L$: the characteristic length; $\lambda_c$: the capillary length.) (magnetic control of deformation of a ferrofluid droplet in simple shear flow).

The effect of the concentration of magnetic nanoparticles (SPION; 8–25 vol%) and surfactant (SDS; 0–17 mmol l$^{-1}$) on the deformation behaviors of FDR is investigated systematically (Figure 6.19a) [49]. When no surfactant exists, the splitting of FDR cannot occur at a low SPION concentration of 8 vol%, the FDR is split into four flattened droplets arranged in a labyrinthine pattern as the SPION concentration increases to 12 vol%, and the FDR is further split to form discrete and conical droplets array by increasing SPION concentration to 17 vol%. However, the addition of SDS surfactant strongly enhances the splitting capability of FDR even at a low SPION concentration of 8 vol%. The splitting numbers for FDR with a high SPION concentration (17 vol%) increase sevenfold as the SDS concentration increases from 0 to 17 mmol $l^{-1}$. These droplets spontaneously flatten to adopt a dumbbell-like shape. The type of surfactant also has an obvious influence on the deformation of FDR. When the surfactant $C_{12}E_5$ with very low interfacial tension is used, these droplets become smaller and are elongated in the lateral direction. At the high $C_{12}E_5$ concentration of 17 mmol $l^{-1}$, well-defined droplets are replaced by ribbons and complex shapes with wide size distribution (Figure 6.19b).

### 6.3.3 The Influence of Contacting Substrate on the Deformation Behaviors of FDR

Both the surface roughness and wetting behaviors of contacting substrate influence the deformation behaviors of FDRs [27]. The surface roughness is mainly aimed at the biological tissue, which has been stated above. The wetting behaviors of contacting substrate have been demonstrated to determine the initial state and magnetic response of FDR. It is necessary to understand and apply the wetting dynamics of FDRs with different substrates. The contacting angle of FDRs on different substrates as a function of time is characterized to reveal the wetting dynamics (Figure 6.20a).

Without applying an external magnetic field, the FDR nearly maintains a spherical shape and cannot stick to the low wetting hydrogel substrate, while it can adhere to high wettability substrates of glass, resin, polymethyl methacrylate, metal, and silicon. Apart from the silicon, the contacting angles of FDR on other substrates nearly

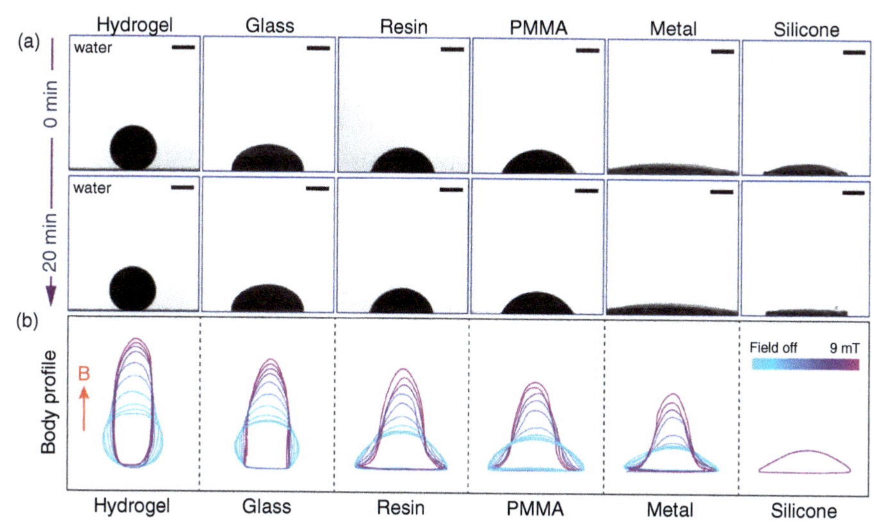


**Figure 6.20** (a) The optical images of FDRs on different substrates at 0 and 20 minutes. (b) The overlapped body profiles of FDRs on different substrates as a function of uniform magnetic field strength ranging from 0 to 9 mT. Source: Sun et al. [27]/Reproduced from Springer Nature/CC BY 4.0.

remain identical as the time increases from 0 to 20 minutes. The contacting angle of FDR changes significantly on the silicon substrate, and its height decreases from 2.0 to 0.5 mm after 20 minutes. This is because the liquid carrier medium is adsorbed by the silicon, forming a layer of magnetic nanoparticles on the silicon substrate. The deformation morphology of FDRs on these different substates changes with the magnetic field ranging from 0 to 9 mT (Figure 6.20b), while the silicon substrate is excluded. The magnetic nanoparticles layer cannot be changed at the low strength magnetic field, which could be detached under a magnetic field strength of 300 mT. Therefore, the choice of substrate is critical for the research of FDRs.

## 6.4 Applications of FDRs

The programmable deformation and motion capabilities of FDRs have enabled them to achieve various applications, including cargo capturing and delivering, object sorting, as well as liquid mixing and transporting. Furthermore, the biocompatibility and small sizes of FDRs allow them to be mainly applied in exploring biomedical applications based on artificial models and biological tissue.

### 6.4.1 Cargo Capturing and Delivering

The FDRs can capture solid and liquid cargo stably with physical confinement even though they pass through complex environments (biological tissue, tortuous ducts). After reaching the target locations, these cargoes are delivered and released on demand. Fan et al. adjust the position and orientation of an external ring-shaped permanent magnet to switch the ring-shape and C-shape of FDR, achieving the capturing, transporting, and releasing of multiple delicate objects efficiently (Figure 6.21a) [29]. The translation and rotation of the permanent magnet are controlled manually. The FDR appears as a ring shape when the magnet is horizontally placed below, while it appears as a C-shape when the magnet is inclined. The FDR is applied for capturing multiple objects if it maintains a C-shape (Figure 6.21b).

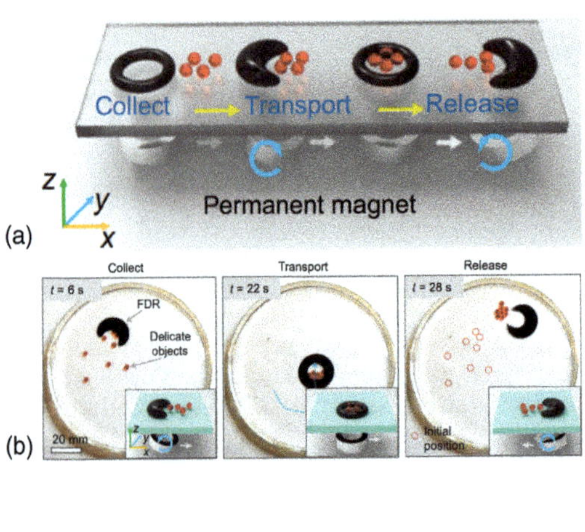


Figure 6.21 (a) The schematic diagram of a ring-shaped FDR for capturing and delivering multiple hydrogel balls; (b) The collecting, transporting, and releasing process of multiple hydrogel balls. Source: Fan et al. [29]/Reproduced from National Academy of Science.

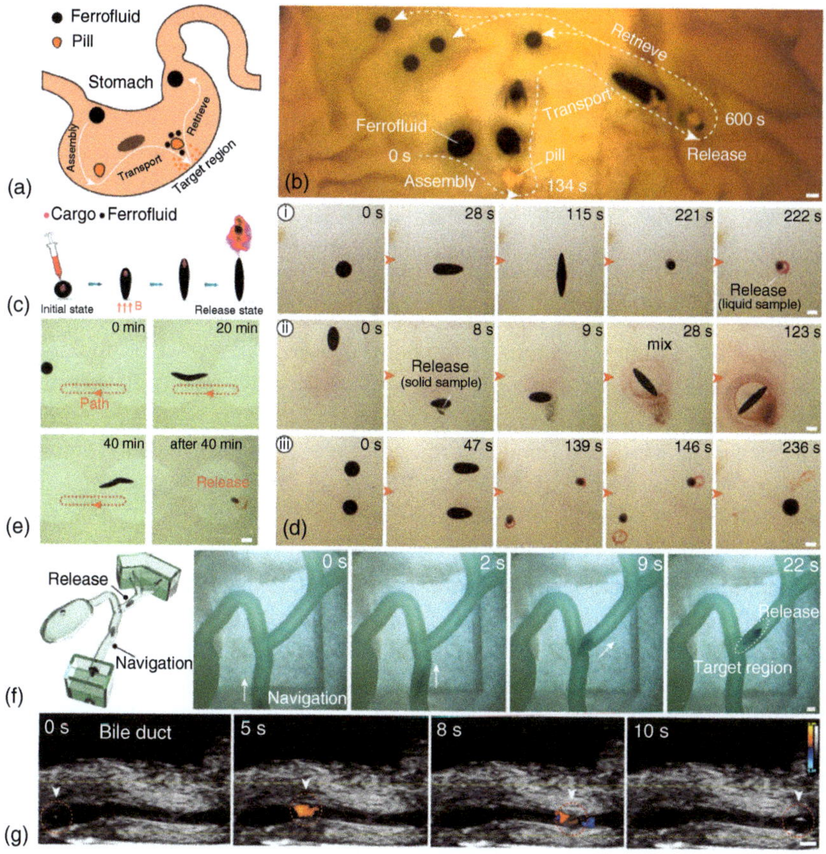


**Figure 6.22** (a) The schematic diagram of transforming a traditional pill into an active drug delivery system with FDRs; (b) The controllable construction, locomotion, and release of a solid pill in an ex vivo animal stomach; (c) The schematic diagram of a liquid capsule that ejects red dyes by applying a magnetic field; (d) Liquid capsules maneuver in various motion modes to achieve multiple delivery; (e) The liquid capsule shuttles back and forth in a narrow channel; (f) The liquid capsule moves in a bile duct phantom and releases the liquid cargo on demand; (g) Ultrasound Doppler-guided locomotion of the FDR in an ex vivo animal bile duct. Source: Sun et al. [27]/Reproduced from Springer Nature/CC BY 4.0.

When the FDR recovers to a ring shape, these objects are trapped, transported, and released simultaneously (Figure 6.21b). Additionally, the release of liquid drugs in a vascular phantom model with narrow branches is achieved.

Sun et al. achieve the capturing and releasing of solid and liquid cargoes. The solid cargoes (passive pills, solid food dye) are captured by either active adhesion of FDR or injecting into the inner space of FDR (Figure 6.22a,b) [27]. The liquid cargo (edible dyes) is captured by injecting drugs inside the FDR (Figure 6.22c). In a pig stomach, the FDR is controlled to approach and actively adhere to a solid pill, endowing it with high mobility capability. When the pill is transported to the infectious area, it is released by the splitting of FDR, rather than being randomly distributed. The FDRs are also controlled to transport and release liquid edible dyes and solid food dyes on a plane. In the transporting process, these cargoes exist stably within the FDRs due to their high surface tension. When an oscillating magnetic field is vertically applied, the FDR is elongated to contract in the lateral direction, which increases the ejection of cargoes from the FDR due to increasing internal pressure (Figure 6.22d). To accelerate the dissolution speed of the released solid food dye, the FDR is further used as a stirring bar when it rotates quickly. Clearly, most of the solid cargo is dissolved within 102 seconds.

In addition, two FDRs are simultaneously controlled for releasing liquid edible dyes on demand, and then they are merged for retrieval. This cooperative strategy allows the delivery of higher volumes of drugs for fast disease therapy. The long-term stability of FDR to transport liquid cargo is also demonstrated when it undergoes repeated large deformations. The FDR filled with red food dye is controlled to navigate through narrow channels back and forth for 40 minutes, and no dye is observed to leak during the locomotion and deformation process (Figure 6.22e). The large deformation and shape adaptation of FDRs also make them suitable for delivering cargo even in the bile duct phantom and natural bile duct. Under the guidance of X-ray fluoroscopy imaging, the FDR moves forward and adapts to the sharp turn of the bile duct phantom, releasing the carried cargo at the target position after applying an oscillating magnetic field (Figure 6.22f). Similarly, under the real-time tracking of ultrasound imaging, controllable transport and delivery of edible dye is achieved in the natural bile duct (Figure 6.22g). The softness and biocompatibility of FDR result in minimal damage to the surrounding cells and tissues, which may be promoted for the therapy of cancer, inflammation, and other diseases by replacing edible dye with potential drugs.

### 6.4.2 Objects Sorting

The deformation (splitting, merging, assembly) and motion capabilities of FDRs allow them to sort different objects from their mixture. Under a vertical oscillating magnetic field, the FDR is split into multiple smaller FDRs, which are then assembled into a vertical collective. The vertical collective can be recognized as a filter screen, which is stabilized in the middle of the microfluidic channel to filter some solid particles dispersed in the dark green flowing fluid [28]. The dark green fluid can escape from the filter screen through holes in the vertical collective, while the solid particles are blocked. Clearly, these particles are sorted on the left side of the vertical collective (Figure 6.23a). Furthermore, an ellipse-shaped FDR could sort two kinds of delicate objects (hydrogel balls) in a clustered environment by pushing them to two sides of a plane (Figure 6.23b) [29]. The large length of the ellipse-shaped FDR allows it to push single or multiple objects. When the ellipse-shaped FDR hits an obstacle or manipulated objects, it can split itself to cross over the obstacle, stay safe to the manipulated objects, and then recover quickly through merging to further complete the sorting task. The splitting appears if the pushing force applied on the obstacle or objects is larger than a threshold. The threshold of pushing force can be adjusted by controlling the viscosity of FDRs, making it appealing for manipulating various delicate objects. With the increasing viscosity, the hydrogel ball will

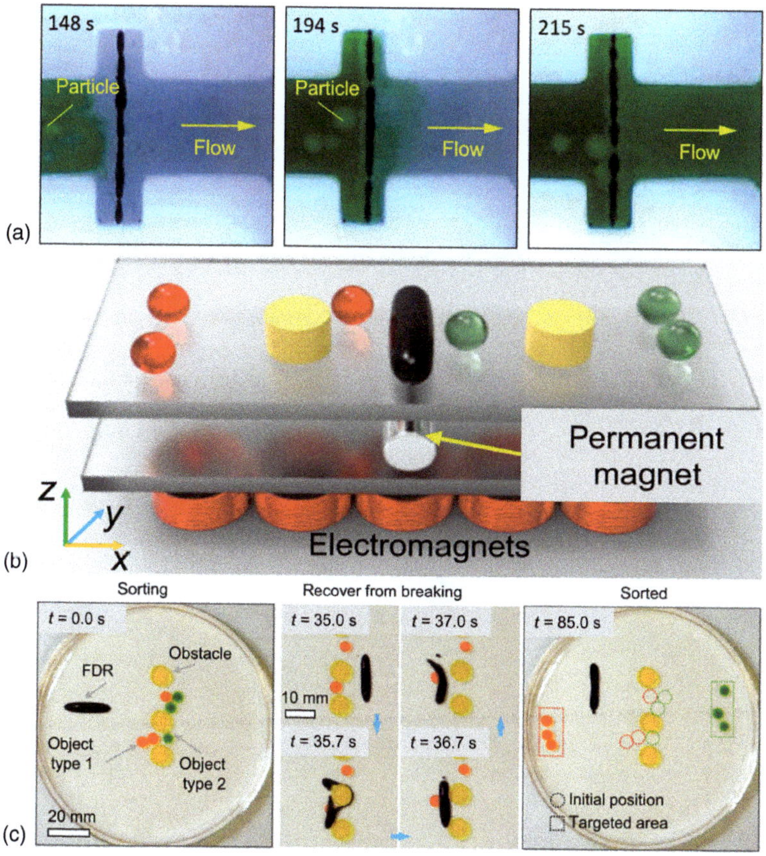


**Figure 6.23** (a) Vertical collectives combined with microfluidic chips to achieve particles sorting. Source: Sun et al. [28]/Reproduced from John Wiley & Sons, Inc. (b) The schematic diagram of an ellipsoidal FDR for sorting delicate objects in a clustered environment. Source: Fan et al. [29]/Reproduced from National Academy of Science. (c) The sorting process of delicate objects by the ellipsoidal FDR. Source: Fan et al. [29]/PNAS.

get a larger velocity (equal to the pushing force) to move forward. As shown in Figure 6.23c, red and green hydrogel balls are quickly pushed to the left and right sides, respectively.

### 6.4.3 Liquid Pumping and Mixing

With coordinated deformation and motion of multiple FDRs, liquid pumping and mixing have been developed to explore their possible biomedical applications. Inspired by the biological cilia that can induce hydrostatic flow by generating nonreciprocal motion, the liquid cilia based on FDRs is successfully investigated for liquid pumping [27]. The FDRs are anchored into the semicircular pit made of resin to form liquid cilia due to their high wettability, and a segmented magnetic field $\boldsymbol{B}_{\text{Cilia}}(t)$ is applied to induce the behavior of liquid cilia. In this magnetic field,

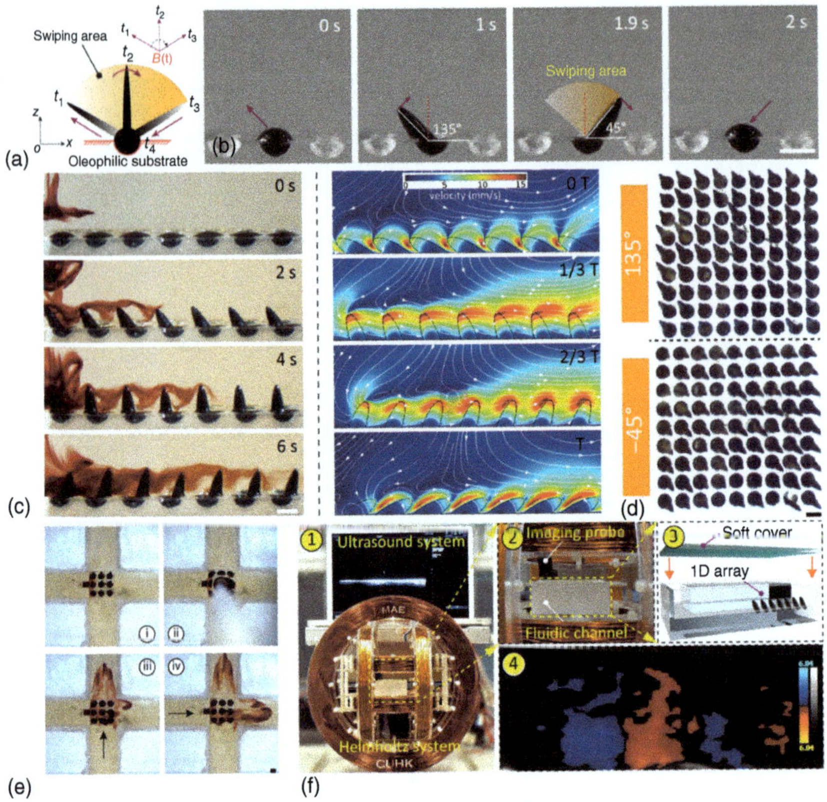


**Figure 6.24** (a) The schematic diagram for 2D nonreciprocal motion of single liquid cilium. (b) Nonreciprocal dynamics of single liquid cilium. (c) Experimental and simulated results of the fluid flow distribution of the liquid cilia array. (d) Top views of liquid cilia matrix with different oscillating direction. (e) Pumping red food dye with a liquid cilia matrix. (f) Ultrasound doppler imaging visualization of liquid cilia array pumping porcine blood in enclosed channels. Source: Sun et al. [27]/Reproduced from Springer Nature/CC BY 4.0.

$B_{\mathrm{m}}$ is the maximum field strength, $\alpha = 2\pi ft$ is the rotating angle, f is the frequency, and $\theta$ is the maximum rotation angle.

$$\boldsymbol{B}_{\mathrm{cilia}}(t) = \begin{cases} B_{\mathrm{m}} \cos(\alpha)\boldsymbol{e}_x - \sin(\alpha)\boldsymbol{e}_z & \text{if } \alpha < \theta \\ 0 & \text{if } \alpha \geq \theta \end{cases} \tag{6.4}$$

The nonreciprocal motion mechanism of FDRs is explained by their programmable deformation from three stages (Figure 6.24a). In the first stage ($t_1$), the liquid cilium is elongated to a spiky shape from the initial elliptical shape, and its elongation length increases with the amplitude of the magnetic field. In the second stage ($t_1$-$t_2$-$t_3$), the spiked liquid cilium rotates along the changing magnetic field direction. In the third stage ($t_3$), the spiked liquid cilium recovers to an ellipse after turning off the magnetic field. By repeating these three stages for controlled oscillation, liquid cilia exhibit nonreciprocal motion and net fluid flow. In the rotating process of liquid cilia, its swiping area determines the degree of nonreciprocal motion, which increases with the swiping area, leading to enhanced fluid flow. The swiping area is controlled by the angle of rotation and the amplitude of the magnetic field. Based on the above mechanism, 1D and 2D cilia are arranged in an array ($1 \times 7$) and matrix ($9 \times 9$) forms, respectively. The tracer edible red dye solution is pumped from the left side to the right side by the 1D cilia when the dye solution is added to the left side (Figure 6.24b), demonstrating the directional liquid pumping capability of 1D cilia. The simulation results of fluid field distribution along the oscillation direction are identical to the experimental results (Figure 6.24c). In addition, the oscillation direction is adjusted arbitrarily with the $9 \times 9$ omnidirectional liquid cilia matrix. Oscillating along 135° or −45° direction and 360° rotating along the z-axis is representatively tuned onto a plane (Figure 6.24d). A $3 \times 3$ liquid cilia is also fixed in the center of the cross-shaped flow channel to control the direction of liquid pumping. When the edible dye is added to the liquid cilia matrix, it is continuously pumped into two channels (Figure 6.24e). The pumping of fresh porcine whole blood in a confined narrow channel is also realized by assisting with wirelessly magnetic actuation and ultrasound imaging tracking (Figure 6.24f). Compared to other elastomers-based artificial cilia, the length of liquid cilia can be dynamically reconfigured in situ to allow more flexible fluid pumping. The special property of liquid cilia will benefit various fluidic applications such as wireless microfluidic devices and robotic platforms.

The FDRs have worked as pistons or valves cooperatively for liquid mixing in tubes by employing their ability of independent motions and adaptive confinement (Figure 6.25a) [29]. The FDR with a lower viscosity (8 mPa s) is first injected into a fluidic channel, and then it is split into six smaller FDRs. After that, these smaller FDRs are controlled to reach the specific positions, where preprogrammed motions are performed to complete the fluid mixing task cooperatively. Meanwhile, six reference points of the fluid-mixing device are set to quantify the preprogrammed motions. The FDRs in the lateral branches of the device are recognized as the valves, while they are recognized as the pistons in the main branches. The closing and opening of valves are controlled by moving away from or close to these reference points in the lateral branches. The injection and drawing back of the fluids are conducted by moving away from or close to these references in the main branch (Figure 6.25b). As shown in Figure 6.25c, red and blue dyes are pumped into the main branches by controlling the opening and closing of valves and the forward and backward motions of pistons. The displacements of these six smaller FDRs within a pumping cycle are plotted to quantitatively describe the state of each FDR in the device (Figure 6.25d,e). These coordinated states of valves and pistons can be arranged in a programmable order to sequentially mix multiple dyes (Figure 6.25f) and allow programmable chemical reactions. This fluid mixing strategy allows more flexible and reconfigurable fluidic functions in situ.

### 6.4.4 Liquid Skin of Soft Robots

Magnetically actuated soft robots made of soft materials (silicon, polydimethylsiloxane) and magnetic particles are capable of achieving programmable structural

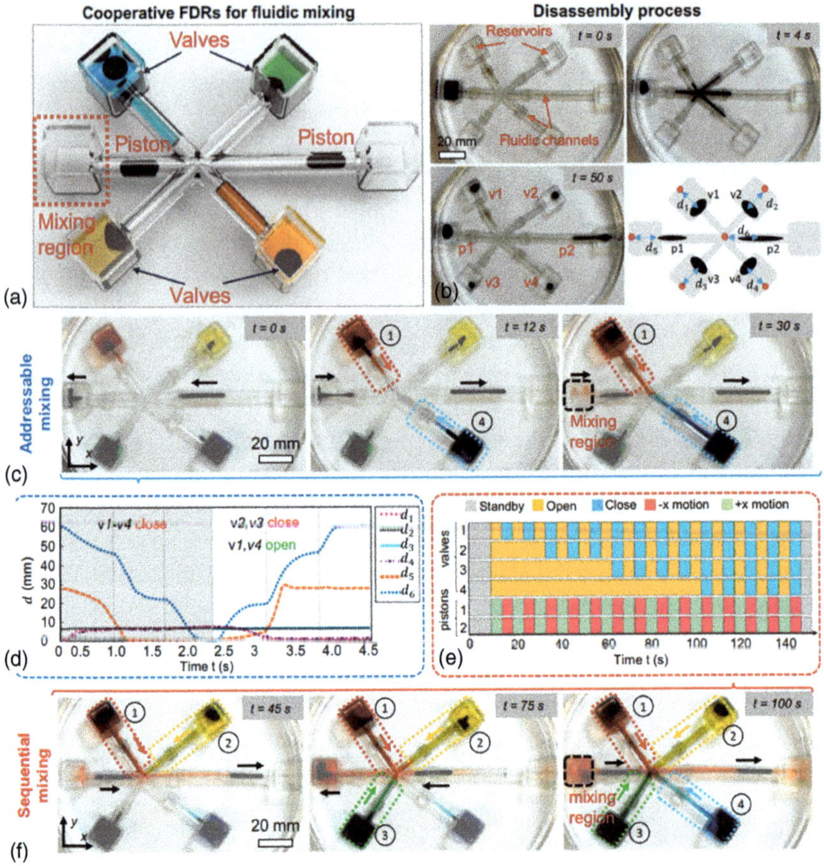


**Figure 6.25** (a) Concept of the programmable fluidic mixing device. (b) The splitting of FDR and coordinated reaching to the specific positions into a fluidic channel. (c) Snapshots of selectively mixing the red and blue dyes. (d) The sequential relative displacements of all FDRs in a pumping period. (e) Sequential states of the valves and pistons in a specified order. (f) Snapshots of sequentially mixing dyes in four channels. Source: Fan et al. [29]/Reproduced from National Academy of Science.

changes. However, the shape-morphing is usually fixed due to the unchangeable magnetic elements when these soft robots are produced, limiting their adaptability to unstructured environments and tasks. Thus, soft robots with reconfigurable ability and dynamically changeable magnetic elements have been developed based on FDRs. The high wettability of oil-based FDR on the silicon elastomer-based soft robots allows it to be used as magnetic liquid skin to coat the soft robots, endowing them with magnetic properties [27, 34]. The magnetic skin can be reused to make FDRs for building soft robots, achieving their reconfiguration. In the coating process, there are three key stages involved. In the first stage, the FDR is split into multiple smaller FDRs, which are then driven to approach and wet the target silicon elastomer thoroughly. In the second stage, ferrofluid penetrates the surface

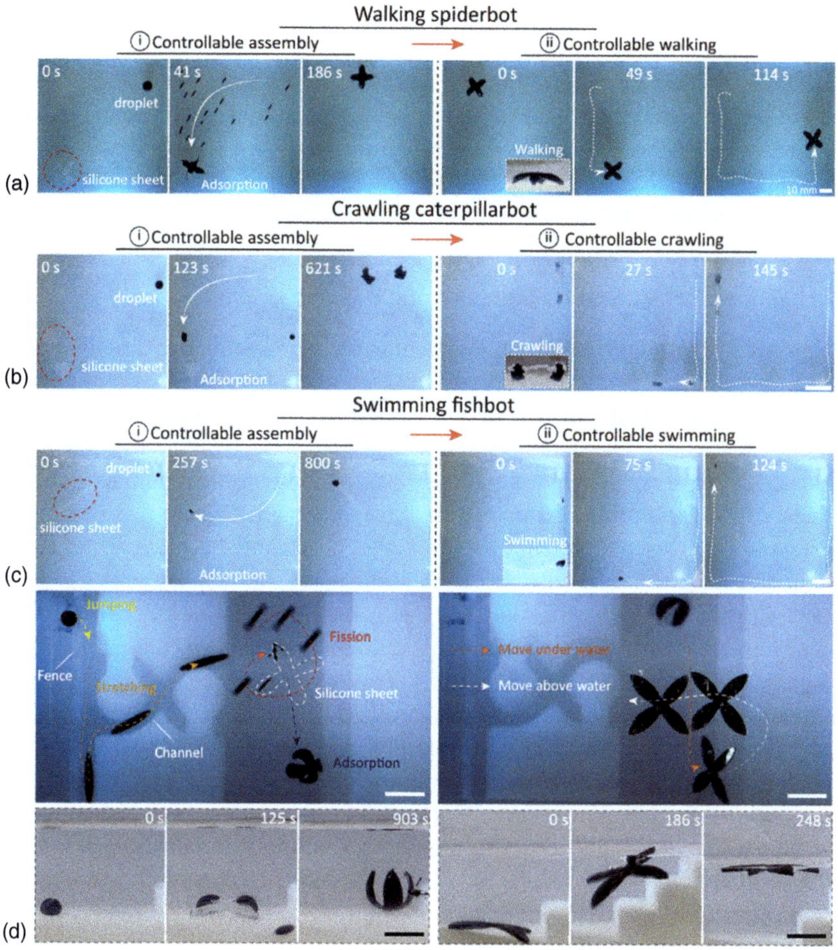


**Figure 6.26** The liquid skin transforms silicone sheet into (a) walking spiderbot, (b) crawling caterpillarbot, and (c) swimming fishbot. (d) The active FDR passes through a complex maze to coat and activate the silicone elastomer-based soft robots. Source: Sun et al. [27]/ Reproduced from Springer Nature/CC BY 4.0.

layer of the elastomer, and the upper part of the elastomer is swollen and inflated due to oil-based carrier fluid. In the third stage, ferrofluid penetrates the body of the elastomer and a continuous rigid layer of magnetic nanoparticles is formed. Multiple miniature soft robots are built through the wetting behavior of FDRs. A petal-shaped elastomeric sheet is transformed into a spider robot with wireless liquid skin, navigating in a walking mode along a predefined way (Figure 6.26a). Additionally, the elastomer sheets are also transformed into a caterpillar robot and a fish robot. The FDR is first split into two smaller FDRs, which are then controlled to adhere to each end of the predesigned elastomer sheet, forming a caterpillar robot. The robot crawls forward under an oscillating magnetic field (Figure 6.26b).

The FDR is adhered to one end of another elastomer sheet to form a fish robot. The robot can swim like a fish under the oscillating magnetic field (Figure 6.26c). Apart from the direct coating, the FDR also can pass through a complex channel to coat itself onto the target silicone elastomer sheet, transforming it into a mobile soft robot. The robot with hydrophobic properties can move from the bottom to the water surface. The robot is first controlled to stand by a vertical magnetic field, and then it will emerge from the water surface (Figure 6.26d).

### 6.4.5 Phase Transitional Metallic Ferrofluid-Based Gripper

The magnetic control method occupies the outstanding penetrability and high controllability, so it has broad application prospects in the fields of micro-manipulation and cargo delivery. Manipulation of a wide range of non-magnetic objects using magnetic fields is a research hotspot. A common strategy is to use magnetic substances to bind target cargoes, such as electroplating or direct doping. But this physical or chemical combination is irreversible, making it difficult to separate the magnetic substances and target cargoes. Wang et al. proposed a strategy of manipulating non-magnetic cargo through ferrofluid, which can realize the grasping of objects with arbitrary shapes, as shown in Figure 6.27 [57]. The developed ferrofluid contains liquid metal gallium and iron particles, which have phase transition capability and magnetic properties, respectively. In high-concentration HCL solution (>3 mol $l^{-1}$), iron particles can be dispersed into molten gallium metal. Ferrofluids can be transformed between solid and liquid states by temperature regulation. When in a liquid state, ferrofluid can fully contact the object being manipulated. After being transformed into a solid state, it can exhibit high stiffness (modulus stiffness: 3.6 GPa), enabling a tight grasp of the manipulated object. Due to the existence of magnetic particles, the goods can be delivered to the designated location through the control of the external magnetic field, and then the separation of the ferrofluid and the goods can be realized by adjusting the temperature.

As shown in Figure 6.28, ferrofluid with phase change ability can manipulate objects of different shapes. In the liquid and solid states, the locking force between the ferrofluid and the manipulated object can reach 0.01 and 1168 N, respectively. To demonstrate the control of objects of different shapes, ferrofluids were controlled by external magnets to adhere and transfer objects such as eggs and apples. In addition, the connection of circuits can be realized through the deformable property and conductivity of ferrofluid in liquid state, which is expected to be used in the construction of smart electronic devices.

## 6.5 Summary

In this chapter, we mainly focus on the deformation behaviors and applications of FDRs. The deformation behaviors include the shape changes and the motions of

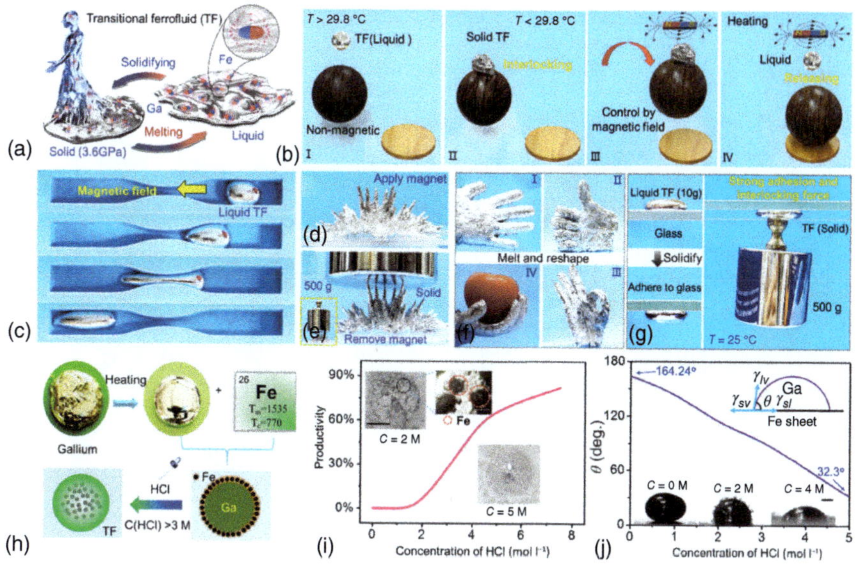


**Figure 6.27** Preparation and application of ferrofluid with phase change ability. (a) Schematic illustration of the phase transition process of ferrofluid. (b) Grasping, manipulation, and release of nonmagnetic spherical objects by ferrofluid. (c) Ferrofluid in the liquid state passes through a narrow channel under the control of a magnetic field. (d) and (e) The ferrofluid transforms into a spine-like structure under the stimulation of a magnetic field. (f) Ferrofluid exhibits gestures of various shapes by melting and reshaping. (g) The ferrofluid can firmly adhere to the glass substrate after transitioning from liquid to solid state. (h) Fabrication of translational ferrofluids. (i–j) The impacts of HCL concentration on the productivity and contact angle of ferrofluid. Source: Wang et al. [57]/ Reproduced from John Wiley & Sons, Inc.

FDR, which are determined by programming external magnetic fields. A series of control devices and systems based on permanent magnets and electromagnetic coils have been developed to modulate the external magnetic fields on demand. Current deformation behaviors of stretching, fission, fusion, and squeezing, as well as jumping, tumbling, rotating, transporting, kayaking, and wobbling motions have been achieved. By employing these deformation behaviors, FDRs have been successfully applied for cargo capturing and delivering, object sorting, liquid pumping and mixing, and activating silicon elastomer-based soft robots. Among current progress, the programming of external magnetic fields has been widely investigated, but it may have reached a high level and a bottleneck. Additionally, these applications have been achieved in some ideal environments, deviating from the original intention to achieve real biomedical applications. Therefore, more attention may be put into exploring or optimizing biomedical applications in the future. For example, the synergetic delivery of different drugs with a group of FDRs to strongly enhance the therapy effect needs to be further explored. Besides, avoiding the adhesion of FDRs to complex biological environments by smart strategies may promote the real biomedical applications of FDRs.

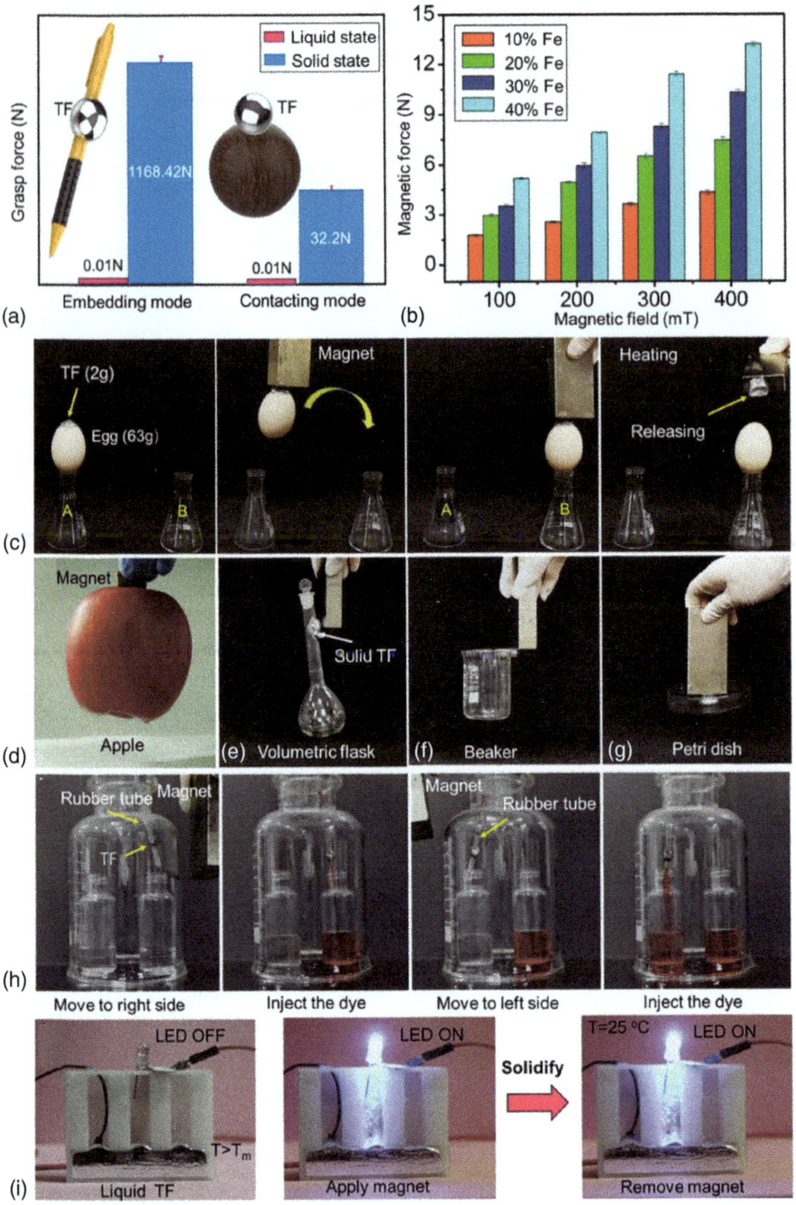


**Figure 6.28** Demonstration of the manipulation of objects of different shapes by the ferrofluid with the ability to change phases. (a) The grasping capability of ferrofluid for objects of different shapes in solid and liquid states. (b) The impacts of the concentration of iron particle and magnetic field strength on the magnetic interaction force of ferrofluid. (c) Grasp, transfer, and release of eggs by ferrofluid with phase transition ability. (d)–(g) Ferrofluid used for grasping of apples, volumetric flasks, beakers, and Petri dishes. (h) Utilizing ferrofluid to manipulate rubber tubes for the delivery of liquid cargo. (i) The connection of the circuit is achieved by the deformation of the ferrofluid in the liquid state activated by a magnetic field. Source: Wang et al. [57]/ Reproduced from John Wiley & Sons, Inc.

## Abbreviations

| | |
|---|---|
| FDR | ferrofluid droplet robot |
| FR | ferrofluid robot |
| SDS | sodium dodecyl sulfate |
| $M^3RA$ | multiscale magnetic miniature robot actuation |
| SPION | superparamagnetic iron oxide nanoparticle |
| TEM | transmission electron microscopy |

## References


1 Kim, Y., Yuk, H., Zhao, R. et al. (2018). Printing ferromagnetic domains for untethered fast-transforming soft materials. *Nature* 558: 274–279.

2 Ze, Q., Kuang, X., Wu, S. et al. (2020). Magnetic shape memory polymers with integrated multifunctional shape manipulation. *Advanced Materials* 32: 1906657.

3 Deng, H., Satti, K., Xie, Y. et al. (2020). Laser reprogramming magnetic anisotropy in soft composites for reconfigurable 3D shaping. *Nature Communications* 11: 6325.

4 Rich, S.I., Wood, R.J., and Majidi, C. (2018). Untethered soft robotics. *Nature Electronics* 1: 102–112.

5 Dong, Y., Wang, J., Guo, X. et al. (2019). Multi-stimuli-responsive programmable biomimetic actuator. *Nature Communications* 10: 4087.

6 Dong, Y., Wang, L., Xia, N. et al. (2021). Multi-stimuli-response programmable soft actuators with site-specific and anisotropic deformation behavior. *Nano Energy* 88: 106254.

7 Ji, X., Liu, X., Cacuccio, V. et al. (2019). An autonomous untethered fast soft robotic insect driven by low-voltage dielectric elastomer actuators. *Science Robotics* 4: eaaz6451.

8 Mitchell, S.K., Wang, X., Acome, E. et al. (2019). An easy-to-implement toolkit to create versatile and high-performance HASEL actuators for untethered soft robots. *Advanced Science* 6: 1900178.

9 Huang, X., Kumar, K., Jawed, M. et al. (2018). Chasing biomimetic locomotion speeds: creating untethered soft robots with shape memory alloy actuators. *Science Robotics* 3: eaau7557.

10 Xu, T., Zhang, J., Salehizadeh, M. et al. (2019). Millimeter-scale flexible robots with programmable three-dimensional magnetization and motions. *Science Robotics* 4: eaav4494.

11 Alapan, Y., Karacakol, A.C., Guzelhan, S.N. et al. (2020). Reprogrammable shape morphing of magnetic soft machines. *Science Advances* 6: eabc6414.

12 Wang, X. et al. (2020). Untethered and ultrafast soft-bodied robots. *Communications Materials* 1: 67.

13 Hines, L., Petersen, K., Lum, G.Z., and Sitti, M. (2017). Soft actuators for small-scale robotics. *Advanced Materials* 29: 1603483.

**14** Dong, Y., Wang, L., Xia, N. et al. (2022). Untethered small-scale magnetic soft robot with programmable magnetization and integrated multifunctional modules. *Science Advances* 8: eabn8932.

**15** Zhang, J., Ren, Z., Hu, W. et al. (2021). Voxelated three-dimensional miniature magnetic soft machines via multimaterial heterogeneous assembly. *Science Robotics* 6: eabf0112.

**16** Yang, X., Shang, W., Lu, H. et al. (2020). An agglutinate magnetic spray transforms inanimate objects into millirobots for biomedical applications. *Science Robotics* 5: eabc8191.

**17** Sitti, M., Ceylan, H., Hu, W. et al. (2015). Biomedical applications of untethered mobile milli/microrobots. *Proceedings of the IEEE* 103: 205–224.

**18** Roh, Y., Kim, M., Won, S.M. et al. (2021). Vital signal sensing and manipulation of a microscale organ with a multifunctional soft gripper. *Science Robotics* 6: eabi6774.

**19** Jin, D., Yuan, K., Du, X. et al. (2021). Domino reaction encoded heterogeneous colloidal microswarm with on-demand morphological adaptability. *Advanced Materials* 33: 2100070.

**20** Yu, J., Wang, B., Du, X. et al. (2018). Ultra-extensible ribbon-like magnetic microswarm. *Nature Communications* 9: 3260.

**21** Yu, J., Jin, D., Chan, K.F. et al. (2019). Active generation and magnetic actuation of microrobotic swarms in bio-fluids. *Nature Communications* 10: 5631.

**22** Yang, Z., Yang, L., Zhang, M. et al. (2021). Ultrasound-guided catheterization using a driller-tipped guidewire with combined magnetic navigation and drilling motion. *IEEE/ASME Transactions on Mechatronics* 27: 2829–2840.

**23** Yang, Z., Yang, L., Zhang, M. et al. (2021). Magnetic control of a steerable guidewire under ultrasound guidance using mobile electromagnets. *IEEE Robotics and Automation Letters* 6: 1280–1287.

**24** Ding, Y., Zeng, M., and Fu, L. (2019). A bright future for liquid functional materials? *Matter* 1: 1099–1103.

**25** Whitesides, G.M. (2018). Soft robotics. *Angewandte Chemie International Edition* 57: 4258–4273.

**26** Wang, Q., Yang, L., and Zhang, L. (2021). Micromanipulation using reconfigurable self-assembled magnetic droplets with needle guidance. *IEEE Transactions on Automation Science and Engineering* 19: 759–771.

**27** Sun, M., Hao, B., Yang, S. et al. (2022). Exploiting ferrofluidic wetting for miniature soft machines. *Nature Communications* 13: 7919.

**28** Sun, M., Yang, S., Jiang, J., and Zhang, L. (2023). Horizontal and vertical coalescent microrobotic collectives using ferrofluid droplets. *Advanced Materials* 2300521.

**29** Fan, X., Dong, X., Karacakol, A.C. et al. (2020). Reconfigurable multifunctional ferrofluid droplet robots. *Proceedings of the National Academy of Sciences* 117: 27916–27926.

**30** Fan, X., Sun, M., Sun, L., and Xie, H. (2020). Ferrofluid droplets as liquid microrobots with multiple Deformabilities. *Advanced Functional Materials* 30: 2000138.

31 Fan, X., Jiang, Y., Li, M. et al. (2022). Scale-reconfigurable miniature ferrofluidic robots for negotiating sharply variable spaces. *Science Advances* 8: eabq1677.

32 Ji, Y., Bai, X., Sun, H. et al. (2022). Biocompatible ferrofluid robot with photothermal property for targeted tumor therapy. *IEEE Robotics and Automation Letters* 7: 11517–11522.

33 Zhang, X., Sun, L., Yu, Y., and Zhao, Y. (2019). Flexible ferrofluids: design and applications. *Advanced Materials* 31: 1903497.

34 Zhang, S., Ke, X., Jiang, Q. et al. (2021). Programmable and reprocessable multifunctional elastomeric sheets for soft origami robots. *Science Robotics* 6: eabd6107.

35 Chen, D., Yang, Z., Ji, Y. et al. (2021). Deformable ferrofluid-based millirobot with high motion accuracy and high output force. *Applied Physics Letters* 118: 134101.

36 Ji, Y., Gan, C., Dai, Y. et al. (2022). Deformable ferrofluid microrobot with omnidirectional self-adaptive mobility. *Journal of Applied Physics* 131: 064701.

37 Deb, R., Sarma, B., and Dalal, A. (2022). Magnetowetting dynamics of sessile ferrofluid droplets: a review. *Soft Matter* 18 (12): 2287–2324.

38 Ahmed, R., Ilami, M., Bant, J. et al. (2020). A shapeshifting ferrofluidic robot. *Soft Robotics* 8: 687–698.

39 Berger, P., Adelman, N.B., Beckman, K.J. et al. (1999). Preparation and properties of an aqueous ferrofluid. *Journal of Chemical Education* 76: 943.

40 Genc, S. and Derin, B. (2014). Synthesis and rheology of ferrofluids: a review. *Current Opinion in Chemical Engineering* 3: 118–124.

41 Dunne, P., Adachi, T., Dev, A.A. et al. (2020). Liquid flow and control without solid walls. *Nature* 581: 58–62.

42 Serwane, F., Mongera, A., Rowghanian, P. et al. (2017). In vivo quantification of spatially varying mechanical properties in developing tissues. *Nature Methods* 14: 181–186.

43 Mongera, A., Rowghanian, P., Gustafson, H.J. et al. (2018). A fluid-to-solid jamming transition underlies vertebrate body axis elongation. *Nature* 561: 401–405.

44 Yu, W., Lin, H., Wang, Y. et al. (2020). A ferrobotic system for automated microfluidic logistics. *Science Robotics* 5: eaba4411.

45 Kole, M. and Khandekar, S. (2021). Engineering applications of ferrofluids: a review. *Journal of Magnetism and Magnetic Materials* 537: 168222.

46 Roger, J., Pons, J., Massart, R. et al. (1999). Some biomedical applications of ferrofluids. *The European Physical Journal - Applied Physics* 5: 321–325.

47 Tari, A., Chantrell, R., Charles, S.W., and Popplewell, J. (1979). The magnetic properties and stability of a ferrofluid containing Fe3O4 particles. *Physica B+ C* 97: 57–64.

48 Zelazo, R.E. and Melcher, J.R. (1969). Dynamics and stability of ferrofluids: surface interactions. *Journal of Fluid Mechanics* 39: 1–24.

49 Latikka, M., Backholm, M., Baidya, A. et al. (2020). Ferrofluid microdroplet splitting for population-based microfluidics and interfacial Tensiometry. *Advanced Science* 7: 2000359.

**50** Komaee, A. and Shapiro, B. (2010). Steering a ferromagnetic particle by magnetic feedback control: algorithm design and validation. In: *Proceedings of the 2010 American Control Conference*, –6543, 6548. IEEE.

**51** Probst, R., Lin, J., Komaee, A. et al. (2011). Planar steering of a single ferrofluid drop by optimal minimum power dynamic feedback control of four electromagnets at a distance. *Journal of Magnetism and Magnetic Materials* 323: 885–896.

**52** Ody, T., Panth, M., Sommers, A.D., and Eid, K.F. (2016). Controlling the motion of ferrofluid droplets using surface tension gradients and Magnetoviscous pinning. *Langmuir* 32: 6967–6976.

**53** Katsikis, G., Breant, A., Rinberg, A., and Prakash, M. (2018). Synchronous magnetic control of water droplets in bulk ferrofluid. *Soft Matter* 14: 681–692.

**54** Varma, V.B., Ray, A., Wang, Z. et al. (2016). Control of ferrofluid droplets in microchannels by uniform magnetic fields. *IEEE Magnetics Letters* 7: 1–5.

**55** Chen, C.-Y., Hsueh, H.-C., Wang, S.-Y., and Li, Y.-H. (2015). Self-assembly and novel planetary motion of ferrofluid drops in a rotational magnetic field. *Microfluidics and Nanofluidics* 18: 795–806.

**56** Cunha, L.H.P., Siqueira, I.R., Oliveira, T.F., and Ceniceros, H.D. (2018). Field-induced control of ferrofluid emulsion rheology and droplet break-up in shear flows. *Physics of Fluids* 30: 122110.

**57** Wang, H., Chen, S., Li, H. et al. (2021). A liquid gripper based on phase transitional metallic ferrofluid. *Advanced Functional Materials* 31: 2100274.

# 7

# Conclusions and Future Prospects

## 7.1 Introduction

Miniaturized soft robots have attracted extensive attention in the field of robotics and soft matter, and are continuously considered as research hotspots. Yang et al. reported a review article in the journal *Science Robotics* and pointed out that the development of related advanced materials and fabrication technologies is one of the major challenges for future robots [1]. We review the development of the past 10 years and find that many major advances have been made in the construction, actuation, and control of miniaturized soft robotic systems. In this book, we take the key work done by the author's research group in the field of miniaturized soft robots, covering materials, preparation, control, and application, and introduce the important achievements in the field of soft robots. In terms of fundamental study, we introduce different types of responsive materials and demonstrate different actuation mechanisms combined with theoretical models. In terms of applications, we highlight representative examples of soft robots in the fields of biomedicine, environmental monitoring, flexible electronics, and micromanipulation. In this chapter, we summarize the major advances in miniaturized soft robots, discuss other functional materials, multi-material integration strategies, and the construction of soft robots with perception capabilities, as well as provide insights into the development of future miniature soft robots and outlook for challenges.

## 7.2 Other Functional Materials Used for Miniature Soft Robots

### 7.2.1 Shape-Memory Materials

As stimuli-responsive polymers, shape-memory materials have attracted growing attention among the applications for soft robots and flexible electronics due to their shape-morphing capability and programmability (Figure 7.1a,b) [6–14]. Upon external stimulation such as the increase of temperature, shape-memory polymers (SMPs) could transform from a glassy state to a rubbery state with the softening of the polymer matrix [15–21]. Both deformation and shape fixation of

*Untethered Miniature Soft Robots: Materials, Fabrications, and Applications*, First Edition.
Li Zhang, Jiachen Zhang, Neng Xia, and Yue Dong.

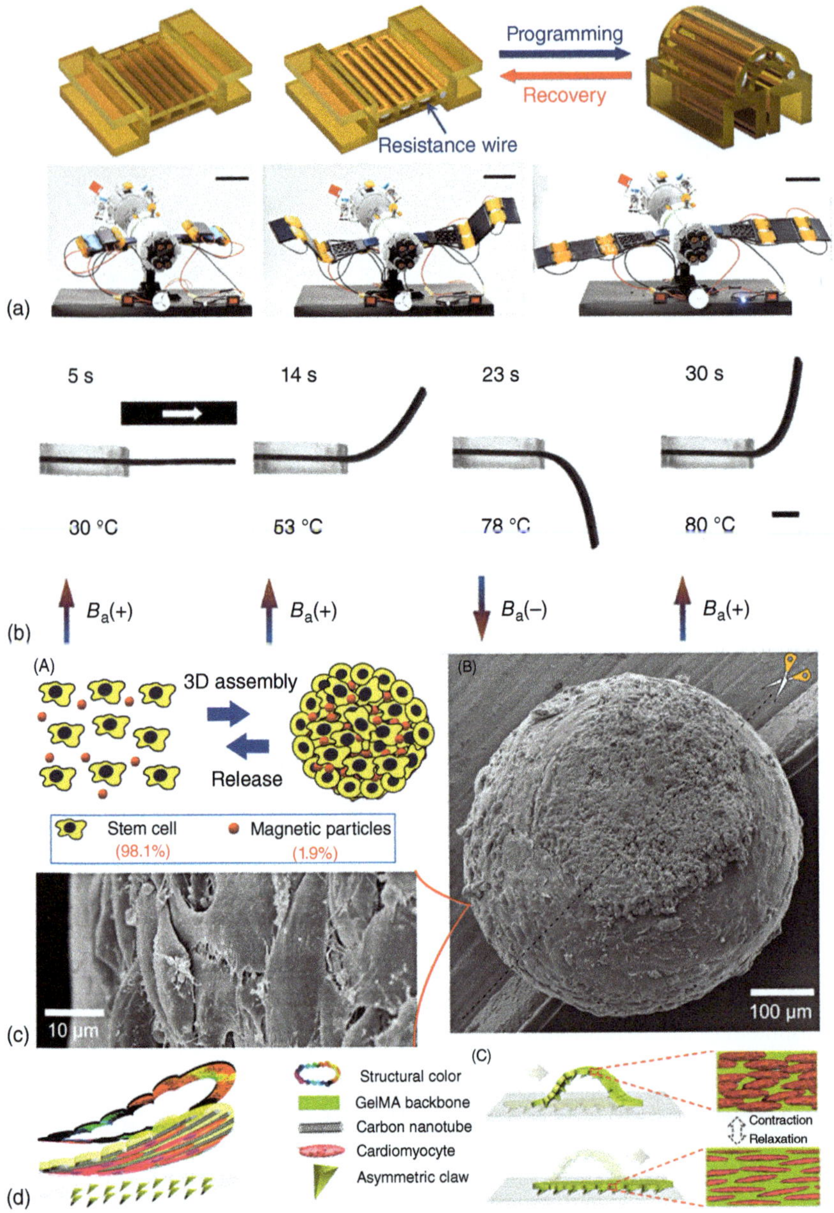


**Figure 7.1** Shape-memory materials and biohybrid materials. (a) 4D-printed SMP structures can be reconfigured from a permanent shape to temporary shapes and applied to the deployment of solar panel. Source: Zhang et al. [2]/Reproduced from John Wiley & Sons, Inc. (b) Shape locking and morphological transformation of magnetic SMP structures by the superimpose of alternating and actuation magnetic fields. Source: Ze et al. [3]/Reproduced from John Wiley & Sons, Inc. (c) Construction of biohybrid microrobots via self-assembly of magnetic particles and stem cells. Source: Wang et al. [4]/American Association for the Advancement of Science. (AAAS). (d) Caterpillar-like soft robot powered by cardiomyocytes. Source: Sun et al. [5]/Reproduced from John Wiley & Sons, Inc.

SMPs are programmable [22–29]. The heated state will reduce the energy barrier of the polymer and the obstruction of molecular chain movement, which will make the polymer easy to deform. When the SMPs transform into the targeted temporary structure, this structure is fixed after cooling and does not require the continuous application of external stress. This temporary structure formed during the programming process is not stable, although it can be maintained for a long time. When the material is stimulated by external heat again, the molecular chains of the polymer will regain their mobility, and then its stored internal stress will be released, allowing the structure to return to its original permanent shape. At the same time, it can also be programmed again into a completely different temporary shape. For example, Ding et al. proposed the 4D printing of a composite material consisting of the SMPs and the elastomer [30]. Through programmable internal strain as well as cooling or heating, the fabricated structure transformed between permanent structure and multiple temporary structures. Apart from thermal activation, other actuation strategies, e.g. light actuation, magnetic actuation, and electric actuation, have been applied to SMPs. Ze et al. doped hard magnetic particles and paramagnetic particles into the matrix of SMPs [3]. Under the action of an external high-frequency magnetic field, the paramagnetic particles would produce a magnetocaloric effect and induce the softening of SMPs. Due to the existence of hard magnetic particles, the structure will produce reconfigurable deformation under the action of low-frequency or static magnetic field. In addition, using the photothermal effect, SMPs-based robots can also be actuated under the combined action of light and magnetic fields, which is expected to achieve selective manipulation.

### 7.2.2 Biohybrid Materials

Compared with traditional robots, biohybrid robots integrating biological composition and soft materials have obvious advantages in energy conversion and environmental adaptability (Figure 7.1c,d) [31–39]. Commonly used biological components include microorganisms, mammalian cell lines, insect dorsal vascular tissue, mammalian primary cells, as well as other biological sources such as sperm [32, 40–50]. Among these biological sources, cardiomyocytes are widely adopted, such as primary cardiomyocytes from rat hearts and cardiomyocytes differentiated from stem cells. The cultured cells can beat under the action of electrical signals, and attach to the flexible material to realize the deformation and movement of the structure. Biocompatible flexible matrix materials provide a suitable environment for the growth of biological components in robots. Widely used matrix materials include bioactive hydrogels, extracellular matrices, as well as bioinert polymers. Among them, hydrogel materials have attracted a lot of attention due to their widely adjustable mechanical properties and excellent biological characteristics, and a variety of artificial hydrogel materials have been developed, such as gelatin methacryl (GelMA), alginate hydrogel, and PEGDA. Sun et al. proposed a crawling robot inspired by caterpillars and snakes, which uses cardiomyocytes as the actuation source and GelMA as the matrix material [5]. The hydrogel matrix deforms under the contraction or elongation of cardiomyocytes, and the fabricated soft robot could crawl on rough substrates under the effect of friction. Wang et al. developed magnetic soft robots for precise *in vivo* delivery via self-assembly of cells [4]. The co-culture of stem cells and magnetic particles coated with polydopamine forms biohybrid microspheres, which can efficiently deliver stem cells driven by an external magnetic field. The biohybrid robot is remarkably soft and exhibits a modulus comparable to that of human brain tissue, allowing it to navigate through tiny and tortuous regions. The viability, proliferation, and differentiation capabilities of stem cells were not affected by the magnetic particles, and meanwhile, the resulting biohybrid robot could suppress the immune response.

## 7.3 Multi-Material Integration Strategies

Complex application scenarios or tasks often require the integration of multiple material compositions (e.g. communication and sensing modules) in the miniature soft machines, which still is a challenge for the development of multifunctional miniature robots. In this respect, multi-material 3D printing techniques including multi material digital light processing (DLP) printing, direct laser writing (DLW), and direct ink writing (DIW) provide promising strategies, which allow the combination of different types of materials (e.g. different stiffnesses, responsive or non-responsive) in one robotic system [51–57] (Figure 7.2). For instance, a multi-nozzle DIW system was developed by Skylar-Scott et al. to fabricate a centimeter-scale pneumatic walking robot consisting of both soft and rigid silicone elastomers [61]. Truby et al. utilized multi-material embedded 3D printing to fabricate a somatosensitive actuator which integrates multiple sensing units and a pneumatic network [62]. The developed pneumatic actuator is capable of providing both proprioceptive and tactile feedback while deforming. At the microscale, Ma et al. combined the DLW printing with a microfluidic system to continuously fabricate a 3D robotic structure with different polymer precursors [58]. Owing to this method, a pH-responsive musculoskeletal system was constructed, while the skeleton and muscle structures were developed by SU-8 resin and bovine serum albumin, respectively. Upon the variation of environment pH, the fabricated microsystem could actively deform to trap a 10 μm microcargo. 3D printing methods based on projection-based photocurings, such as DLP, could quickly form high-precision 3D structures. However, utilizing this approach to integrate different types of materials such as hydrogels and elastomers into 3D robotic structures remains challenging due to the need for frequent replacement of the liquid tank and the issue of interface delamination. Recently, Ge et al. used an air-jet method to remove residual precursors during the DLP printing, while constructing covalent bonding between hydrogels and other polymers through photoinitiator-induced incomplete polymerization [63]. The proposed multi-material DLP printing technique was applied to fabricate pneumatic grippers with perceptive capability and shape-memory-hydrogel stents with drug-delivery functions. Cui et al. used a DLP-based multi-material 3D printing technique to prepare 3D metamaterial structures, which consist of piezoelectric units (piezoelectric ceramics), conductive units, and structural units [64]. This novel metamaterial robot enables multimodal locomotion and active feedback control.

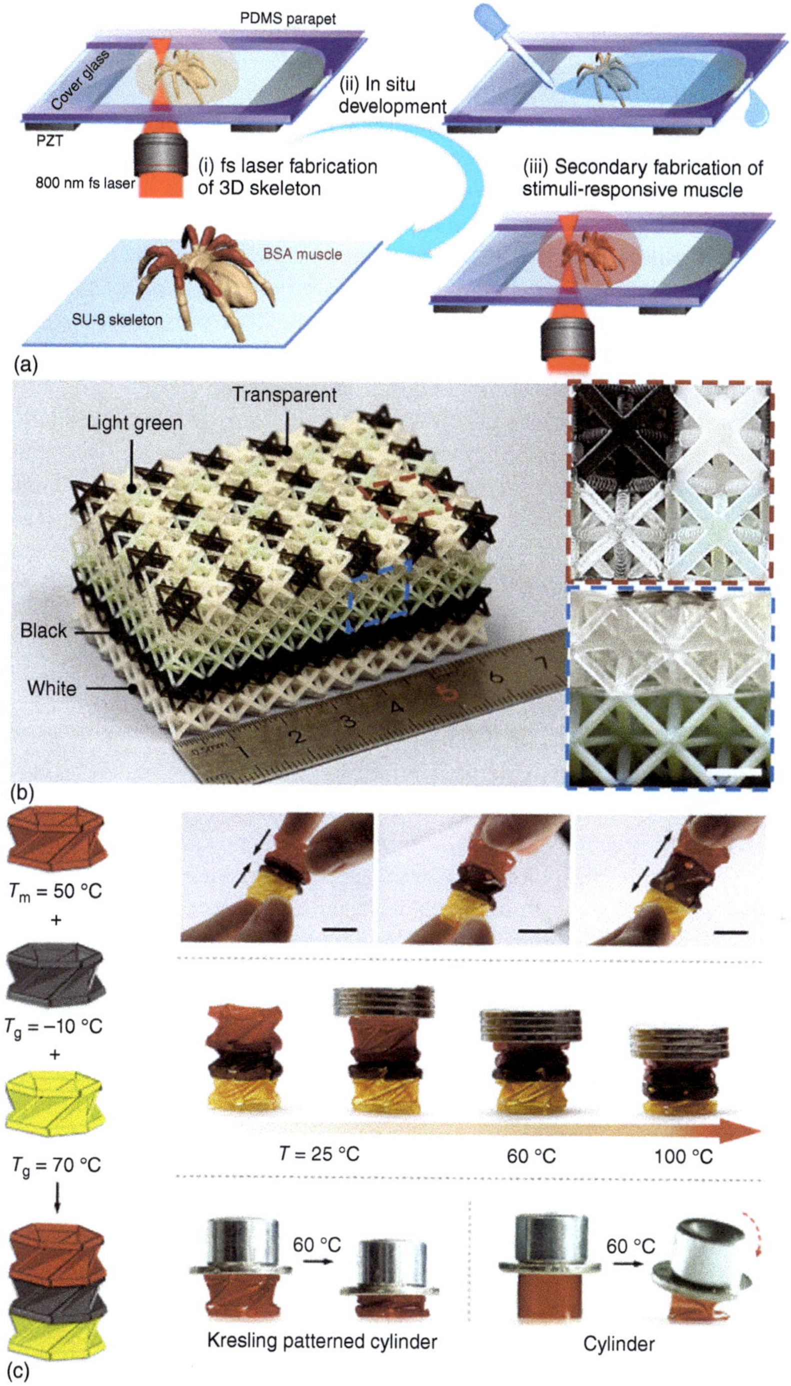


**Figure 7.2** Multi-material integration. (a) Fabrication of a spider-like musculoskeletal system using multi-material DLW printing. Source: Ma et al. [58]/Springer Nature /CC BY 4.0 /Public domain. (b) Multi-material 3D printing fabricates lattice structures with varied colors. Source: Cheng et al. [59]/Reproduced from Springer Nature/CC BY 4.0. (c) Modular assembly of Kresling-patterned cylinders via interfacial bonding. Source: Fang et al. [60]/Reproduced from Elsevier.

In addition to the continuous improvement in manufacturing methods, researchers have also made a lot of contributions to robotic materials [65]. For example, smart materials can endow robots with the ability to weld or heal, thereby promoting the development of modular robots. Non-covalent networks and dynamic covalent bonding have been developed for stable interfacial healing of various materials under temperature, photo-stimulation, or in an automated manner. Fang et al. proposed modular reconfigurable structures based on SMPs with dynamic covalent networks [60]. Three deformable samples with Kresling patterns were prepared by 4D printing and welded together by an interfacial healing mechanism. Each module has different thermal transition properties, so each region in the assembled structure exhibits sequentially compressed deformation behaviors under external loads. Combining the material's self-healing properties with the 4D printing technique significantly expands the complexity and scalability of deformable devices. In addition, Zhang et al. proposed liquid crystal elastomer (LCE)-based multi-material robots with weldable properties [66]. The developed LCE material utilizes the thermal polymerization of acrylate groups to form robust material connections. Therefore, this strategy allows the integration of LCE materials with different physical or chemical properties into robotic systems, enabling complex deformation modes such as multi-stimuli responsiveness, selective actuation, and multi-level actuation.

## 7.4 Multifunctional Integration for Miniature Soft Robots with Perception Capabilities

Organisms in nature could perceive their own movement and changes in the environment through receptors, and then provide somatosensory feedback to regulate the locomotion. Imitating biological intelligence and constructing autonomous robots with proprioception and environmental perception capabilities is a continuing goal in the field of soft robotics [67–76] (Figure 7.3). A variety of advanced sensors such as temperature sensors, pressure sensors, luminescent units, and optical fibers have been embedded in soft robotic systems through physical integration or material design. On the other hand, in order to improve the flexibility of the sensing module, advanced materials such as conductive fillers (Ag flakes, graphene), ionic conductors, and liquid metals are widely used [77–82]. For example, Ohm et al. doped silver flakes in polyacrylamide and alginate interpenetrating networks [83]. The developed conductive hydrogel material maintains high electrical conductivity and softness simultaneously, and can be used as an energy transfer unit for soft swimming robots and wearable devices. Due to the limited size of miniaturized soft robots, their signal transmission and sensing capabilities are limited. To address this challenge, Lu et al. integrated magnetic materials, piezoelectric energy conversion devices (piezoelectric ceramics), near-field communication devices (NFC), and temperature sensors (thermistors) in a layer-by-layer manner [84]. Manufactured robots with an overall size of $10 \times 30$ mm can exhibit different motion patterns under a programmable magnetic field. Using piezoelectric composite materials, the mechanical energy generated during motion can be

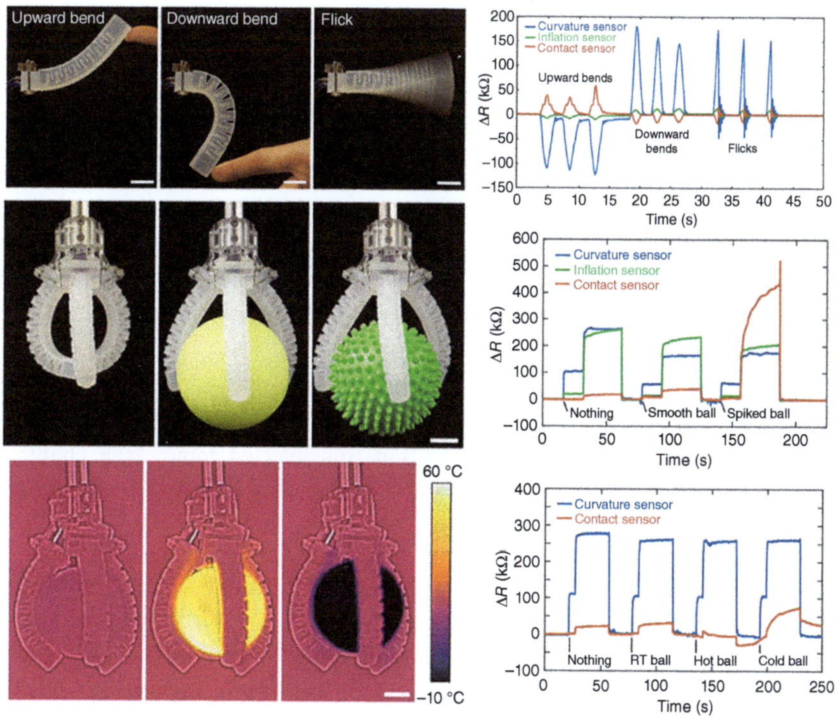


**Figure 7.3** Soft robots with perception capabilities. Soft robotic grippers for handling objects with varying surface roughness and varying surface temperatures. Source: Truby et al. [62]/Reproduced from John Wiley & Sons, Inc.

collected, and the contact interface and ambient temperature can be monitored through its own sensors. This simple stacking method is difficult to integrate actuation and sensing functions on the 3D robot structure. In this regard, multi-material printing technologies including embedded 3D printing, and multi-material DLP printing are used to build soft robotic structures with complex sensor networks. For example, Truby et al. used embedded 3D printing to print curvature sensors, expansion sensors, and contact sensors made of conductive ionic gels while building the pneumatic network (elastomer matrix) required for actuation, which can endow the printed actuators with proprioception and tactile feedback [62].

Integrating actuation and sensing functions at the polymer network level is very attractive for the development of soft robots, because this strategy could avoid delamination caused by mismatched material properties or interference between different functional modules. Inspired by the neuromuscular system of organisms, Zhao et al. proposed a dual-network hydrogel actuator with interpenetrating conductive polymer (polyaniline, PANI) and actuation unit (poly(*N*-isopropylacrylamide), PNIPAAm) [85]. Soft robots adopting this special molecular design could integrate piezoresistive sensing and photothermal actuation functions. To enable the uniform distribution of the responsive and sensing components throughout the robotic structure, a one-pot polymerization method was adopted. With the assistance of an ice template, PANI and PNIPAAm were subjected to low-temperature polymerization and UV irradiation polymerization, respectively. The prepared hydrogel material has good stretchability (stretch at break: 170%) and high electrical conductivity (24.6–36.8 mS $cm^{-1}$). Soft robots based on the developed double-network hydrogel could shrink or bend under the action of near-infrared light, and the resistance change of the hydrogel body can accurately monitor the mechanical deformation generated by itself, and then provide sensory feedback and realize active control.

## 7.5 Perspectives Toward Intelligent and Autonomous Soft Robots

This book focuses on one particularly interesting topic of miniature robotics, i.e. untethered miniature soft robots. Within this selected topic, each chapter presents extensive discussions on one representative category of robots made of a different material (Figure 7.4). Not only is the general landscape of this highly dynamic and quickly evolving research area depicted, but the unique advantages and constraints of each category are also revealed. The silicone elastomer matrix-based robots exhibit 3D programmability in morphology and fast responsibility under magnetic control. The carbon-based robots display dexterous structural changes once triggered by multiple potential stimuli. The hydrogel-based

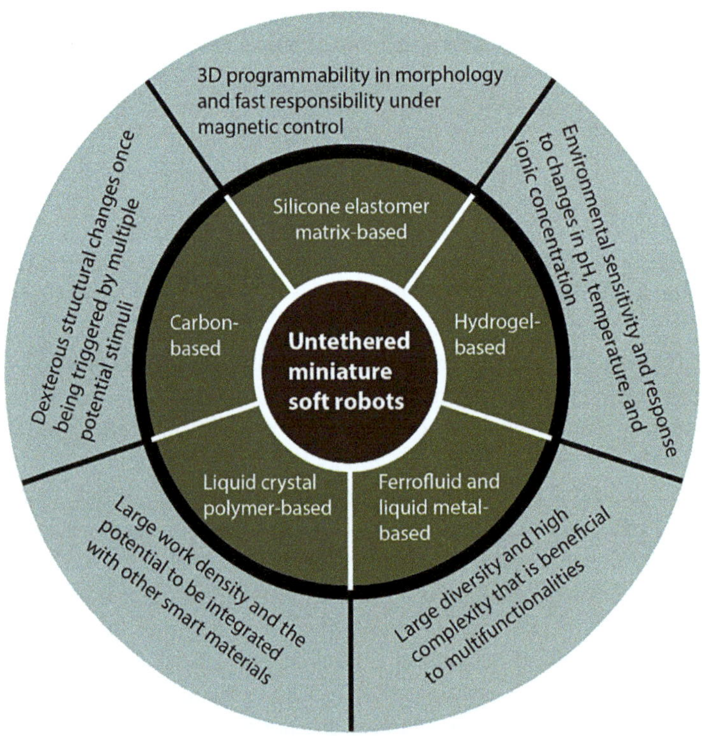


**Figure 7.4** Comparison and summary of smart materials for untethered miniature soft robots.

robots possess environmental sensitivity and could respond to changes in pH, temperature, and ionic concentration. The liquid crystal polymer-based robots showcase large work density and the potential to be integrated with other smart materials. The ferrofluid and liquid metal-based robots demonstrate the possible large diversity and high complexity that is beneficial to multifunctionalities. The distinct features of each category of robots make them uniquely suitable for one or more specific kinds of applications. And the large variety of candidate categories leads to a vast design and search space to find auspicious solutions to meet demands posed by different real-world tasks.

It is impressive to reflect on how much has been achieved so far in this field. The stage of preliminary investigations has passed, which aimed at exploring and validating novel ideas with proof-of-concept results. Current focuses are being placed upon the realization of specific functionalities on these robots, while each robot is only capable of one or a limited number of capabilities. Moving into the next 10 years or so, a new phase is anticipated in which complete robotic systems and platforms will be developed for targeted genuine applications, delivering this new robotic technique from laboratories to the real world. Thus, much more improvement still needs to be made in order to further advance miniature soft robotics and fully materialize its promised potential (Figure 7.5). The next generation of such robotic systems should be reliable, multifunctional, cost-effective, standalone, and application-orientated.

For this field to move forward, there are several grand challenges that need to be fully or at least partially addressed. **First, untethered miniature soft robots suffer from low output forces and torques**. Being untethered, small, and compliant is the most significant advantage of these robots, but at the same time, it is also the most severe limitation regarding the capability of outputting forces and torques. Many applications require the robots to be capable of exerting forces and torques that are higher than a threshold value to achieve specific functionalities. For example, these robots could access small and constrained workspaces buried deep inside human bodies to conduct surgical operations for healthcare purposes, which is arguably its most important motivation. But these diagnostic and therapeutic operations, such as puncturing, cutting, and suturing, demand a force and torque value that is often several orders of magnitude larger than the achievable values of

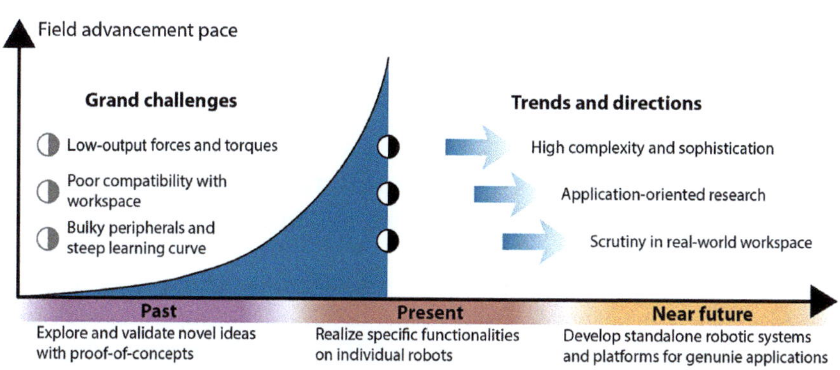


**Figure 7.5** Perspective for future soft robotic systems.

existing robots. This giant gap needs to be narrowed and filled for these untethered miniature soft robots to achieve their promised potential.

**Second, untethered miniature soft robots suffer from low compatibility with the targeted real-world workspace and working conditions**. So far, the compatibility of a such robot with its targeted real-world workspace has been seldom considered in the investigation, as a well-controlled, known laboratory environment is used as the testbench. This ignorance of workspace constraints damages the real-world applicability of these robots. For robots aiming to work inside human bodies, they could become a potential hazard to the patient without a thorough consideration of their biocompatibility and biodegradability in the specific physiological environment inside the targeted intracorporeal workspace. Unfortunately, biocompatibility and biodegradability are often ignored during the discussion of these novel kinds of devices. Even when they are mentioned, the problem is often over-simplified to lose realistic relevance. Biocompatibility and biodegradability are not on–off properties and cannot be generically defined. They are dependent on the targeted workspace and environment, since the internal environment in the human body varies dramatically from region to region, such as the change in pH values from the stomach to intestine. Thus, biocompatibility and biodegradability need to be placed at the central place for the design consideration of a robotic system at a stage as early as possible, for this consideration will restrain the pool of feasible design choices such as the constituent materials. In addition, for robots aiming to work in places without a direct line-of-sight, its compatibility with other imaging modalities, such as X-ray and ultrasound in clinical scenarios, will play a pivotal role in determining its real-world feasibility.

**Third, untethered miniature soft robots suffer from bulky peripherals and steep learning curves**. Although existing miniature robots have a characteristic length way smaller than conventional robots, they can only work with support from bulky peripherals. So far, the robots reported in this field are mostly proof-of-concept and require expertise to run them. Next-generation miniature robotic systems need to be developed as standalone and intuitive systems that could be incorporated into the existing work routines of the specific targeted field. For example, these robots need to work reliably and with efficacy in a stressful and time-sensitive clinical environment, operated by healthcare practitioners who are less experienced in engineering than robotic researchers. Thus, the control, actuation, fail-safe mechanisms, imaging modalities, graphic user interfaces, and others are all important factors that need to be properly addressed, even though they may seem trivial compared with the robot itself.

To address the aforementioned challenges, and also to further expand the horizon of this exciting scientific field, researchers have explored several directions that are right now under intensive investigation. **The first trend is designing multifunctional miniature robotic systems with increasing complexity and sophistication**. It used to be a commonly seen statement that miniature robots cannot carry onboard components and thus novel actuation and control strategies are needed. But as the microfabrication technique advances, complex robot designs made of multiple materials and sophisticated 3D geometries become achievable. The groundbreaking stratagems in microfabrication have extended the design space of these robots, enabling the physical realization of unprecedented designs that were previously too complex to fabricate. The increased complexity and tunability in robot design allow for multifunctionalities and a better capability to perform demanding tasks.

**The second trend is framing the research tightly around a specific application**. An increasing percentage of new miniature soft robotic systems are developed with thorough consideration of the requirements and constraints of their targeted real-world tasks. The designated tasks, workspace, and applications are clearly defined at the start and at the center of the research, in order to better tailor the robot to the specific conditions to improve feasibility and efficiency. This trend helps these application-oriented robotic systems to attain better real-world applicability and relevance, making them superior solutions to address true and well-acknowledged needs unmet by existing solutions.

**The third trend is scrutinizing the developed robots in the realistic workspaces**. For example, robots developed for biomedical applications are now more commonly tested in *ex vivo* and *in vivo* experiments. The interactions between these robots and biological cells, tissues, and organs are being investigated. Exposing these robots to the targeted biological workspace is advantageous to gain a better and more realistic understanding of the challenges and the problems. Specifically, a direct line-of-sight is often available in a laboratory environment for easy observation, which is not the case for a clinical workspace. In a realistic scenario, medical imaging modalities, e.g. X-ray, ultrasound imaging, MRI, and fluorescent imaging, need to be employed to replace optical cameras which are impractical in scenarios without a direct line-of-sight, to further enhance the real-world relevance of the developed robots.

The recent development of untethered miniature soft robotics is encouraging. This field is attracting more and more attention not only from researchers, but also from industrial companies and government agencies, amassing resources to be invested to push its advancement. The potential of miniature soft robots is prominent and well acknowledged, and this potential is being materialized at a breakneck pace. Admittedly, each of the aforementioned grand challenges is arduous to tackle and could take decades to be fully addressed, assuming future technological breakthroughs or disruptive inventions are needed. Nevertheless, with diligent researchers bravely tackling the grand challenges, it is envisioned that they will be steadily and surely addressed, sustaining this field along its exponential growth curve toward real-world applications, bringing transformative positive impact on multiple aspects of our socio-economic life, particularly the modern healthcare, in the near future.

## Abbreviations

| | |
|---|---|
| CAL | computed axial lithography |
| CLIP | continuous liquid interface production |
| DIW | direct ink writing |
| DLP | digital light processing |

| | |
|---|---|
| FDM | fused deposition modeling |
| FFF | fused filament fabrication |
| GelMA | gelatin methacryloyl |
| LCE | liquid crystal elastomer |
| NFC | near-field communication |
| PNIPAAm | poly(*N*-isopropylacrylamide) |
| SLA | stereolithography |
| SLS | selective laser sintering |
| SMP | shape-memory polymer |
| TPP | two-photon polymerization |

## References


1 Yang, G.-Z., Bellingham, J., Dupont, P.E. et al. (2018). The grand challenges of *Science Robotics*. *Science Robotics* 3: eaar7650.

2 Zhang, B., Li, H., Cheng, J. et al. (2021). Mechanically robust and UV-curable shape-memory polymers for digital light processing based 4D printing. *Advanced Materials* 33: 2101298.

3 Ze, Q., Kuang, X., Wu, S. et al. (2020). Magnetic shape memory polymers with integrated multifunctional shape manipulation. *Advanced Materials* 32: 1906657.

4 Wang, B., Chan, K.F., Yuan, K. et al. (2021). Endoscopy-assisted magnetic navigation of biohybrid soft microrobots with rapid endoluminal delivery and imaging. *Science Robotics* 6: eabd2813.

5 Sun, L., Chen, Z., Bian, F., and Zhao, Y. (2020). Bioinspired soft robotic caterpillar with cardiomyocyte drivers. *Advanced Functional Materials* 30: 1907820.

6 Lendlein, A. (2018). Fabrication of reprogrammable shape-memory polymer actuators for robotics. *Science Robotics* 3: eaat9090.

7 Jin, B., Song, H., Jiang, R. et al. (2018). Programming a crystalline shape memory polymer network with thermo-and photo-reversible bonds toward a single-component soft robot. *Science Advances* 4: eaao3865.

8 Mattmann, M., De Marco, C., Briatico, F. et al. (2022). Thermoset shape memory polymer variable stiffness 4D robotic catheters. *Advanced Science* 9: 2103277.

9 Yang, Q., Peng, C., Ren, J. et al. (2019). A near-infrared photoactuator based on shape memory semicrystalline polymers toward light-fueled crane, grasper, and walker. *Advanced Optical Materials* 7: 1900784.

10 Chen, T., Bilal, O.R., Shea, K., and Daraio, C. (2018). Harnessing bistability for directional propulsion of soft, untethered robots. *Proceedings of the National Academy of Sciences of the United States of America* 115: 5698–5702.

11 Reeder, J., Kaltenbrunner, M., Ware, T. et al. (2014). Mechanically adaptive organic transistors for implantable electronics. *Advanced Materials* 26: 4967–4973.

12 Sekitani, T., Zschieschang, U., Klauk, H., and Someya, T. (2010). Flexible organic transistors and circuits with extreme bending stability. *Nature Materials* 9: 1015–1022.

13 Li, W., Liu, Y., and Leng, J. (2017). Programmable and shape-memorizing information carriers. *ACS Applied Materials & Interfaces* 9: 44792–44798.

14 Zhang, G., Peng, W., Wu, J. et al. (2018). Digital coding of mechanical stress in a dynamic covalent shape memory polymer network. *Nature Communications* 9: 4002.

15 Liu, W., Chen, H., Ge, M. et al. (2018). Electroactive shape memory composites with $TiO_2$ whiskers for switching an electrical circuit. *Materials & Design* 143: 196–203.

16 Zhang, F., Xia, Y., Wang, L. et al. (2018). Conductive shape memory microfiber membranes with core–shell structures and electroactive performance. *ACS Applied Materials & Interfaces* 10: 35526–35532.

17 Testa, P., Style, R.W., Cui, J. et al. (2019). Magnetically addressable shape-memory and stiffening in a composite elastomer. *Advanced Materials* 31: 1900561.

18 Guo, Q., Bishop, C.J., Meyer, R.A. et al. (2018). Entanglement-based thermoplastic shape memory polymeric particles with photothermal actuation for biomedical applications. *ACS Applied Materials & Interfaces* 10: 13333–13341.

19 Toncheva, A., Khelifa, F., Paint, Y. et al. (2018). Fast IR-actuated shape-memory polymers using in situ silver nanoparticle-grafted cellulose nanocrystals. *ACS Applied Materials & Interfaces* 10: 29933–29942.

20 Wang, W., Shen, D., Li, X. et al. (2018). Light-driven shape-memory porous films with precisely controlled dimensions. *Angewandte Chemie* 130: 2161–2165.

21 Zhang, F., Xiong, L., Ai, Y. et al. (2018). Stretchable multiresponsive hydrogel with actuatable, shape memory, and self-healing properties. *Advanced Science* 5: 1800450.

22 Liu, R., Kuang, X., Deng, J. et al. (2018). Shape memory polymers for body motion energy harvesting and self-powered mechanosensing. *Advanced Materials* 30: 1705195.

23 Liu, Z., Li, Q., Bian, W. et al. (2019). Preliminary test and analysis of an ultra-light lenticular tube based on shape memory polymer composites. *Composite Structures* 223: 110936.

24 Li, F., Liu, L., Lan, X. et al. (2019). Ground and geostationary orbital qualification of a sunlight-stimulated substrate based on shape memory polymer composite. *Smart Materials and Structures* 28: 075023.

25 Lendlein, A. and Langer, R. (2002). Biodegradable, elastic shape-memory polymers for potential biomedical applications. *Science* 296: 1673–1676.

26 Lin, C., Lv, J., Li, Y. et al. (2019). 4D-printed biodegradable and remotely controllable shape memory occlusion devices. *Advanced Functional Materials* 29: 1906569.

27 Shin, Y.C., Lee, J.B., Kim, D.H. et al. (2019). Development of a shape-memory tube to prevent vascular stenosis. *Advanced Materials* 31: 1904476.

28 Neffe, A.T., Hanh, B.D., Steuer, S., and Lendlein, A. (2009). Polymer networks combining controlled drug release, biodegradation, and shape memory capability. *Advanced Materials* 21: 3394–3398.

29 Wu, P., Shen, X., Schäfer, C.G. et al. (2019). Mechanochromic and thermochromic shape memory photonic crystal films based on core/shell nanoparticles for smart monitoring. *Nanoscale* 11: 20015–20023.

30 Ding, Z., Yuan, C., Peng, X. et al. (2017). Direct 4D printing via active composite materials. *Science Advances* 3: e1602890.

31 Mazzolai, B. and Laschi, C. (2020). A vision for future bioinspired and biohybrid robots. *Science Robotics* 5: eaba6893.

32 Morimoto, Y., Onoe, H., and Takeuchi, S. (2018). Biohybrid robot powered by an antagonistic pair of skeletal muscle tissues. *Science Robotics* 3: eaat4440.

33 Ricotti, L., Trimmer, B., Feinberg, A.W. et al. (2017). Biohybrid actuators for robotics: a review of devices actuated by living cells. *Science Robotics* 2: eaaq0495.

34 Tetsuka, H., Pirrami, L., Wang, T. et al. (2022). Wirelessly powered 3D printed hierarchical biohybrid robots with multiscale mechanical properties. *Advanced Functional Materials* 32: 2202674.

35 Park, S.-J., Gazzola, M., Park, K.S. et al. (2016). Phototactic guidance of a tissue-engineered soft-robotic ray. *Science* 353: 158–162.

36 Kim, T.H., Kwon, C.H., Lee, C. et al. (2016). Bio-inspired hybrid carbon nanotube muscles. *Scientific Reports* 6: 1–8.

37 Raman, R., Cvetkovic, C., Uzel, S.G. et al. (2016). Optogenetic skeletal muscle-powered adaptive biological machines. *Proceedings of the National Academy of Sciences of the United States of America* 113: 3497–3502.

38 Vizsnyiczai, G., Frangipane, G., Maggi, C. et al. (2017). Light controlled 3D micromotors powered by bacteria. *Nature Communications* 8: 15974.

39 Pagan-Diaz, G.J., Zhang, X., Grant, L. et al. (2018). Simulation and fabrication of stronger, larger, and faster walking biohybrid machines. *Advanced Functional Materials* 28: 1801145.

40 Sun, L., Yu, Y., Chen, Z. et al. (2020). Biohybrid robotics with living cell actuation. *Chemical Society Reviews* 49: 4043–4069.

41 Wang, Q., Tian, Y., Du, X. et al. (2021). Magnetic navigation of collective cell microrobots in blood under ultrasound doppler imaging. *IEEE/ASME Transactions on Mechatronics* 27: 3174–3185.

42 Yan, X., Zhou, Q., Vincent, M. et al. (2017). Multifunctional biohybrid magnetite microrobots for imaging-guided therapy. *Science Robotics* 2: eaaq1155.

43 Zhang, Y., Zhang, L., Yang, L. et al. (2019). Real-time tracking of fluorescent magnetic spore-based microrobots for remote detection of *C. diff* toxins. *Science Advances* 5: eaau9650.

44 Aydin, O., Zhang, X., Nuethong, S. et al. (2019). Neuromuscular actuation of biohybrid motile bots. *Proceedings of the National Academy of Sciences of the United States of America* 116: 19841–19847.

45 Yoon, J., Eyster, T.W., Misra, A.C., and Lahann, J. (2015). Cardiomyocyte-driven actuation in biohybrid microcylinders. *Advanced Materials* 27: 4509–4515.

46 Shang, Y., Chen, Z., Fu, F. et al. (2018). Cardiomyocyte-driven structural color actuation in anisotropic inverse opals. *ACS Nano* 13: 796–802.

47 Sakar, M.S., Neal, D., Boudou, T. et al. (2012). Formation and optogenetic control of engineered 3D skeletal muscle bioactuators. *Lab on a Chip* 12: 4976–4985.

48 Yasa, O., Erkoc, P., Alapan, Y., and Sitti, M. (2018). Microalga-powered microswimmers toward active cargo delivery. *Advanced Materials* 30: 1804130.

49 Akiyama, Y., Hoshino, T., Iwabuchi, K., and Morishima, K. (2012). Room temperature operable autonomously moving bio-microrobot powered by insect dorsal vessel tissue. *PLoS One* 7: e38274.

**50** Xu, H., Medina-Sánchez, M., Magdanz, V. et al. (2018). Sperm-hybrid micromotor for targeted drug delivery. *ACS Nano* 12: 327–337.

**51** Zhang, Y.F., Zhang, N., Hingorani, H. et al. (2019). Fast-response, stiffness-tunable soft actuator by hybrid multimaterial 3D printing. *Advanced Functional Materials* 29: 1806698.

**52** Sachyani Keneth, E., Kamyshny, A., Totaro, M. et al. (2021). 3D printing materials for soft robotics. *Advanced Materials* 33: 2003387.

**53** Bastola, A.K., Paudel, M., and Li, L. (2018). Development of hybrid magnetorheological elastomers by 3D printing. *Polymer* 149: 213–228.

**54** Xia, K., Chen, X., Shen, X. et al. (2019). Carbonized chinese art paper-based high-performance wearable strain sensor for human activity monitoring. *ACS Applied Electronic Materials* 1: 2415–2421.

**55** Bodkhe, S., Noonan, C., Gosselin, F.P., and Therriault, D. (2018). Coextrusion of multifunctional smart sensors. *Advanced Engineering Materials* 20: 1800206.

**56** Ren, L., Song, Z., Liu, H. et al. (2018). 3D printing of materials with spatially non-linearly varying properties. *Materials & Design* 156: 470–479.

**57** Mao, Y., Yu, K., Isakov, M.S. et al. (2015). Sequential self-folding structures by 3D printed digital shape memory polymers. *Scientific Reports* 5: 1–12.

**58** Ma, Z.-C., Zhang, Y.L., Han, B. et al. (2020). Femtosecond laser programmed artificial musculoskeletal systems. *Nature Communications* 11: 4536.

**59** Cheng, J., Wang, R., Sun, Z. et al. (2022). Centrifugal multimaterial 3D printing of multifunctional heterogeneous objects. *Nature Communications* 13: 7931.

**60** Fang, Z., Song, H., Zhang, Y. et al. (2020). Modular 4D printing via interfacial welding of digital light-controllable dynamic covalent polymer networks. *Matter* 2: 1187–1197.

**61** Skylar-Scott, M.A., Mueller, J., Visser, C.W., and Lewis, J.A. (2019). Voxelated soft matter via multimaterial multinozzle 3D printing. *Nature* 575: 330–335.

**62** Truby, R.L., Wehner, M., Grosskopf, A.K. et al. (2018). Soft somatosensitive actuators via embedded 3D printing. *Advanced Materials* 30: 1706383.

**63** Ge, Q., Chen, Z., Cheng, J. et al. (2021). 3D printing of highly stretchable hydrogel with diverse UV curable polymers. *Science Advances* 7: eaba4261.

**64** Cui, H., Yao, D., Hensleigh, R. et al. (2022). Design and printing of proprioceptive three-dimensional architected robotic metamaterials. *Science* 376: 1287–1293.

**65** Su, L., Jin, D.D., Pan, C.F. et al. (2021). A mobile magnetic pad with fast light-switchable adhesion capabilities. *Bioinspiration & Biomimetics* 16: 055005.

**66** Zhang, Y., Wang, Z., Yang, Y. et al. (2020). Seamless multimaterial 3D liquid-crystalline elastomer actuators for next-generation entirely soft robots. *Science Advances* 6: eaay8606.

**67** Roh, Y., Kim, M., Won, S.M. et al. (2021). Vital signal sensing and manipulation of a microscale organ with a multifunctional soft gripper. *Science Robotics* 6: eabi6774.

**68** Yang, X., Chang, L., and Pérez-Arancibia, N.O. (2020). An 88-milligram insect-scale autonomous crawling robot driven by a catalytic artificial muscle. *Science Robotics* 5: eaba0015.

**69** Wehner, M., Truby, R.L., Fitzgerald, D.J. et al. (2016). An integrated design and fabrication strategy for entirely soft, autonomous robots. *Nature* 536: 451–455.

**70** Acome, E., Mitchell, S.K., Morrissey, T.G. et al. (2018). Hydraulically amplified self-healing electrostatic actuators with muscle-like performance. *Science* 359: 61–65.

**71** Hubbard, J.D., Acevedo, R., Edwards, K.M. et al. (2021). Fully 3D-printed soft robots with integrated fluidic circuitry. *Science Advances* 7: eabe5257.

**72** Joyee, E.B., Szmelter, A., Eddington, D., and Pan, Y. (2022). 3D printed biomimetic soft robot with multimodal locomotion and multifunctionality. *Soft Robotics* 9: 1–13.

**73** Zhang, S., Wang, B., Jiang, J. et al. (2019). High-fidelity conformal printing of 3D liquid alloy circuits for soft electronics. *ACS Applied Materials & Interfaces* 11: 7148–7156.

**74** Ke, X., Zhang, S., Chai, Z. et al. (2021). Flexible discretely-magnetized configurable soft robots via laser-tuned selective transfer printing of anisotropic ferromagnetic cells. *Materials Today Physics* 17: 100313.

**75** Ma, K.Y., Chirarattananon, P., Fuller, S.B., and Wood, R.J. (2013). Controlled flight of a biologically inspired, insect-scale robot. *Science* 340: 603–607.

**76** Tang, W., Zhang, C., Zhong, Y. et al. (2021). Customizing a self-healing soft pump for robot. *Nature Communications* 12: 2247.

**77** Chen, J., Han, K., Luo, J. et al. (2020). Soft robots with self-powered configurational sensing. *Nano Energy* 77: 105171.

**78** Kim, S.Y., Choo, Y., Bilodeau, R.A. et al. (2020). Sustainable manufacturing of sensors onto soft systems using self-coagulating conductive Pickering emulsions. *Science Robotics* 5: eaay3604.

**79** Thuruthel, T.G., Shih, B., Laschi, C., and Tolley, M.T. (2019). Soft robot perception using embedded soft sensors and recurrent neural networks. *Science Robotics* 4: eaav1488.

**80** Van Meerbeek, I., De Sa, C., and Shepherd, R. (2018). Soft optoelectronic sensory foams with proprioception. *Science Robotics* 3: eaau2489.

**81** Oh, B., Park, Y.G., Jung, H. et al. (2020). Untethered soft robotics with fully integrated wireless sensing and actuating systems for somatosensory and respiratory functions. *Soft Robotics* 7: 564–573.

**82** Chen, J., Chen, B., Han, K. et al. (2019). A triboelectric nanogenerator as a self-powered sensor for a soft–rigid hybrid actuator. *Advanced Materials Technologies* 4: 1900337.

**83** Ohm, Y., Pan, C., Ford, M.J. et al. (2021). An electrically conductive silver–polyacrylamide–alginate hydrogel composite for soft electronics. *Nature Electronics* 4: 185–192.

**84** Lu, H., Hong, Y., Yang, Y. et al. (2020). Battery-less soft millirobot that can move, sense, and communicate remotely by coupling the magnetic and piezoelectric effects. *Advanced Science* 7: 2000069.

**85** Zhao, Y., Xuan, C., Qian, X. et al. (2019). Soft phototactic swimmer based on self-sustained hydrogel oscillator. *Science Robotics* 4: eaax7112.

# Index



*Untethered Miniature Soft Robots: Materials, Fabrications, and Applications*, First Edition.
Li Zhang, Jiachen Zhang, Neng Xia, and Yue Dong.